INTRODUCTION TO WILDLIFE MANAGEMENT

McGraw-Hill Series in Forest Resources

Avery and Burkhart: Forest Measurements
Brockman and Merriam: Recreational Use of Wild Lands
Brown and Davis: Forest Fire: Control and Use
Dana and Fairfax: Forest and Range Policy
Daniel, Helms, and Baker: Principles of Silviculture
Davis: Forest Management
Davis: Land Use
Dykstra: Mathematical Programming for Natural Resource Management
Harlow, Harrar, and White: Textbook of Dendrology
Heady: Rangeland Management
Knight and Heikkenen: Principles of Forest Entomology
Nyland, Larson, and Shirley: Forestry and Its Career Opportunities
Panshin and De Zeeuw: Textbook of Wood Technology
Panshin, Harrar, Bethel, and Baker: Forest Products
Sharpe and Hendee: An Introduction to Forestry
Shaw: Introduction to Wildlife Management
Stoddart, Smith, and Box: Range Management

Walter Mulford was Consulting Editor of this series from its inception in 1931 until January 1, 1952.

Henry J. Vaux was Consulting Editor of this series from January 1, 1952, until July 1, 1976.

INTRODUCTION TO WILDLIFE MANAGEMENT

James H. Shaw
Associate Professor of Wildlife Ecology
Oklahoma State University

McGRAW-HILL BOOK COMPANY
New York St. Louis San Francisco Auckland Bogotá
Hamburg Johannesburg London Madrid Mexico Montreal New Delhi
Panama Paris São Paulo Singapore Sydney Tokyo Toronto

INTRODUCTION TO WILDLIFE MANAGEMENT

1 2 3 4 5 6 7 8 9 0 D O C D O C 8 9 8 7 6 5 4

ISBN 0-07-056481-7

This book was set in Times Roman by Black Dot, Inc. (ECU).
The editors were Marian D. Provenzano and Barry Benjamin;
the production supervisor was Leroy A. Young.
The drawings were done by ECL Art.
R. R. Donnelley & Sons Company was printer and binder.

Library of Congress Cataloging in Publication Data

Shaw, James H.
Introduction to wildlife management.

(McGraw-Hill series in forest resources)
Includes index.
1. Wildlife management. I. Title. II. Series.
SK355.S46 1985 639.9 84-9713
ISBN 0-07-056481-7

To the memory of
IVENS PINTO FRANQUEIRA
Delegado
Instituto Brasilero de Desenvolvimento Florestal
Minas Gerais, Brazil (1967–1980),
an outstanding conservation administrator
of exceptional vision and energy

and
Ralph M. Wetzel,
the discoverer of the Chacoan Peccary
and a scientist of extraordinary
dedication, generosity, warmth, and humor

CONTENTS

PREFACE

The field of wildlife management has grown enormously in the half century since Aldo Leopold's *Game Management* first appeared. New areas of expertise have blossomed and flourished, including endangered species preservation, nongame management, socioeconomic aspects of wildlife management, and an increasing emphasis upon management of natural communities to preserve species diversity. No longer is single-species, production-oriented game management by itself sufficient to meet either the magnitude of environmental changes or the greater expectations and interest of the public as a whole.

Much of the expansion in wildlife management stems from pressures of human population growth, depletion of natural resources in general, and pollution. Concern over environmental problems has attracted a broader segment of the public to the problems of conserving and managing wild species. Once the rural and small-town hunters were the primary proponents of wildlife conservation in the United States. Today hunters are joined by an increasing number of urban dwellers comprised of animal protectionists, outdoor recreation enthusiasts, bird watchers, photographers, campers, and hikers. Although these diverse groups often express serious differences of opinion, all are concerned with the future of the world's wildlife and all feel that they have some stake in that future. As a result professional wildlife managers now serve a much larger constituency than they did a generation ago.

This textbook grew out of nine years of teaching an upper-division, undergraduate course entitled, "Principles of Wildlife Ecology." In such a rapidly expanding discipline, I was unable to find an up-to-date textbook with sufficient breadth. Most books available were either sadly out of date or confined to detailed treatments of their author's primary research interests. This book attempts a different approach by providing an introduction to virtually the entire field of terrestrial wildlife management.

Chapter 1 introduces the basic conflict between short-term human interests and the longer-term needs of wildlife. It includes a history of wildlife management in the United States, organizing events into "eras" based upon prevailing attitudes and practices. Some of the more important people in this history receive mention as well. Other human-related aspects of wildlife management such as the values of wildlife, the legal status of wildlife, and sources of funding for wildlife management also appear.

Wildlife habitat, the subject of Chapter 2, begins with basic terms and elementary concepts including energy transfers, trophic levels, niche, carrying capacity, succession, and the edge effect. Conservationists and wildlife managers must often evaluate different areas in terms of habitat value to establish priorities for acquisition and improvement. A substantial portion of Chapter 2 is therefore devoted to habitat evaluation. Once evaluation is completed, the next step is habitat manipulation to fulfill management objectives. The chapter concludes with a presentation of mitigation, a potentially very important form of wildlife habitat management.

Chapter 3 reviews the primary components of wildlife populations and the most general model for population growth, the logistic growth equation. Cyclic and irruptive populations are also treated. Some relatively new additions to population ecology appear in this chapter. The concept of r and K selection illustrates how different population characteristics are typically linked together. This concept helps explain why, in the wake of human disturbance, some species become pests while others are relegated to endangered lists. Wildlife managers have learned that changes in population sizes and changes in the genetic makeup of populations are intricately related. Chapter 3 thus concludes with an introduction to applied population genetics.

Abundance, including estimates of actual numbers and measurements of indexes or population trends, is the subject of Chapter 4. The actual numbers can be calculated from such methods as mark-and-recapture, transect surveys, selective additions or removals, or pellet group surveys. Although indexes cannot furnish estimates of actual numbers, they are generally faster and easier to apply over wide areas, making them more suitable for extensive management needs. Some of the more common methods include roadside counts, road kills, and scent station surveys. Methods of population analysis for such crucial characteristics as rate of increase, mortality, and survival also appear in Chapter 4.

The application of animal behavior to wildlife management forms the basis for Chapter 5. This topic, rarely incuded in most other wildlife management texts, is important both from the standpoint of its influence upon wildlife populations and for dealing with practical problems resulting from human-induced alterations in animal behavior. The best-known example of a behavioral problem in wildlife management is the grizzly bear controversy in the national parks of both the United States and Canada.

Chapter 6 describes two major types of interactions between populations of different species: competition and predation. Once thought of primarily in terms of a violent struggle for existence, competition between species has been found to have been largely resolved through natural selection. Practical problems do arise, though they are typically the result of habitat disturbances or from the introduction of exotic species. Predation has likewise been widely misunderstood. The effects of predation on prey abundance vary and can be resolved only on a case-by-case basis by examining the components. For example, the important matter of age-specific selection of prey by a predator depends largely upon the relative sizes of the predator and the prey and upon whether the predator relies upon ambush or upon chases.

Much of wildlife management is aimed at providing hunter demands while guarding against excessive harvest. This traditional form of game management is brought into the computer age by questioning whether hunting mortality is additive or compensatory. In practice, much of the mortality inflicted by hunters on big-game population is additive, while most of that incurred by small game is compensatory. Chapter 7 also introduces harvest models and discusses maximum sustained yield and other harvest schemes.

A newer and more pressing form of wildlife management, aimed at preserving natural diversity, is the subject of Chapter 8. First it addresses the problem of endangered species, including causes of endangerment, the general management strategy, captive propagation, and some case studies. The chapter then shifts to an ecosystem approach, concentrating upon preserving representative natural communities rather than single species. It concludes with application of management for natural diversity under two quite different sets of conditions, urban-suburban wildlife and wildlife management in wilderness.

Four special management problems are featured in Chapter 9. Exotic introductions have been justified on the grounds of esthetics, economics, and improved hunting opportunities and, more recently, to provide new homes for endangered species. Most ecologists, however, cite disappointing and sometimes tragic consequences of exotic introductions and many governments have taken action to regulate importation of foreign wildlife. Predator control is an even more controversial topic, though a distinction should always be made between predator control aimed at protecting livestock and predator control intended to benefit game. The former results primarily from political and economic pressures and the latter could result from increased pressures for game harvests. Wildlife diseases are not so controversial as exotic introductions or predator control, but they are less well understood ecologically. They remain a special problem in part because of the risks posed by exotic diseases and in part because some modern wildlife management tends to concentrate wildlife, thereby increasing the risks and the consequences of diseases. Chapter 9 concludes with a review of the status of wetlands, extremely productive and important wildlife habitats that are being lost to agriculture and other land uses.

Chapter 10 reviews the crucial relationship between wildlife populations and changes in land use. As land use intensifies, the natural communities become simpler and populations of both game and nongame wildlife decline. The chapter examines land use trends on commercial forests, rangelands, and croplands. Recommendations regarding the integration of wildlife management into other forms of land use appear in Chapter 10, together with a discussion and update of "the land ethic" as proposed by Leopold.

The final chapter presents a global perspective of wildlife management. As our planet grows smaller, more populous, and more interdependent, the needs for international cooperation in wildlife management also increase. The chapter addresses the controversial Convention on International Trade in Endangered Species as the first major step. It then reviews the status of wildlife management in various parts of the world. The most important need in international wildlife conservation will be the establishment of adequate habitat reserves, particularly in the developing nations of the tropics. The chapter concludes with some suggestions regarding shared responsibilities for international wildlife management.

The losses of species and the general declines in wildlife diversity throughout the world are at present not encouraging. According to some informed projections, as many as 25 percent of the world's wild species will become extinct by the end of the twentieth century. Yet despite these trends and projections, there is nonetheless cause for optimism. Within the developed world, particularly the United States, human population growth has been substantially lower than projected even as recently as the 1960s. The technology that often looms menacingly over patches of wildlife habitat has brought new tools for conservation such as remote sensing, radio telemetry, and computer technology. The United Nations has increased its commitment to the world's wildlife through its UNESCO Biosphere Reserve Programme. Effective international treaties have been forged to regulate trade in endangered species and establish precedents for further agreements. Meanwhile private conservation organizations such as the National Wildlife Federation, The Nature Conservancy, National Audubon Society, the World Wildlife Fund, and the International Union for Conservation of Nature and Natural Resources continue to grow in both membership and effectiveness.

So the race goes on between conservation programs and the loss of wild species and natural communities. Under the very best conditions, not all the species alive today will be with us in the twenty-first century. But with diligent planning, education, hard work, and not just a little luck, a quite large diversity of wild organisms will be preserved and managed for centuries to come. I hope that this book proves useful in attaining such a goal.

Such a broadly based text would not have been possible without the dedicated assistance of many people. Dave Balph, John Kadlec, Tracy Carter, Robert Chambers, Bruce Coblentz, and Tom Townsend carefully reviewed entire drafts, making many useful suggestions. Others who helped with various outlines, portions of manuscripts, or both, include John Barclay, John Bissonette, Wayne King, Dave Kitchen, Tom Logan, Dale McCullough, Bart O'Gara, Robert Rolley, Scott Shalaway, Larry Talent, and Jim Teer. Tracy Carter also edited and typed the final manuscript. I thank them all for their help and accept final responsibility for any errors that may have slipped through.

James H. Shaw

CHAPTER 

PEOPLE AND WILDLIFE

PREHISTORIC PEOPLE AND WILDLIFE

Wildlife got along quite well without human intervention for millions of years. The decline of wild species in more recent times has been almost exclusively due to various human activities. Just as people are an integral part of the problem, so must they be a critical part of the solution. Wildlife managers often remark that the most difficult part of wildlife management is the management of people, not wildlife. Thus, an understanding of this delicate relationship between people and wildlife is fundamental to understanding wildlife management.

Prehistoric people subsisted as hunters and gatherers. Undoubtedly, game animals, including some very large species, provided important components of human diets. What effects did millennia of hunting by humans have upon wild species? The evidence is, and probably always will be, incomplete, so interpretations are bound to vary. One important controversy centering on the question of human effects on prehistoric wildlife concerns the sudden disappearance of so many species of large animals at or near the end of the Pleistocene epoch. Most paleontologists have suspected that abrupt changes in climate led to the mass extinctions. Paul Martin and his associates, however, have concluded that prehistoric people drove many of those species to extinction through overhunting. In their "Pleistocene overkill hypothesis" they cite what seems to be a remarkable coincidence between the arrival of prehistoric peoples in Europe, parts of Asia, and North and South America and the time during which mammoths (*Elephas* spp.), giant ground sloths (*Megatherium americanum*), the giant bison (*Bison latifrons*), and numerous other large mammals became extinct. Perhaps our species was driving others to extinction long before the dawn of history (Martin and Wright, 1967).

Hunter-gatherers may have contributed to Pleistocene extinctions in more indirect ways. Besides overhunting, at least three other kinds of effects have been suggested: direct competition, imbalances between competing game species, and early agricultural practices (Krantz, 1970). Direct competition may have brought about the demise of large carnivores such as the saber-toothed cats (*Machairodus* and *Smilodon*). These animals simply may have been unable to compete with the more sophisticated hunting skills of Pleistocene people.

Human hunters could have caused competitive imbalances among game species, leading to the extinctions of species less able to compete. When other predators such as the gray wolf (*Canis lupus*) prey upon large mammals, they generally take high proportions of each year's crop of young. Human hunters, in contrast, tend to take the various age classes of large mammals in proportion to their actual occurrence. If humans first outcompeted many of the larger predators and then replaced them, they may have allowed more young to survive each year, gradually increasing the populations of favored game. As these populations expanded, they may have outcompeted other species, forcing the latter into extinction (Krantz, 1970). This theory suggests that human hunters played an indirect role in the Pleistocene extinctions by actually favoring the hunted species over those not hunted.

Agriculture developed during the Neolithic period of the Pleistocene epoch, some 10,000 years ago. Farmers of this period cleared forests, drained swamps, and implemented irrigation. The natural plant communities that had once supported wide ranges of animal species were simplified and were rendered capable of supporting far fewer wild species. Humans may have been causing extinctions through habitat changes as early as the Neolithic period.

Human hunters were probably modifying wildlife habitat even before the Neolithic. They presumably made widespread use of fire, first to drive game toward waiting hunters, and later to attract game back into a burned area in search of rich, new plant growth. In prehistoric times, people lacked the ability to start fires quickly and reliably, so they kept glowing embers as fire starters. Such fire starters could have led to frequent and widespread accidental fires, significantly altering wildlife habitat, especially in drier regions (Eckholm, 1976).

HISTORIC PEOPLE AND WILDLIFE

As humans gradually shifted from hunter-gatherers to agrarians, so too did their impact on wildlife. Agriculture required a more sedentary way of life, one in which people grew more possessive of the land; wild animals were regarded as intruders and competitors upon this land. As agriculture grew, forests and other natural areas shrunk. One United Nations study in 1963 estimated that between one-third and one-half of the world's forests had been destroyed and that the principal cause of this destruction was land clearing for agriculture (Eckholm, 1976). More recent reports (Ehrlich et al., 1977) place the worldwide deforestation figures at close to 50 percent, with more occurring each year. Along with agriculture came domestication. Early farmers confined a few species of birds and mammals, raising them for food, for use in religious rites, as pets, and to provide ornaments and other products. At first these animals were allowed to breed freely with their wild counterparts. Gradually, breeding among these confined animals was limited to those animals providing more or "better" products. Over the centuries, this process of domestication took its toll on wildlife. Artificial selective breeding requires that undesirable subjects be barred from reproducing, and in the early stages of domestication wild stock was deemed undesirable. Thus many of the wild counterparts of these early domesticates were slaughtered or simply driven away, a practice that may explain why so many progenitors of domestic species, such as the horse, the camel, and the cow, became extinct. In addition, many of the grazing or browsing domesticated animals were kept at densities much higher than their wild ancestors could ever have achieved under natural conditions. Excessive grazing or browsing reduced the natural vegetation and altered the environment, making it even less suitable for those wild animals that were left.

As selective breeding caused domesticated animals to become more desirable to their owners, it also contributed to the animals' increasing dependence. One way in which they became more dependent was for protection against wild predators. Thus, one of the earliest forms of wildlife management was predator control, which generally meant eradication whenever possible. The larger predatory species such as wolves, lions (*Panthera leo*), and brown bears (*Ursus arctos*), which had managed to survive the Pleistocene, declined or disappeared wherever agriculture and animal husbandry developed.

Humanity continues its pressures upon wild nature, fed by the sheer increase in human numbers. As of 1975 there were about eight times as many people on earth as there were in 1650 (see Figure 1-1). If present growth rates continue, the human population will double again by the early

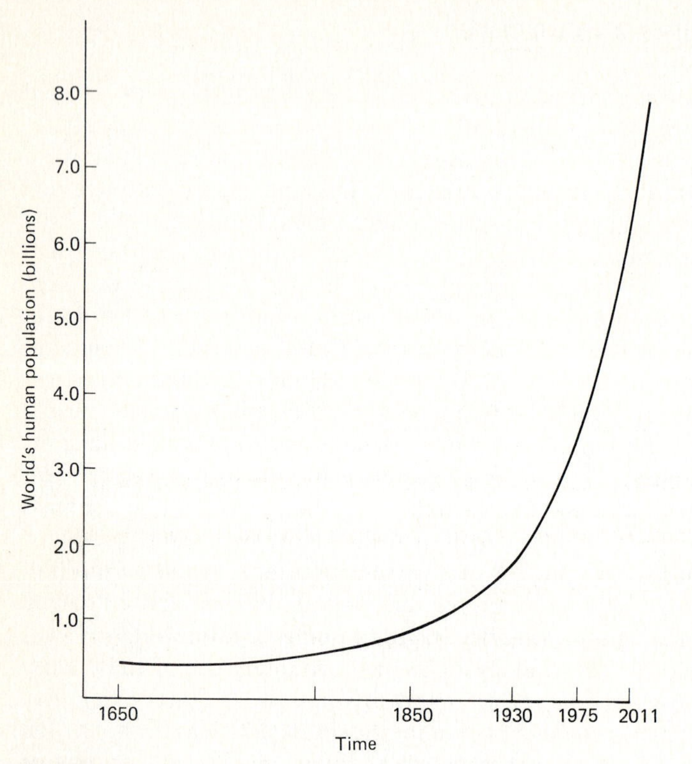

FIGURE 1-1
Human population growth from 1650 to 1975 and projected through 2011. Note that each point on the curve represents a doubling from the previous point. It took 200 years for the population to grow from 0.5 to 1.0 billion, 80 years to expand from 1.0 to 2.0 billion, and only 45 years to grow from 2.0 to 4.0 billion (Ehrlich et al., 1977, by permission W. H. Freeman & Co.).

twenty-first century. Along with rising human populations comes more stress on land to produce food, shelter, and other necessities. People also aspire for higher standards of living, resulting in increased demands for energy, minerals, and other resources. All these forces combine to advocate further reductions in wildlife habitat.

HISTORY OF WILDLIFE MANAGEMENT IN THE UNITED STATES

Wildlife management in the United States and Canada had roots in English common law. The king and his nobles claimed formal legal ownership of game and exclusive hunting rights until the signing of the

Magna Charta in A.D. 1215. That document transferred ownership of wildlife from the crown to the people. Since then wildlife has remained public property tended by the government as a "sacred trust."

This custom of public ownership crossed the Atlantic to the eastern shores of North America, where colonists found what must have seemed an inexhaustible bounty of wildlife and other natural resources. The fate of that bounty is summarized in the next few pages. Dates of the various "eras" are only approximations designed to help organize material according to attitudes, conditions, and management strategies (see Table 1-1).

TABLE 1-1
IMPORTANT EVENTS IN AMERICAN WILDLIFE CONSERVATION

Year	Event
Era of abundance (1600–1849):	
1630	Massachusetts Bay Colony established the first bounty system for large predators (wolves at one penny each).
1646	The first closed hunting season on deer in the Portsmouth, Rhode Island, Colony.
1708	The first closed seasons on birds; in some New York counties heath hens, grouse, quail, and turkey were protected.
1848	Rhode Island passed laws against spring hunting of migratory game birds (later repealed).
Era of exploitation (1850–1899):	
1852	The first salaried game wardens were established in Maine.
1864	New York became the first state to require hunting licenses.
1872	Yellowstone became the world's first national park.
1878	The first state game and fish commission was established in California.
1878	Iowa was the first state to establish bag limits on game birds (25 prairie chickens per day).
1896	The Supreme Court ruled in *Geer v. Connecticut* that wildlife is the property of the state, not the landowner.
Era of protection (1900–1929):	
1900	The Lacey Act curbed market hunting and limited importation of exotic wildlife.
1903	Pelican Island, Florida, became the first federal wildlife refuge.
1913	The Weeks-McLean Act established federal control over migratory birds and ended spring waterfowl hunting.
1915	The first federal funds were appropriated for predator control.
1916	Management cooperation with Canada was established through the Migratory Bird Treaty Act between the United States and Great Britain.
Era of game management (1930–1965):	
1930	Report of the Committee on North American Game Policy was published.
1933	Aldo Leopold published *Game Management*.
1934	The Federal Bird Hunting Stamp Act required waterfowl hunters to purchase a duck stamp, with proceeds devoted to wetland preservation.
	The Taylor Grazing Act regulated livestock grazing on federal rangelands.
1935	The first Cooperative Wildlife Research Units were established at state universities to provide graduate training in wildlife conservation.
1936	The first North American Wildlife Conference was held.

TABLE 1-1
IMPORTANT EVENTS IN AMERICAN WILDLIFE CONSERVATION

1937	The Federal Aid to Wildlife Restoration Act (Pittman-Robertson Act) placed a federal excise tax on sporting arms and ammunition, with proceeds to be reapportioned to states for wildlife research and restoration.
	The Migratory Bird Treaty Act was amended to include Mexico.
1950	The Dingell-Johnson Act established a federal excise tax on sport fishing equipment, with proceeds to be reapportioned to states for fisheries projects.
1958	The Outdoor Recreation Resources Review Committee (ORRRC) was created to inventory America's outdoor recreation needs and make management recommendations.
1960	The Multiple Use Act revised management policy on national forests.
1961	The Wetlands Loan Act passed, authorizing Congress to appropriate $105 million to purchase critical waterfowl habitat. The loan was to be repaid from future sales of duck stamps.
1964	The Land and Waters Conservation Act authorized federal agencies to charge recreation use fees and established excise tax on motor boat fuels.
	The wilderness bill established a National Wilderness Preservation System on selected federal lands.
Era of environment management (1966–1984):	
1966	The Endangered Species Act authorized the Secretary of the Interior to maintain a list of native endangered species.
1969	The National Environmental Policy Act (NEPA) established the Council on Environmental Quality (CEQ).
	The Endangered Species Act curbed importation into the United States of species (or parts thereof) threatened abroad.
1970	The Environmental Protection Agency (EPA) was established as an independent federal regulatory agency.
1973	The Endangered Species Act superceded earlier legislation and provided federal protection to native endangered species.
1976	The National Forest Management Act regulated clear-cutting on National Forests.
	Congress extended the 1961 Wetlands Loan Act and authorized up to $200 million to purchase waterfowl habitat.
	The Bureau of Land Management Organic Act facilitated resource management on 182 million hectares (450 million acres) of federal land.
1978	The Cooperative Training Agreement was signed between United States and Mexico to provide cooperative wildlife management training.
	The Endangered Species Act of 1973 was amended to establish a committee to review cases of conflict between development projects and critical habitat.
1980	The Nongame Wildlife Act passed Congress to provide $20 million for 4 years to aid state wildlife agencies in developing programs for the benefit of wildlife species not normally hunted.
	The Alaska National Interest Lands Act established federal protection in the forms of National Parks, National Monuments, and National Wildlife Refuges to more than 40.5 million hectares (101 million acres) of land in Alaska.
1982	Congress passed a 3-year extension of the 1973 Endangered Species Act.

Era of Abundance (1600–1849)

Wherever settlers established farms, homes, and villages, local wildlife populations changed. A few species benefited, but many declined. There was little interest in these changes, aside from active efforts to eradicate predators to protect livestock. Local exterminations aroused no concern, so long as there were wild places to the west presumed to be full of wild animals. The notion that entire species could permanently disappear was not generally accepted during this era. Even Thomas Jefferson believed a trapper's account of a mammoth heard trumpeting in the Virginia woods (Eiseley, 1957). In general, wild animals were treated as though they would always remain abundant somewhere and never become extinct.

Wildlife diversity was probably quite high prior to settlement, and people then presumed that wild animals were abundant in unsettled areas. For some larger animals, including bison (*Bison bison*), grizzly bear (*Ursus arctos*), elk (*Cervus elaphus*), and pronghorn (*Antilocapra americana*), along with ducks, geese, cranes, and eagles, abundance was quite high. Many of today's more popular game species, however, including white-tailed deer and bobwhite quail, were originally either rare or absent from regions in which they became abundant following settlement.

Predatory mammals were persecuted in North America just as they had been in England because of the threat they posed to livestock. Thus, the earliest record of wildlife management in the colonies was the payment of bounties of one penny each on wolves killed in the Massachusetts Bay Colony beginning in 1630.

There were a few efforts to boost local game populations through legal protection. The first was a closed hunting season on white-tailed deer around the Portsmouth, Rhode Island, Colony in 1646, followed by similar action from the Massachusetts Bay Colony in 1718. Some New York counties closed seasons on upland game birds as early as 1708 (Trefethen, 1975). Modest by modern standards, these protective efforts did little to offset wildlife declines and were completely insufficient to stem the pressures brought about during the second half of the nineteenth century.

Era of Overexploitation (1850–1899)

Perhaps the destruction of the vast buffalo or bison herds best exemplifies this era of conservation history. In 1850, tens of millions of bison roamed the Great Plains in herds exceeding those known for any other mammal ever. By the century's close, the bison was all but extinct. Market hunting and hunting just for sport had flourished before 1850, but these activities increased greatly during the late 1800s. With the completion of the railroad came people who shot bison from the windows of the trains with no thought of ever using the carcasses. The decline in game abundance was accelerated because settlement and technological improvements (such as repeating firearms and more efficient rail transportation) greatly increased without corresponding increases in conservation measures.

Most conservation attempts of the time were, like those of the previous era, designed to curb harvest of game species. In 1852 Maine became the first state to employ salaried game wardens. Twelve years later New York made conservation history by implementing the first state hunting license requirement. The first daily bag limits restricted Iowa hunters to what must have seemed an austere 25 prairie chickens (*Tympanachus cupido*) a day in 1878.

A major step in American conservation occurred in 1872, when Yellowstone became the world's first national park. Although never intended exclusively for wildlife, the park became our first federal preserve and as such benefited wildlife by eliminating local hunting and preserving natural habitat.

Era of Protection (1900–1929)

The most obvious cause of American wildlife's precipitous decline in the late nineteenth century was direct overexploitation. The most obvious remedy was direct legal protection. Thus legal protection became the primary means of wildlife conservation during the first three decades of the twentieth century. The era began with the Lacey Act, which helped curb market hunting by making interstate transportation of illegally killed game a federal offense. It also began formal regulation of imported (exotic) wildlife species. Federal protection for migratory waterfowl began in 1913 with the Weeks-McLean Act, followed five years later by the Migratory Bird Treaty Act, which established formal cooperation between the United States and Canada.

During these years, state game and fish departments emerged to assume jurisdiction over what was, after all, state-owned resident wildlife. Commissions, composed of interested laypeople appointed by state governors, were created to direct departmental policy and oversee spending. State game and fish departments received most of their funding through sales of hunting and fishing licenses and used these funds to hire wardens to enforce state laws.

The generally negative attitude toward predators led to one noteworthy departure from the era's protectionist theme. In 1915 the first federal funds became available for the destruction of wildlife deemed noxious to agriculture and animal husbandry.

Era of Game Management (1930–1965)

Although legal protection is an important part of wildlife conservation, alone it is insufficient. By 1930 it was evident that biological research into the habits and habitat requirements of wild animals was generally lacking and that federal-state agency cooperation needed improvement. These were some of the major points emphasized by the Committee on North American Game Policy. Three years later the Committee's chair, Aldo Leopold, published his classic *Game Management* (Leopold, 1933a), an event that signaled the birth of professional wildlife conservation.

Wildlife conservation of that era was largely confined to game animals. This emphasis stemmed from conservation leadership by Americans active in hunting for sport. Their license fees provided some of the first (and often the only) financial support for conservation programs. To the legal protection of previous eras were added programs with emphasis on restoration of natural food and cover as well as on understanding population fluctuations in relation to environmental conditions.

The era's most important achievements were in increased public funding and more effective conservation administration. In 1934 the Duck Stamp Act required all American waterfowl hunters to purchase a federal stamp annually. Proceeds have been used ever since to secure critical wetlands for breeding, migration stopover, and wintering of waterfowl. Perhaps the single important event of the era was the 1937 passage of the Pittman-Robertson Act. This act placed a 10 percent excise tax on the manufacture of sporting arms and ammunition. Funds were reapportioned to state fish and game agencies on a matching basis for use in wildlife research and restoration. The Pittman-Robertson Act proved to be such a reliable fund raiser that it was followed 13 years later by the Dingell-Johnson Act, similar legislation to generate funds for fisheries projects.

Other federal legislation directed management policy. The Multiple Use Act of 1960 required National Forest policy to include forms of land use besides timber management. Another federal excise tax, this time on motor boat fuels and offshore oil drilling, resulted from the Land and Waters Conservation Act to fund outdoor recreation programs. A National Wilderness Preservation System was established in 1964 with passage of the Wilderness Act.

Era of Environmental Management (1966–1984)

Just as legal protection alone proved inadequate for game management, game management proved inadequate to assure perpetuation of the vast majority of wild vertebrates not ordinarily considered game species. This era arrived in the late 1960s, beginning with the 1966 Endangered Species

Act, a mild piece of legislation authorizing the Secretary of the Interior to maintain a list of rare and endangered species. No legal protection or federal aid resulted from the 1966 act, but it began a commitment that would grow and flourish in the decade to come. A few legal teeth appeared in the 1969 Endangered Species Act, prohibiting importation of foreign endangered species or parts thereof. That same year Congress passed the National Environmental Policy Act (NEPA), establishing the Council on Environmental Quality (CEQ) and requiring environmental impact statements for federally assisted projects likely to have environmental effects. The Environmental Protection Agency (EPA), an independent federal agency, came into being in 1970 when Congress authorized a merger and expansion of several previously separate air- and water-quality agencies. A third Endangered Species Act granted federal protection to native endangered species in 1973. Meanwhile, state fish and game departments began broadening their programs. In keeping with their new responsibilities, many changed their names to "wildlife" departments.

The era saw more attempts to improve the performance of resource management agencies. The Bureau of Land Management (BLM) received a boost in 1976 with passage of an organic act authorizing expanded resource conservation on the agency's 182 million hectares (450 million acres) of public land. An agreement to provide cooperative training for wildlife managers was signed in 1978 between wildlife agencies of the United States and Mexico. The Endangered Species Act of 1973 was amended in 1978, establishing a committee to review conflicts between development projects and endangered species.

Each era retained the principal elements of those before it (Figure 1-2). The first closed seasons of the era of abundance led to more extensive state protection and institution of bag limits during the era of overexploitation. Federal protective measures were added in the era of protection. The era of game management introduced professional training and expanded funding sources. Finally, concern over deteriorating environments and accompanying declines in nongame species resulted in broader wildlife management programs during the era of environmental awareness.

Notable Figures

Wildlife management in the United States has been influenced by many people, the most notable of whom are described here in chronological order. The first two were also closely associated with forestry. Wildlife management, a younger profession, inherited forestry's principal philosophical division.

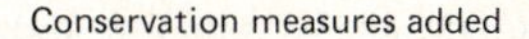

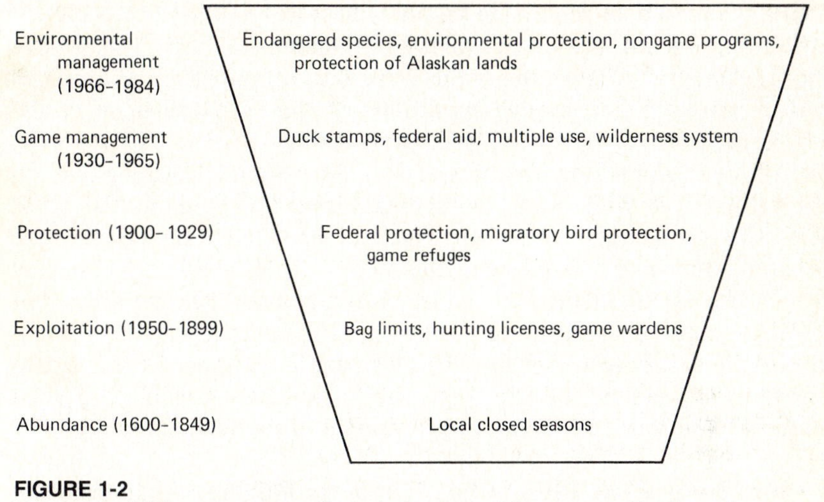

FIGURE 1-2
The major wildlife management efforts during each of the conservation "eras."

John Muir (1838–1914) A native of Scotland, John Muir immigrated to the United States with his family in 1849. Muir had considerable talent as an inventor and might have had a long and profitable career in industry had he not suffered a freak accident that left him temporarily blinded. This loss of sight caused Muir to reconsider his personal values and goals and, after his vision returned, he embarked on a vastly different way of life. Amid the social disorder following the Civil War, Muir walked alone and unarmed from Indianapolis to Florida. The next year, 1868, found Muir in California, where he took quickly to the Sierra Nevadas and the mountaineering life with which he was to become forever associated. During Muir's time in the mountains, California was experiencing serious destruction of its natural resources; Muir's concern for their protection led him in 1892 to form the Sierra Club, now a leading conservation organization. Muir is best remembered as the founder of the preservationist school of conservation. Preservationists maintain that the most effective conservation strategy is complete protection of the resource, be it timber or wildlife, from exploitation.

Gifford Pinchot (1865–1946) Educated in European forestry, Gifford Pinchot is credited with having brought professional forest management to America. In the 1890s American forestry was limited largely to complete protection of governmental reserves, sometimes enforced by the military. The only alternatives seemed to be unregulated exploitation or complete protection. Pinchot was, in the words of Trefethen (1975:98), "the Moses who led American forestry out of this wilderness of confusion." He did this by demonstrating that through application of proper management principles, forests could be harvested indefinitely without wholesale destruction. Pinchot founded the utilitarian school of conservation, advocating that the best means of assuring future resources is through wise use on a sustained yield basis. An energetic organizer, Pinchot helped found the first forestry school in America, at Yale University in 1899. He established the Society of American Foresters the following year and became the first director of the Department of Agriculture's Division of Forestry. He even invented the word "conservation" in 1907. But his most important contribution to conservation politics came about through his association with a big-game hunter who became the twenty-fifth President of the United States.

Theodore Roosevelt (1858–1919) The twenty-fifth President is probably best remembered for his "big stick" diplomacy. But if Roosevelt had not had such a consuming interest in natural resources in general and wildlife in particular, wildlife management in America would have followed a far less effective course. Thrust into office with McKinley's assassination in 1901, Roosevelt quickly took measures to ensure vigorous enforcement of the Lacey Act. Two years later, he signed an executive order making Pelican Island, Florida, a federal bird reservation. This action established a precedent that eventually gave rise to a National Wildlife Refuge System. Roosevelt helped Pinchot strengthen the staff of the young Division of Forestry and set aside another 59 million hectares (148 million acres) of new national forest. The number of national parks doubled under his administration. Roosevelt convinced Congress to pass the Monuments and Antiquities Act in 1906, allowing the President to create new national monuments by proclamation. Before his term of office was over, Roosevelt had used this authority to establish 23 national monuments, including the Grand Canyon. Thus the Roosevelt administration bolstered wildlife conservation through improved federal law enforcement, a National Wildlife Refuge System, more national parks, and a greatly expanded system of national forests (Clarke, 1979; Trefethen, 1975).

Aldo Leopold (1887–1948) A professional forester, Aldo Leopold is widely regarded as the founder of American wildlife management. After graduating from the Yale School of Forestry in 1909, Leopold joined the U.S. Forest Service in New Mexico and there developed an interest in game species of the southwest. This interest led him to integrate ecological principles into practical methods of aiding game populations, culminating in his textbook, *Game Management,* in 1933.

More than 30 years later, the principles presented in *Game Management* were carefully reevaluated by Kelker (1964), who found them still applicable and enjoying widespread acceptance by professional wildlife biologists. The same year that he published his textbook, Leopold published an article entitled "The Conservation Ethic" in the *Journal of Forestry* (Leopold, 1933b); republished several times since, this article remains a classic.

Soon after publishing his text Leopold visited Germany to learn that country's intensive methods of forestry and wildlife management. The trip was a turning point in his thinking because Leopold realized that artificial methods of intensive management required sacrifice of important esthetic values (Flader, 1974).

Leopold chaired the committee on North American Game Policy and became the nation's first professor of game management at the University of Wisconsin. His graduate students went on to become leading wildlife professionals. Leopold's best-known work is *A Sand County Almanac,* published posthumously in 1949. This timeless classic is a beautifully written statement of Leopold's dedication and vision.

Aldo Leopold left more than just a written legacy. Of his five children, all became naturalists of one kind or another. Four earned Ph.D.s, and three have been elected to the National Academy of Sciences, an achievement unequalled by any other family in American history (Carter, 1980).

J. N. "Ding" Darling (1876–1962) A political cartoonist from Iowa, Ding Darling served along with Aldo Leopold on the waterfowl restoration committee in 1934. That same year President Franklin Roosevelt appointed him director of the Bureau of Biological Survey, the forerunner of today's U.S. Fish and Wildlife Service. Darling quickly gained public support for waterfowl restoration. When the bureau needed funds to purchase wetland areas for waterfowl, Darling came up with the idea of a duck stamp and persuaded Congress to pass the Duck Stamp Act. He even contributed the artwork for the first stamp himself.

Darling was the driving force behind development of the cooperative wildlife research units, first in his native state of Iowa and later as a network of units, primarily at land-grant colleges throughout the United States. Through his energies, the Wildlife Management Institute, the National Wildlife Federation, and the North American Wildlife and National Resources Conference all came into being (Trefethen, 1975).

VALUES OF WILDLIFE

Most of today's endangered species reached their precarious positions because of human activity of one form or another. Do we have an ethical obligation to guard them from extinction? What good are they anyway? If we exploit game animals, have we then any moral responsibilities to ensure their perpetuation for future human generations? Ethics and morals are, of course, constructions of society (Klein, 1973). All of which brings up the important matter of values. If wildlife had no value, its continued existence would largely be limited to those species whose life history needs happened to be compatible with human activities.

Wildlife management is carried on through the application of biological principles toward maintaining, increasing, or reducing wildlife abundance according to specific management objectives. But such applications cost money and quite often their successes are achieved only by sacrificing or modifying land uses that would, in the short run anyway, yield greater profits. Wildlife values, both economic and otherwise, are essential to justify wildlife management. So before delving into the "hows" of wildlife management, it is necessary to understand the "whys."

Wildlife values have been classified in several ways. Ralph King (1947) recognized six overlapping classes that have been widely accepted (cf. Steinhoff, 1978), especially for game species. Describing the values for what he termed "nonresource" wildlife, David Ehrenfeld (1976) elaborated on more subtle though equally important values. More recently, Norman Myers (1979, 1983) made an excellent case for utilitarian values, a category often granted too little importance by conservationists. The classification used here is based upon King's, with modifications from Ehrenfeld and Myers. Some negative values, including accidents, crop and livestock depredations, and diseases, are also mentioned.

Positive Values

Esthetic Values Esthetic values are those relating to inherent natural beauty and artistic appreciation. Inevitably, these values overlap with those in other categories since virtually all animals have some esthetic values. Esthetics are difficult to measure or compare, but they are critically important and play major roles in conservation movements such as the one during the late 1960s and early 1970s. Most Americans wlll never see a polar bear (*Thalarctos maritimus*), a resplendent quetzal (*Pharomachrus mocinno*), or a markhor (*Capra falconeri*) in the wild. Yet to many of these people there is considerable value in merely knowing that they still exist free from human dependency somewhere. These "irrational" values may indeed be the most popular and widespread category.

Recreational Values Values relating to sports and hobbies are termed recreational values, and are often subdivided into consumptive and nonconsumptive categories. Hunting is the oldest and most common form of consumptive recreation. In 1975 over 20 million hunters in the United States spent nearly half a billion days afield in pursuit of game. Since the late 1960s, though, the most rapid increases in recreational values of wildlife have been in nonconsumptive forms. Nearly 50 million Americans spent more than 1.5 billion days observing birds and other wildlife and 15 million photographers passed 160 million days snapping pictures of wildlife during 1975 (U.S. Fish and Wildlife Service 1977). Unlike hunting, these nonconsumptive activities require little regulation and will accommodate an almost unlimited number of participants.

Ecological Values Every species plays its own role within the larger biotic community. Beavers (*Castor canadensis*), for example, build dams to form ponds that in turn stabilize water flow. Hawks reduce small mammal populations by daylight and owls provide similar services after sundown. Such activities can help maintain stability and sustain natural diversity.

Other ecological values are more subtle than those of beavers, hawks, or owls and frequently they are less well understood. Leopold (1953) declared, "To keep every cog and wheel is the first precaution of intelligent tinkering." Through this analogy he advocated retention of all wild species on the grounds that, although we cannot completely understand their function, neither can we predict the long-range consequences of their disappearance. Wild species should be conserved to avoid irreversible ecological changes.

Educational and Scientific Values Values that serve in teaching and learning about wildlife belong in this category. Examples include population characteristics, habitat requirements, and social organization of wild animals. With their greater popular appeal, larger species of birds and mammals may serve more effectively to illustrate scientific principles. Social behavior of the gray wolf is more likely to capture public attention than will the social behavior of the snail. Population dynamics of ringneck pheasants (*Phasianus colchicus*) mean more to most people than do those same phenomena in fruit flies (*Drosophila melanogaster*).

Only a small fraction of the world's species of wild vertebrates have been thoroughly studied under natural conditions. Detailed field investigations can add enormously to our knowledge of life on earth, but in the future such studies will be possible only if adequate measures are taken to ensure preservation of these species in their native habitats.

Utilitarian Values Perhaps the most important values of wildlife are the utilitarian or practical ones. As our breeds of livestock become genetically more specialized through selective breeding, they become suited for an increasingly narrow range of environmental conditions. Whenever such conditions change, new bloodlines must be found. Wild sheep (*Ovis* spp.), for example, are interfertile with domestic sheep, and wild goats (*Capra* spp.) can interbreed with their domestic counterparts. Sometimes new genetic material can be found in what are recognized as different genera. Cattle (*Bos taurus*) have been successfully interbred with American bison. The offspring, when backcrossed with cattle, produce meat with higher protein and lower fat than meat from purebred cattle. These "cattalo" or "beefalo," as they are called, subsist on rougher forage, too.

Wild animals also have important medical uses. Several species of wild primates including the rhesus monkey (*Macaca mulatta*) and the chimpanzee (*Pan troglodytes*) have long played important roles in biomedical research. A small South American monkey, the cotton-topped marmoset (*Saguinus oedipus*), is being used to develop a potent vaccine against lymphatic cancer (Myers, 1976, 1983). The nine-banded, or common long-nosed, armadillo (*Dasypus novemcinctus*) is playing an important role in research against leprosy (*Mycobacterium leprae*) because it can be infected with this ancient scourge.

Larger mammals also offer excellent opportunities for biomedical research. The African elephant (*Loxodonta africana*) may serve as a model for research on atherosclerosis and thrombosis, and the black bear (*Ursus americanus*) is playing a similar role in studies of kidney disease (Myers, 1983). The American bison is highly resistant to all forms of cancer and some evidence suggests that people who eat bison show increased cancer resistance (Myers, 1983; Rorabacher, 1970). Perhaps the bison, after its narrow escape from extinction, will help medical science understand the causes of cancer and its effective treatment.

Even venomous snakes are important. Venom from various new world vipers is being used, with modifications, as nonaddictive painkillers and in the treatment of hypertension and thrombosis (Myers, 1983). Ironically snake venoms may soon save far more human lives than they claim.

Another practical value provided by wild species is service as environmental indicators to warn of contamination. The best-known examples are predatory birds such as the osprey (*Pandion haliatus*) and the peregrine falcon (*Falco peregrinus*), whose population declines helped call attention to the buildup of DDT and related persistent pesticides in the environment.

Commercial Values Commercial values are those in which wildlife yields economic returns. Furbearing species in North America and other parts of the world serve as a controversial, but valid, example of direct commercial values derived from wildlife. Commercial game ranching in parts of Africa, though still plagued with some practical problems, offers promise as a means of producing protein for profit. Hunting leases in Europe and parts of North America provide less direct commercial returns but yield good income with low overhead.

A less fortunate example of commercial value is the flourishing international trade in wild animals as pets. Although such practices are undeniably profitable to animal dealers and traffickers, they can pose serious threats to species survival. Ten wild animals may die in capture or transport for every one offered for sale in a pet store. Ironically, many of the more expensive and most sought after wild animals, such as most species of monkeys, prove to be troublesome and destructive pets.

Negative Values

Accidents Every year there are numerous collisions between large mammals and automobiles, resulting in injury and property damage. A very few wild animal species such as the grizzly bear occasionally pose direct threats to human safety. Although the headlines reporting such rare tragedies usually refer to them as "unprovoked," they ordinarily should be regarded as accidents. Their occurrence remains quite low and the application of new studies of the behavior of such species should allow managers to reduce risks even further.

A less dramatic, but even more deadly form of accident can be brought about by flocks of birds. They have been sucked into jet aircraft engines in sufficient numbers to cause the engines to stall, with fatal results for all. Sixty-two people died in a Boston air crash in 1960 after starlings (*Sturnus vulgaris*) entered three of the craft's four engines (Hawthorne, 1980).

Crop and Livestock Damage Livestock and crop losses to wildlife amount to millions of dollars annually. Coyotes (*Canis latrans*) kill sheep, perhaps not so often as some who work among the stock claim, but frequently enough to cause serious losses to some owners. Ducks, geese, and sandhill cranes (*Grus canadensis*) damage an estimated $25 million worth of grain crops annually (Schwilling, 1975), and weaver birds (*Quelia* spp.) inflict severe crop depredation in Africa. Rats damage crops throughout most of the world both in the field and in storage facilities.

Disease Reservoirs Wildlife can act as reservoirs of diseases transmissible to humans and domestic animals. Serious zoonotic diseases (those transmissible from wild animals to people) include rabies, anthrax, and tularemia. Some evidence suggests that brucellosis, found in many wild mammals, may be transferable to livestock. Sleeping sickness, carried from wild ungulates to livestock by the tsetse fly (*Glossina* spp.), makes parts of Africa uninhabitable for cattle and leads to severe conflict between people and wildlife. Hoof-and-mouth disease, now more common in Europe and South America than in North America, can be transferred from wild animals to livestock.

The transfer of disease between wildlife and livestock works both ways, with the most severe problems taking place when either livestock or wildlife is introduced into a new environment, bringing with it the disease. Rinderpest, a viral disease, arrived in Somalia with domestic cattle in 1889. Through the next decade it swept through much of Africa, severely reducing populations of Cape Buffalo (*Syncerus caffer*) for decades thereafter (Sinclair, 1977).

Measuring Wildlife Values

Wildlife managers often need measurements of their resource's value, particularly when confronted with yet more habitat losses in favor of what appear to be more profitable forms of land use. In 1955, the U.S. Fish and Wildlife Service began collecting basic data on numbers of hunting licenses sold, number of hunter-days afield, and similar activities at five-year intervals. The 1975 survey was expanded to measure various nonconsumptive activities such as wildlife observation and wildlife photography (U.S. Fish and Wildlife Service, 1977). The 1970s brought an infusion of sociological and economic survey techniques, providing wildlife managers with more sensitive and sophisticated tools. Most of these tools fall into one of two broad categories: economic surveys and attitude surveys.

Economic Surveys Aside from the small proportion of species such as furbearers, few wild animals yield direct economic returns. Most economic benefits accruing from wildlife are indirect, and surveys are commonly undertaken to measure them. Hunters, bird-watchers, and nature photographers spend money on travel, lodging, meals, and equipment in pursuit of wildlife, expenditures important both to the economy as a whole and to local communities in particular. To measure these expenses, studies usually employ some form of questionnaire surveys and then use the results to illustrate economic benefits of wildlife. One important variation of this approach determines "consumer surplus," or the difference between what people actually spend and the amount they are willing to spend (Charbonneau and Hay, 1978; Steinhoff, 1978). Other economic surveys have measured wildlife values through assessment of alternative opportunities lost, diseconomies (costs sacrificed for alternative values), and the cost of wildlife management.

A common economic survey, the cost-benefit analysis, has been applied to evaluate a particular wildlife area in Alaska (Steinhoff, 1971). Martin and Gum (1978) compared indirect economic values of cattle ranching, hunting, and general rural outdoor recreation in seven regions of Arizona. Guidelines in selecting economic tools for use in wildlife-related matters have been summarized by Bart et al. (1979).

Attitude Surveys Conservation programs require public support for both funding and implementation. This support can be gained or lost because of public attitudes toward wildlife management. Measuring public attitudes is thus an important step in wildlife management. If unfavorable attitudes are detected for a particular management program, it may have to be modified, delayed, or even abandoned, no matter how sound its biological basis. In time, unfavorable attitudes may be changed through patience and careful education.

Attitude surveys are most often measured through systematic questionnaires directed either to particular interest groups or to the general public. One example of the former is a survey of Maryland deer hunters to determine their attitudes toward crowding in a public hunting area (Kennedy, 1974). A survey directed at the general public measured the relationship between knowledge of wildlife management principles and attitudes toward conservation issues (Dahlgren et al., 1977). Another survey devised a typology toward animals (Kellert, 1976), for use in more detailed and specific surveys (Kellert, 1978, 1980). When sport hunting became increasingly controversial in the 1970s, several surveys (Applegate, 1979; Kellert, 1978; Rohfling, 1978; Shaw and Gilbert, 1974) measured attitudes on that issue.

THE LEGAL STATUS OF WILDLIFE

Under English common law, wild animals living under natural conditions are considered to be public property. Their legal status is therefore determined by public resource management agencies charged with enforcement of wildlife laws. These agencies have authority to grant a wild species either complete protection throughout their jurisdiction or partial protection during certain seasons, within certain areas, or both. Many songbirds and birds of prey are granted complete, year-round legal protection. Game and furbearing species receive partial protection and may legally be taken only by licensed persons within seasons and in numbers prescribed by law. Many species of wild vertebrates, including most snakes, amphibians, and small mammals, are afforded *no legal protection* and may legally be taken by anyone at any time. The more numerous and unpopular of these unprotected species are called "vermin" or "varmints."

Legal protection emerged as a conservation tool in the United States during a period when direct, deliberate killing was thought to pose the greatest threat to wildlife. Although they are still important, these laws providing direct protection are alone quite inadequate to ensure the survival of wild species. Strict enforcement of protection laws has reduced or eradicated market hunting throughout the developed world. But these measures do nothing to control indirect destruction of wildlife through changes in land use that destroy natural habitat. So subtle is habitat destruction that it rarely meets with public outcry. Wildlife, as public property, repeatedly becomes subjugated to development and exploitation from both public and private sectors.

Private landowners cannot legally take game out of season even on their own property. But they can alter land use practices that in turn affect wildlife populations far more profoundly and permanently than the mere taking of a few out-of-season animals. Private landowners can affect wildlife in yet another way, through limiting access to their lands, an important consideration, addressed in Chapter 10.

State and Federal Jurisdiction

In the United States, states have primary legal authority over resident wildlife. This state jurisdiction was tested in a U.S. Supreme Court decision in 1896. A Connecticut state court convicted a market hunter named Geer of possessing game birds with the intent of shipping them out of the state. Although he took the birds legally, Geer's plans to transport them to another state violated Connecticut law. Geer and his attorneys appealed his conviction on the grounds that the state law interfered with congressional authority to regulate interstate commerce. The Supreme Court ruled in favor of Connecticut, citing the state's "right to control and regulate the common property of game" (Bean, 1977). This in essence confirmed state jurisdiction over resident wildlife.

Migratory birds, of course, regularly cross state lines, making state jurisdiction alone inadequate. So the federal government began assuming jurisdiction over migratory waterfowl and other birds early in this century. Early federal regulations were concerned with commerce in, and means of taking, migratory birds. In more recent years, the federal government's role has expanded, especially in regulations regarding the acquisition and management of critical management areas and the impacts of certain types of development upon wildlife (Bean, 1977). Most federal jurisdiction in practice involves regulations on importation of exotic wildlife, interstate shipments, protection and management of migratory birds and, most recently, protection and management of endangered species. Despite occasional differences of opinion, state-federal cooperation generally remains good in these areas. State wildlife agencies still assume primary jurisdiction over resident or nonmigratory species, with most efforts directed toward resident game and furbearers.

FUNDING FOR WILDLIFE MANAGEMENT

Wildlife management programs cost money. Sources of funding are, of course, critical in maintaining these programs, so much so that they often influence management priorities. Historically, the most successful and reliable funding sources in the United States have been various forms of license fees and taxes paid by those who use wildlife. With users paying the tab, most state wildlife agencies naturally cater to the needs of those species most commonly taken by the users: game and furbearers. But wildlife is owned by the public, user and nonuser alike, so that conflicts sometimes arise between the two groups.

Licensed sporthunters were once the only organized groups to lobby in favor of wildlife management. Since the mid-1960s, however, more and more nonhunting (and even antihunting) groups have banded together in support of wildlife management for reasons other than hunting or trapping. A gradual trend toward other funding sources has followed, with new, untapped sources to be found, no doubt, in the near future.

General Tax Revenues

Most federal agencies and about one-third of the state wildlife agencies in the United States rely upon general tax revenues for their funding. Money is appropriated by Congress at the federal level and by state legislatures for the states. One advantage of general tax revenue is its flexibility, particularly in meeting short-term needs. General tax revenues are more democratic since they originate from such a broad segment of the public. The primary drawback, however, is that such funding becomes unreliable over the long term, since appropriations can be reduced whenever the legislative body so desires. Agencies relying upon appropriations must adapt as best they can to a feast or famine existence.

License Sales

Most state wildlife agencies receive over half their funding from sales of licenses for hunting, fishing, trapping, game breeding, or operation of commercial shooting preserves. Since 1923, an estimated $2.5 billion in revenue has been raised in the United States through license sales (Jahn and Trefethen, 1978). The number of licenses sold does not fluctuate much between years, making license sales more reliable sources of funds than are appropriations. Moreover, state agencies can anticipate their incomes based upon projections of license sales. One drawback, though, is that license sales provide support only from hunters and other users of wildlife, furnishing little direct support for the majority of wild vertebrates not hunted or trapped.

User Excise Tax

The Pittman-Robertson Act and the Dingell-Johnson Act, mentioned earlier in this chapter, have provided well over half a billion dollars in revenues since 1937 (U.S. Fish and Wildlife Service, 1978). The money is derived from a federal excise tax placed upon manufacturers of firearms, ammunition, sport fishing tackle, and related paraphernalia. The U.S. Fish and Wildlife Service receives the revenue and then reapportions it to state wildlife agencies on a three-to-one matching basis. This federal aid money constitutes between 15 and 25 percent of the budgets of most state wildlife departments. Like license sales, these user excise taxes furnish steady levels of revenue and they have the additional advantage of increasing automatically with inflation.

Prior to the late 1930s state legislatures commonly diverted hunting license revenues from state wildlife departments to pay for roads, schools, and other public projects. The Pittman-Robertson Act ended this practice. Before any state could receive "P-R" funds, it had to pass an enabling act. The enabling act had to include: "A prohibition against the diversion of license fees paid by hunters for any purpose other than the administration of said State fish and game departments" (U.S. Fish and Wildlife Service, 1978). All states complied promptly.

Special Use Stamps

The duck stamp, the most successful of the special use stamps, has been required for waterfowl hunters since 1934. In recent years sales of duck stamps have generated more than $10 million annually to help preserve waterfowl habitat. Some states now require hunters to purchase either a state duck stamp or a special habitat stamp to provide funds for state programs.

Nongame wildlife management presents a special problem. Funding should obviously come from sources other than user pay. Several states including Ohio, Colorado, Michigan, and California have produced nongame stamps and offered them for sale on a voluntary basis. These programs have so far fared poorly and sales have been meager. Special use stamps prove to be far better fund raisers when their purchase is required than when it is voluntary.

Funding from the Private Sector

Private organizations provide important financial support for wildlife management. The Nature Conservancy and the National Audubon Society own or lease their own wildlife sanctuaries and natural areas. The Conservancy sometimes serves as an agent in delicate transactions between private landowners and public wildlife agencies, purchasing the land and then deeding it to the agency when the latter is unable to make the purchase directly.

During the drought of the 1930s, with waterfowl numbers at an all-time low, a private organization called Ducks Unlimited was formed in the United States and Canada. The organization has received more than $32 million in contributions from its members for restoration of waterfowl habitat (Stahr and Callison, 1978).

The International Union for the Conservation of Nature and Natural Resources (IUCN) and its sister organization, the World Wildlife Fund (WWF), are based in Gland, Switzerland. Funded by private contributions, these organizations serve wildlife conservation on a global scale. The IUCN publishes a series of *Red Data Books,* continually updated, which furnish complete information on the status of endangered species. In addition, the IUCN provides international guidelines for parks, nature reserves, and biological reserves. The WWF supports endangered species research and restoration projects and publicizes the plight of endangered wildlife.

Private foundations such as the Welder Wildlife Foundation in Texas and the Max McGraw Foundation in Illinois have their headquarters on their own refuges and provide both facilities and funding for wildlife research. Both are supported by private endowments.

Special Provisions of State Sales Tax

In the late 1970s, Missouri passed a constitutional amendment to provide additional funding for conservation programs, including those for wildlife. The amendment specified that one-eighth of 1 percent of the general sales tax be earmarked for conservation. Other states may follow Missouri's lead. This imaginative arrangement draws funding from the general public, adjusts automatically for inflation, and is virtually immune from legislative cutbacks.

State Tax Checkoffs

Colorado in 1978 became the first state to use voluntary state tax checkoffs to fund nongame wildlife management. Simply by checking a box on a state tax form, taxpayers can contribute a part of their tax refund to the nongame program. This imaginative source of revenue has proven to be so successful that by late 1982 at least 19 other states had passed similar measures. This may prove to be the most important means of funding nongame wildlife management. It is voluntary and furnishes support from a very broad segment of the public, not just the users.

Miscellaneous Sources of Funds

The state of Washington receives about $200,000 annually from sales of personalized license plates, with the proceeds going to wildlife programs. A state tax on horse racing nets the California Fish and Game more than half a million dollars each year, and at least six state wildlife departments receive funding from bond issues (Jahn and Trefethen, 1978).

SUMMARY

Prehistoric people relied upon wildlife for food, clothing, and shelter. The impact of these activities upon wildlife populations remains unclear and controversial. One point of view, known as the "Pleistocene overkill hypothesis," concludes that early hunters actually drove many of the larger mammals to extinction at the end of the Pleistocene epoch. Others have contended that prehistoric humans contributed to Pleistocene extinctions in more indirect ways, such as through competition, the alteration of competitive relationships, or through habitat disturbance by Neolithic agriculture. It is possible then, that people could have been contributing to the extinction of some species long before recorded history.

As people began to rely more and more on agriculture, their impact upon wildlife changed. Human effects become less direct and involved changes in the habitat rather than direct exploitation. Domestication of a few wild species further estranged people from wildlife and encouraged them to destroy or drive away wild relatives as they selectively bred these animals. People also began to destroy predators as threats to their livestock. As livestock numbers increased, so did heavy grazing and browsing, thereby altering wildlife habitat. But the greatest change brought about by agriculture was the increase it made possible in human populations, presently threatening wildlife habitat on an unprecedented scale.

Wildlife conservation in the United States can be classified by "eras" according to attitudes and conservation measures of the time. Until the mid-1800s there was little sentiment in favor of conservation, with wild species treated as though they would forever remain abundant. The most severe direct exploitation of wildlife occurred during the late 1800s when market hunters, using new technology, seriously reduced populations of plume birds, big game, and other wildlife. These reductions led to federal involvement and the emphasis on direct legal protection during the first three decades of the 1900s. After 1930 it became evident that direct legal protection was alone inadequate, so to it were added habitat restoration and new sources of funding such as the Migratory Bird Hunting Stamp Act and the Pittman-Robertson Act, with emphasis on game species. Environmental awareness of the late 1960s broadened wildlife management still further, adding programs for endangered species and nongame wildlife.

Wildlife values can be placed into six overlapping categories. Esthetic values are those relating to inherent natural beauty and artistic appreciation and, although hard to quantify, are universal and basic. Recreational values can be subdivided into consumptive values such as hunting and trapping, and nonconsumptive such as bird-watching and wildlife photography. Ecological values include those effects that wild species have upon one another and upon their natural communities. These values emphasize the avoidance of irreversible changes. Educational and scientific values encompass those that serve in teaching and learning about wildlife and illustrating scientific principles. Practical or utilitarian values comprise those associated with such endeavors as improvements in livestock, uses in biomedical research, and service as environmental indicators. Commercial values include trapping of furbearers, game ranching, and commercial hunting leases.

There are also negative values associated with wildlife. Collisions between automobiles and large mammals pose serious hazards as do occasional attacks on people. Birds can be sucked into air intakes of jet aircraft, causing crashes. Wild animals also cause millions of dollars' worth of crop and livestock depredations each year. Some wild animals may serve as reservoirs for parasites or infectious diseases that effect the health of people or their domestic animals.

The social sciences have provided wildlife specialists with means of measuring wildlife values. Two of the more common types of measurements are economic surveys and attitude surveys. Economic surveys evaluate the monetary returns accruing from wildlife, although these are usually indirect. Other surveys address public attitudes toward wildlife management, the basis for such attitudes, and often make recommendations for correcting negative attitudes through information and education programs.

Any wild species may have one of three basic legal classifications. Songbirds and birds of prey often are granted complete protection year-round. Game and furbearer species receive partial protection at certain times of the year, in particular areas, or both. Many species are afforded no legal protection whatsoever.

State wildlife agencies have primary legal authority over resident (nonmigratory) wildlife. Protection and management of migratory birds is a responsibility of the U.S. Fish and Wildlife Service, carried out in cooperation with state wildlife agencies. Federal jurisdiction also includes exotic wildlife, interstate shipments, and management and protection of endangered species.

Funding for wildlife management can be either user-pay or more broadly based. User-pay sources of revenue come from hunters, trappers, and others who exploit wildlife. These sources include license fees, user excise taxes, and duck stamps. More broadly based funding derives from the general public and includes legislative appropriations, special provisions of the state sales tax, state tax checkoffs, contributions from private organizations, and several miscellaneous sources.

REFERENCES

Applegate, J. E. 1979. Attitudes toward deer hunting in New Jersey: a decline in opposition. *Wildl. Soc. Bull.* 7: 127–129.

Bart, J., D. Allee, & M. Richmond. 1979. Using economics in defense of wildlife. *Wildl. Soc. Bull.* 7: 139–144.

Bean, M. J. 1977. The evolution of national wildlife law. Council on Environmental Quality. Washington, D.C.

Carter, L. J. 1980. The Leopolds: a family of naturalists. *Science* 207: 1051–1055.

Charbonneau, J. J., & M. J. Hay. 1978. Determinants and economic values of hunting and fishing. *Trans. N. Am. Wildl. and Natur. Resour. Conf.* 43: 391–403.

Clarke, J. 1979. *The life and adventures of John Muir.* San Francisco, Calif.: Sierra Club Books.

Dahlgren, R. B., Alice Wywialowski, T. A. Bubolz, & V. L. Wright. 1977. Influence of knowledge of wildlife management principles on behavior and attitudes toward resource issues. *Trans. N. Am. Wild. and Natur. Resour. Conf.* 42: 146–155.

Eckholm, E. P. 1976. *Losing ground: environmental stress and world food prospects.* New York: W. W. Norton and Co.

Ehrenfeld, D. W. 1976. The conservation of non-resources. *Am. Scient.* 64: 648–656.

Ehrlich, P., Anne P. Ehrlich, & J. P. Holdren. 1977. *Ecoscience: population, resources, environment.* San Francisco, Calif.: W. H. Freeman and Sons.

Eiseley, Loren. 1957. *The immense journey.* New York: Random House.

Flader, Susan L. 1974. *Thinking like a mountain.* Columbia, Mo. University of Missouri Press.

Hawthorne, D. W. 1980. Wildlife damage and control techniques. In S. D. Schemnitz (Ed.), *Wildlife management techniques manual* (4th ed.). Washington, D.C.: The Wildlife Society.

Jahn, L. R., & J. Trefethen. 1978. Funding wildlife conservation programs. In H. P. Brokaw (Ed.), *Wildlife and America.* Washington, D.C.: Council on Environmental Quality.

Kelker, G. H. 1964. Appraisal of ideas advanced by Aldo Leopold thirty years ago. *J. Wildl. Manage.* 28: 180–185.

Kellert, S. R. 1976. Perceptions of animals in American society. *Trans. N. Am. Wildl. and Natur. Resour. Conf.* 41: 533–546.

Kellert, S. R. 1978. Attitudes and characteristics of hunters and anti-hunters. *Trans. N. Am. Wildl. and Natur. Resour. Conf.* 43: 412–423.

Kellert, S. R. 1980. Americans' attitudes and knowledge of animals. *Trans. N. Am. Wildl. and Natur. Resour. Conf.* 45: 111–124.

Kennedy, J. J. 1974. Attitudes and behavior of deer hunters in a Maryland forest. *J. Wildl. Manage.* 38: 1–8.

King, R. T. 1947. The future of wildlife in forest land use. *Trans. N. Am. Wildl. and Natur. Resour. Conf.* 12: 454–467.

Klein, D. 1973. Ethics of hunting and the anti-hunting movement. *Trans. N. Am. Wildl. and Natur. Resour. Conf.* 38: 256–267.

Krantz, G. S. 1970. Human activities and megafaunal extinctions. *Am. Scient.* 58:164–170.

Leopold, A. 1933a. *Game management.* New York: Charles Scribner's Sons.

Leopold, A. 1933b. The conservation ethic. *J. Forestry.* 31: 634–643.

Leopold, A. 1949. *A Sand County Almanac.* New York: Oxford University Press.

Leopold, A. 1953. *Round River.* New York: Oxford University Press.

Martin, P. S., & H. E. Wright (Eds.), 1967. *Pleistocene extinctions.* New Haven, Conn.: Yale University Press.

Martin, W. E., & R. L. Gum. 1978. Economic value of hunting, fishing, and general rural outdoor recreation. *Wildl. Soc. Bull.* 6: 3–7.

Myers, N. 1976. An expanded approach to the problem of disappearing species. *Science.* 198–202.

Myers, N. 1979. *The sinking ark.* New York: Pergamon Press.

Myers, N. 1983. *A wealth of wild species.* Boulder, Colo.: Westview Press.

Rohlfing, A. H. 1978. Hunter conduct and public attitudes. *Trans. N. Am. Wildl. and Natur. Resour. Conf.* 43: 404–411.

Rorabacher, J. A. 1970. *The American buffalo in transition.* Saint Cloud, Minn.: North Star Press.

Schwilling, M. 1975. Waterfowl damage control. *Proc. Great Plains Animal Damage Control Workshop* 2: 163–169.

Shaw, D., & D. Gilbert. 1974. Attitudes of college students toward hunting. *Trans. N. Am. Wildl. and Natur. Resour. Conf.* 39: 157–162.

Sinclair, A. R. E. 1977. The African buffalo. Chicago: University of Chicago Press.

Stahr, E. and C. Callison. 1978. The role of private organizations. In H. P. Brokaw (Ed.), *Wildlife and America.* Washington, D.C.: Council on Environmental Quality.

Steinhoff, H. W. 1971. Communicating complete wildlife values of Kenai. *Trans. N. Am. Wildl. and Natur. Resour. Conf.* 36: 428–438.

Steinhoff, H. W. 1978. Big game values. In J. Schmidt and D. Gilbert (Eds.), *Big game of North America: ecology and management,* Harrisburg, Pa.: Stackpole Books.

Trefethen, J. B. 1975. *An American crusade for wildlife.* New York: Winchester Press and the Boone and Crockett Club.

U.S. Fish and Wildlife Service. 1977. *1975 national survey of hunting, fishing, and wildlife-associated recreation.* Washington, D.C.: U.S.D.I. Fish and Wildlife Service.

U.S. Fish and Wildlife Service. 1978. *Environmental impact statement.* Federal aid in fish and wildlife restoration program. Washington, D.C.: U.S.D.I. Fish and Wildlife Service.

WILDLIFE HABITAT

Most wildlife management is habitat management. Putting it another way, almost any wildlife management objective from increasing the harvest of a game species to prevention of serious disease outbreaks, through maintenance of maximum species diversity can be met largely, if not exclusively, by maintaining proper habitat conditions.

People generally realize that national parks or biological reserves provide better wildlife habitat than do suburban developments, agricultural monocultures, or golf courses. But most of the earth's land surface falls somewhere between these extremes of protection and development and provides varying degrees of suitability as habitat for wildlife. In addition, species vary in their habitat requirements, so that any given area may provide excellent habitat for one species but may be completely unsuited for others.

This chapter introduces some of the components of habitat and the basic concepts upon which habitat management is based. It then furnishes some principles of habitat evaluation and management. But habitat management, like any other wildlife management, must begin with a clear set of objectives. Applications of habitat management in relation to specific objectives appear in Chapters 7, 8, and 10.

HABITAT COMPONENTS

Think of habitat as a particular place or area. To understand how habitat affects wild populations of animals, one must examine its components. The four basic components of habitat are food, cover, water, and space.

Food

The most obvious component of habitat for any animal is food. Availability of food usually changes with the seasons, particularly in temperate and arctic regions, so food may be plentiful in one season and critically short in another. For carnivorous or predatory species, food availability simply means prey availability. Carnivores expend energy searching for, chasing (or ambushing), capturing, and killing prey, but these extra expenditures are offset by the higher concentration of energy contained in the animal matter of their food. Because animal matter is nutritionally complete and easy to digest, predatory species seldom, if ever, experience qualitative food deficiencies from natural diets.

Herbivores, on the other hand, depend upon foods that require no active pursuit but that are lower in energy and more variable in protein and nutrient composition. Thus, they must spend more time actually eating than do carnivores. For herbivores, food can become critical in two ways: an overall shortage (quantitative food stress) or an unbalanced diet (qualitative food stress), such as one deficient in protein. Browsing and grazing animals do not feed randomly but instead show clear and predictable preferences for certain plants over others. This selective feeding has led biologists to classify plant species according to a given herbivore's preferences. These preferences are assumed to be related to palatability or taste. The usual classifications are preferred (first choice and always taken more frequently than it occurs), staple (second choice but still providing all nutritional needs), emergency (able to furnish only short-term nutritional needs), and stuffers (useless nutritionally and ingested apparently to relieve hunger pangs).

Cover

Any variation in the habitat that provides protection from weather or predators or that offers a better vantage point is termed "cover." A common use of cover is escape from predators, although predators that ambush prey may themselves require cover to get close enough to make a kill. Cover may also furnish important protection against severe weather, providing shade from the heat, relief from wind and precipitation, or protection against radiant heat loss to a cold night sky. Nesting and roosting cover can also be critical for many kinds of birds. Many species use certain types for resting during periods of inactivity, a practice that has led some biologists to describe it as "loafing" cover.

Water

Most species replenish body water by drinking surface water. When surface water is not available, some species can sustain themselves by taking in morning dew or by ingesting the water contained in succulent plants. A very few animals, such as kangaroo rats (*Dipodomys* spp.) can derive all their water needs from that produced during metabolism.

Water can affect wildlife indirectly through changes in the habitat. Some of the most productive, yet vulnerable, wildlife habitats are those classed as wetlands. Marshes, for example, provide important nesting and brood-rearing habitat for waterfowl and other aquatic birds, but only so long as the water level is maintained at fairly shallow depths. Too much or too little rain can reduce productivity markedly by changing water levels. Human activities commonly abolish marshes through drainage, creation of landfills, and pollution (see Chapter 9).

Riparian habitats are those that occur naturally along rivers and streams. These ribbons of woody and herbaceous plant life may constitute the only natural habitats over large areas of midwestern farmlands and rangelands. Riparian habitats act as important travel lanes along which wild animals can move, thus playing a crucial role in sustaining otherwise isolated populations. Unfortunately, riparian habitats diminish through stream channelizations and through pasture "improvement" that clears away native vegetation right up to the stream edge.

Space

Individual animals require varying amounts of space in which to find enough food, cover, water, and to locate mates. Populations of wild animals require even more space and, since wildlife managers are generally concerned with populations, the question naturally arises as to how much space is enough. In each case, the amount of space or suitable habitat depends upon the size of the population desired. This population size depends, in turn, upon the size of the species (generally the larger the animal, the larger the area required), its diet (carnivores require larger areas than do herbivores), and the productivity and diversity of the habitat in relation to the habitat requirements of the species. Methods used to predict minimum sizes for parks and biological reserves appear in Chapter 8.

CONCEPTS RELATED TO WILDLIFE HABITAT

The four basic components of wildlife habitats are rather straightforward and easy to understand. The concepts underlying habitat management involve the ways in which the basic components interact with populations of living organisms occurring there. Some of these concepts also include the element of time.

Trophic Levels

Wild animals live in natural habitats as interwoven parts of ecosystems or communities. Species obtaining food in the same general way are on the same trophic level; they are the same number of steps from the green plants or primary producers that comprise the first trophic level. Herbivores make up the second trophic level, and carnivores the third. Aquatic ecosystems may contain ten or more trophic levels, with all but the first two being carnivores that prey on each other. Terrestrial communities, however, seldom have more than four trophic levels and some may have only three. The species in adjacent trophic levels are diagrammatically linked together to form food chains in very simple communities and food webs in more complex ones.

The Second Law of Thermodynamics

The second law of thermodynamics states that energy cannot be transferred from one form to another without enormous loss in the form of dissipated heat. This law applies to the transfer of energy between trophic levels. Many introductory ecology texts report an average energy transfer between trophic levels of about 10 percent, the remaining 90 percent lost. Recent studies of energy dynamics have uncovered so much variability in rates of energy transfers that the 10 percent figure no longer suffices even for a crude (and extremely liberal) estimate. About the only safe generalization is that a great deal of energy is lost at each step up the food chain or web. Thus, productivity of carnivores within a community is much less than that of herbivores. When the consumers are birds or mammals, the energy transfer rates are particularly low, as these homeothermic animals pay high energetic costs for maintenance of more or less constant body temperatures. For them the efficiency of transfer may be less than 1 percent.

Relationships between trophic levels in a community can be expressed in terms of relative numbers, biomass, or energy transfers (productivity). Table 2-1 shows all three relationships for moose (*Alces alces*) and their predators, gray wolves, at Isle Royale National Park in Michigan. These

TABLE 2-1
A comparison of two adjacent trophic levels in the Isle Royale, Michigan, ecosystem (Jordan et al., 1971, Copyright © 1971, the Ecological Society of America, used by permission)

Comparative data	Moose	Wolf
Numbers	1,000	24
Biomass (kg)	367,000	812
Productivity (kg/yr)	91,000	141

data represent one of the few detailed studies on energy transfers between homeotherms in an undisturbed state, and they are of particular interest because Isle Royale supports only one dominant herbivore with a single predator preying on it. Note especially the enormous difference between the two trophic levels. One kilogram of growth in moose results in less than .002 kilogram of growth in wolves.

The Niche

One of the most important concepts in ecology is that of the niche, or role played by each species in its natural habitat. Because the niche is a role rather than a place, it can be described only in terms of an interaction between the species and the environment. The most obvious portion of a niche is that pertaining to food, although other important niche parameters include ways of using cover, water, or even space. Bobcats (*Lynx rufus*) and cougars (*Felis concolor*) share some of the same habitat but feed on prey of different sizes. Bobcats and coyotes, on the other hand, take similar-sized prey but use cover differently, with bobcats hunting more in heavy cover while coyotes concentrate more on hunting in open areas.

In practice, all the variables that make up a niche have not been measured or defined completely even for widely studied species. Biologists must content themselves with the results of experimental studies (done mostly under controlled conditions in the laboratory) or with partially defined niches gleaned from fortuitous field studies. Nonetheless, basic niche theories are widely accepted and sound.

Some species have broad niches that allow them to exist in a wide variety of habitat types. They are known as generalists. At the other end of the scale are species with quite narrow niches, confined to one or two habitat types, making them specialists. Familiar generalists include the opossum (*Didelphis virginianus*) and the raccoon (*Procyon lotor*), both of which are omnivorous and can live almost anywhere. The black-footed ferret (*Mustela nigripes*) subsists only on prairie dogs (*Cynomys* spp.), and the everglades kite (*Rostrhamus sociabilis*) is so specialized that it feeds only upon the apple snail (*Pomacea caliginosa*). Specialists are more likely to end up on the endangered species lists since they are poorly adapted to environmental changes, whereas generalists are sometimes so successful that they become pests.

Carrying Capacity

One of the most important concepts in wildlife management is that of carrying capacity, or the number of conspecific animals that can be supported within a given area. Various authorities have defined carrying capacity differently. Population ecologists, for example, use the term "K" to denote the upper limit beyond which no further population increase can be sustained by the habitat. Wildlife managers sometimes must rely on more practical criteria that are easier to measure in the field. One of these criteria is the condition of individual animals. In populations below carrying capacity, the greater amounts of food, cover, and space per animal leave animals generally heavier, in better nutritional condition, and with fewer signs of parasites and diseases than animals from populations at carrying capacity. A third way to define carrying capacity is the average number of animals that an area supports over a period of years. Obviously, this definition could only be applied to those habitats for which several years of census data were available for the species in question.

Expressed as the number per unit area or the total number within a given tract, carrying capacity for any given species can vary a great deal between different habitats. These variations are often used as indications of habitat quality.

The usual definitions of carrying capacity imply that it is fixed or static for a given species in a particular habitat. Actually, habitat conditions are almost always changing to some extent, altering carrying capacity. Habitats are disturbed by fires, storms, or human activities. Carrying capacity also changes between seasons, with colder or drier seasons furnishing the "crunch" or "pinch period" that results in lower carrying capacity than could be supported at other times of the year (Fretwell, 1972).

Biotic Succession

Also called plant succession and ecological succession, biotic succession is a predictable change in the species composition of a community through time. This gradual change takes place in overlapping phases known as seral stages. Primary succession is succession occurring for the first time in "new," previously uninhabited sites such as fresh sand dunes along lake shores or on exposed, bare rock.

The more common form of succession follows disturbances such as storms or fires and is called secondary succession. Secondary succession generally advances faster than primary succession, as soil and seed conditions are more favorable.

If, in a wooded region, a disturbance is severe enough to destroy trees, the plants that predominate in the early successional stages will be herbaceous (nonwoody) species that thrive in open areas under direct sunlight. The first trees and shrubs will be either species with light, motile seeds, drifting in from adjacent areas, or else species that sprout readily from surviving rootstocks. These "pioneer" trees grow quickly, capturing more and more of the sunlight, making conditions less favorable for many of the herbaceous species below. During the midsuccessional stage, other tree species appear that tolerate shading better and live longer than the pioneer species. The growth and development of shade-tolerant trees is a slow process, but they gradually outgrow the pioneering species that can no longer survive in the resulting shade. When the community reaches this late-successional stage, the shade-tolerant species are clearly dominant. Since these trees can reproduce and develop in their own shade, they will perpetuate themselves indefinitely until another disturbance begins the process again.

The time required for succession to change a fresh disturbance to a stable, late-successional community varies with climate, habitat type, and frequency of disturbance. Table 2-2 lists the times required to reach late-successional stages in a variety of habitat types. The variations in these estimates result from differences in local conditions. If the soil is removed or otherwise substantially altered, recovery can take much longer.

No species of animal is equally well adapted to all successional stages. Each generally does best and reaches its highest numbers within the span of a few years as succession advances, although species best suited to late-successional conditions may remain at maximum densities until the next disturbance. Most of our familiar game species in North America become most numerous under early- to midsuccessional stages. An example of how successional advance changes carrying capacity is provided by a study of ruffed grouse (*Bonasa umbellus*) in Minnesota. During the spring breeding season, the grouse feed largely on staminate flower buds of two species of aspen, *Populus tremuloides* and *P. grandidentata*. Aspen are pioneer species that reach their own population peak early in the successional advance. Thus, the maximum breeding densities of ruffed grouse occur fairly early, when aspen furnishes an ideal blend of food and cover (see Figure 2-1). Once the shade-tolerant trees begin to replace the aspens, breeding densities of grouse sharply decline, ultimately to zero under late-successional conditions (Gullion, 1970).

TABLE 2-2
Time required to reach a stable, late-successional condition in selected habitat types in Illinois
(Graber and Graber, 1976, by permission, Illinois Natural History Survey)

	Years required for successional completion	
Habitat type	**If soil remains intact**	**If soil is removed or greatly altered**
Bottomland forest		
Oak-gum-cypress	20–600	120–750
Elm-ash-cottonwood	5–99	40–134
Hackberry-gum-oak-hickory	100–500	235–1100
Upland forest		
Elm-oak-hickory	60–99	160–199
Oak-hickory	100–500	200–600+
Maple-beech	35–500+	185–700+
Aspen	5–39	10–44
Pine	10–100+	35–125+
Shrub	3–30	6–33
Natural marsh	600+	1600+
Prairie	10–30+	20–45

The long-term effects of secondary succession can be quite profound. Before settlement by European colonists, most of North America was comprised of pristine, undisturbed habitat. Exceptions occurred wherever the native Americans used fires, either to clear lands for crops, to drive out game, or to attract game as the vegetation recovered from burning. But the popular image of pristine America as a place teeming with wild game and abounding with wildlife is often incorrect.

An example of just how extensively secondary succession can alter wildlife habitat comes from the former town of Remington, Wisconsin (Grange, 1938). Prior to settlement, the area was a wilderness, dominated by peat and spagnum bogs with stands of tamarack (*Larix latricina*) and black spruce (*Picea mariana*). Three species of pine and a scattering of hardwoods inhabited the sandy uplands. Ruffed and spruce grouse (*Pedioecetes phasianellus*) were present, though not very abundant, a few sharp-tailed grouse (*Canachites canadensis*) may have lived there, but prairie chickens were definitely absent. There were snowshoe hares (*Lepus americanus*) but no cottontails (*Sylvilagus floridanus*), and if any white-tailed deer lived in the vicinity, they were extremely rare.

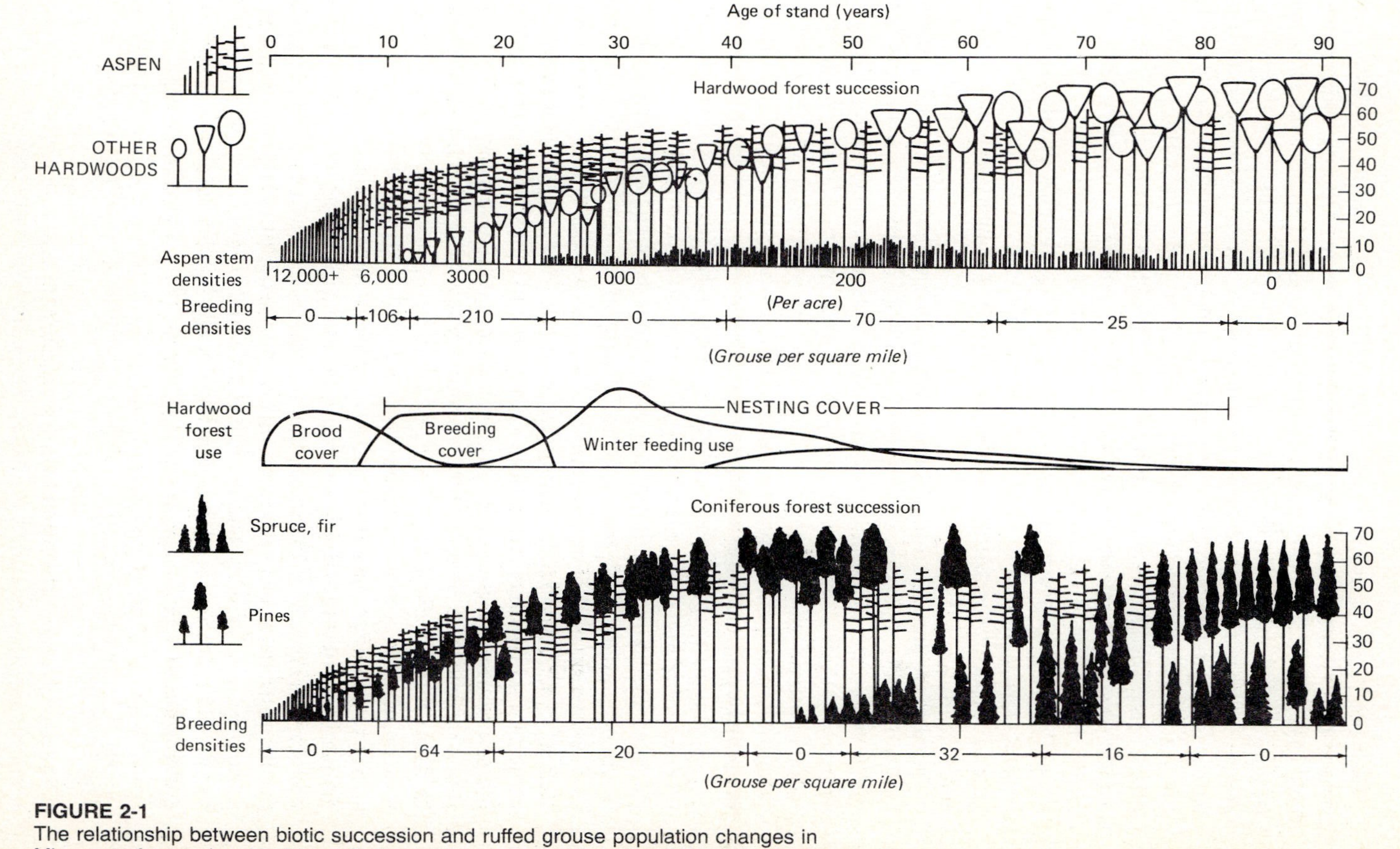

FIGURE 2-1
The relationship between biotic succession and ruffed grouse population changes in Minnesota forests (modified from Gullion, 1970 by permission, The Wildlife Management Institute).

Settlers began arriving in the 1850s and 1860s, clearing the land and setting fires. Settlement increased rapidly between 1870 and 1900, more land was cleared and burned, and habitat conditions changed sufficiently to permit establishment for the first time of both prairie chickens and ruffed grouse. Where there had once been peat bog, there was now extensive wiregrass (*Spartina pectinate*) in stands so thick that settlers harvested it for use in the manufacture of carpets. Fires became still more common, further improving habitat for prairie chickens and sharptails.

The first year of the twentieth century witnessed development of numerous drainage ditches intended to open up still more land to agriculture. As this dredging continued and the lowlands dried out, fires became more frequent and widespread. Various weeds, sedges, and grasses replaced the wiregrass. Prairie chicken populations reached their peak by 1920.

Then came hard economic times which, along with the usual assortment of fires, floods, and droughts, together with heavy assessments to pay for the dredging, led to agricultural depression. Farmers in increasing numbers abandoned their lands during the 1920s. A huge fire raged through the area in the autumn of 1930, destroying much of what peat still remained. Quaking aspen quickly invaded many of the former peat bogs, and others remained bare long after the fire.

For the most part, the secondary succession following the big fire proceeded rapidly. The season of 1932 produced record harvests of prairie chickens and sharptails. Cottontail populations boomed, while the formerly abundant snowshoe hares all but disappeared. As browse increased, deer populations shot up so rapidly that they were plentiful for the first time by 1937 and so numerous by the winter of 1942 to 1943 that they were subject to widespread starvation. Grouse populations headed downward. As sharptails and prairie chickens declined, ruffed grouse populations started to pick up. Despite the starvation, the deer population continued to climb through 1947.

Secondary succession had changed the land from extensive bogs interspersed with thin pine forests to wiregrass pastures, then to extensive cultivation, followed by drainage, fires, and abandonment. Once abandoned, the now drier habitat became extensive tracts of forest. Wildlife populations changed as radically as did the vegetation. Most people upon seeing the region around Remington, Wisconsin, could never guess that the very character of the landscape had changed markedly since the time of earliest settlement.

The Edge Effect

The zone of contact between two or more habitat types such as a forest and a cultivated field is called an edge. The common observation that game species were more abundant along edges led Leopold (1933) to conclude that the more edge per unit area, the higher the production of game. This is the edge effect, or law of interspersion, a major principle in wildlife habitat management. Its acceptance stems from the commonsense notion that more edge must benefit those species with limited mobility and varied habitat requirements, traits typical of many game species.

Baxter and Wolfe (1972) compared indexes of edge with population indexes in bobwhite quail (*Colinus virginanus*) in three study sites. They found that the higher the edge index, the higher the population index, indicating that bobwhite quail are subject to the edge effect.

Perhaps the width and quality of edge are more important than the amount of edge, but these criteria are far more difficult to measure in the field. The relative importance of edge may also change drastically in different habitat types. To complicate matters further, there is both empirical and theoretical evidence that increasing amounts of edge actually decreases some species populations. Some forest dwellers, such as the ovenbird (*Seiurus aurocapillus*), need large, contiguous expanses of undisturbed habitat for nesting, rearing of young, and dispersal.

When Leopold first described the edge effect, he had in mind the relationship between edge and productivity of game species. Since then, a corollary has become established: the greater the edge, the greater the overall species diversity and total abundance of wildlife. This modification of the edge effect is strongly supported by ecological field studies. For example, Strelke and Dickson (1980) studied breeding birds along transects placed perpendicularly to edges of forest clearcuts in east Texas. They found significantly greater species diversity and abundance within the first 25 meters of forest adjacent to the clearcut than either in the clearcut area or deeper into the forest (see Table 2-3).

Within limits, then, species diversity tends to increase with edge. Some species, however, including some of those most difficult to manage like the ovenbird, require large tracts of unbroken habitat. Is it then reasonable to suppose that there is some optimum, some compromise between the amount of edge and the size of the blocks or units of the various habitat types? Jack Ward Thomas and his associates have found that there is. In forested regions, maximum numbers of avian species occur in habitat blocks or forest stands of between 30 and 40 hectares. Stands of less than 30 hectares lose too many area-sensitive species, while stands greater than 40 hectares cannot satisfy the needs of edge species (Thomas et al., 1979).

In conclusion, increases in edge should generally stimulate increases in those populations of animals that have limited mobility and varied habitat requirements. It should also boost species diversity by increasing overall habitat diversity. Edge is also easily measured, especially when recent aerial photographs are available. Researchers will continue to study edge, eventually reaching the point where they can make accurate evaluations of

TABLE 2-3
The number of species of breeding birds in relation to edges between mixed pine-hardwood forests and clearcuts
(Strelke and Dickson, 1980, by permission, The Wildlife Society)

	Habitat type			
Bird habitat association	**Woods, interior**	**Woods, edge**	**Clearcut, edge**	**Clearcut, interior**
Woods	7	7	4	1
Edge	2	3	3	2
Clearcut		3	4	4
All habitats	1	2	2	2
Infrequent observations (<10)	3	12	8	8
Totals	13	27	21	17

the effects of changes in edge on certain species or on overall species diversity. Meanwhile, most wildlife managers will continue to increase edge for most early- to midsuccessional game species.

HABITAT EVALUATION

Habitat evaluations fall into two categories. The first type of evaluation is done for specific management objectives, typically involving an assessment in terms of the needs of a single species. The second type of evaluation is broader, aimed primarily at determining habitat values for several species or even for entire biological communities.

Evaluation for Specific Management Objectives

Wildlife managers often evaluate habitat for specific objectives such as estimating habitat quality for a particular game species. In such cases, food production, availability of cover, and amount of edge are three important measurements. Experienced field-workers can often estimate the population of herbivores in relation to the carrying capacity by checking the condition of key plants as indicator species. Other methods estimate habitat quality from blood chemistry, parasite loads, bone marrow condition in individual animals, and reproductive success.

Food Production All plant materials consumed by wildlife are collectively known as forage (see Table 2-4). These include grasses, the narrow-leafed herbaceous members of the botanical family Graminae.

TABLE 2-4
A simple classification of plant foods eaten by wildlife
(Examples from Martin, Zim, and Nelson, 1951, by permission, Dover Publications)

	Examples	
Forage class	**Plant name**	**Consuming species**
Herbaceous plants		
Grasses	Fescuegrass (*Festuca*)	Wild turkey
Forbs	Ragweed (*Ambrosia*)	Ringneck pheasant Hungarian partridge
Sedges	Bulrush (*Scirpus*)	Canvasback & avocet
Woody plants (many produce both browse and mast)		
Hardwoods	Oak (*Quercus*)	Squirrels & wild turkey
Conifers	Pine (*Pinus*)	Pinon jay & mule deer
Shrubs	Blackberry (*Rubus*)	Ruffed grouse Cottontail rabbit

Forage also includes broad-leafed herbaceous plants called forbs. Browse, an important forage class for many large mammals, consists of leaves, buds, and newly grown twigs from shrubs and trees. Seeds consumed by wild animals are termed mast, with fruits called soft mast and nuts referred to as hard mast.

Some habitat evaluations require sampling of all forage available to the animals. To do this, biologists clip and weigh all grass, forbs, and browse within small sample plots as high as the species under study can reach [about 1.5 meters for deer, 2.75 meters for moose, and perhaps 6 meters for giraffe (*Giraffa camelpardalis*)]. This is an expensive and laborious procedure and should not be done unless management objectives can be met within the time and the limits of the workforce available (cf. Harlow, 1977). Faster forage inventories can be made through visual estimates of forage weights, confirmed by clipping and weighing forage from a small fraction of the plots, randomly selected after the visual estimate is made (Blair, 1959).

Mast production is estimated in two basic ways. Hard mast from trees and larger shrubs is generally measured through the use of mast traps, catchment devices open at the top to catch mast as it falls. Soft mast from shrubs can be measured by clipping and weighing.

Cover The distribution of cover types can be measured by cover mapping. Depending upon the region, cover types might include trees, shrubs, pasture, cultivated fields, rocky outcroppings, marsh, or alpine tundra. Cover types are determined through aerial photographs, field surveys, or both; they are then delineated on a map of the study area.

Various cover types are indicated on the map by reference codes (cf. DeVos and Mosby, 1971; Mosby, 1980; Gysel and Lyon, 1980). The area within each cover type can then be resolved with a dot grid or a compensating polar planimeter.

Within each cover type, cover density can be estimated in a variety of ways. Vertical density is estimated on a rating scale (e.g., 1 through 10) but this simple method requires careful, experienced field personnel to assure reasonable consistency. A more objective method employs a freestanding, vertical optical density board painted black and white at 0.5-meter intervals (Nudds, 1977). The observer sets up the board, moves a standard distance from it, and estimates visually the proportion of each 0.5-meter interval blocked by cover.

Photoelectric devices are used to estimate horizontal cover (DeVos and Mosby, 1971). These tools measure the amount of light admitted through the forest or shrub canopy cover in comparison with that recorded in full sunlight.

Edge Patton (1975) devised a simple and effective technique that related the amount of edge in a given area to the perimeter (circumference) of a circle of the same size. Since the circle has the least amount of edge per unit area of any geometric figure, it serves as a convenient standard. Patton's formula is:

$$DI = \frac{TP}{2\sqrt{A \cdot \pi}}$$

where *DI* represents the total amount of edge, *TP* the total perimeter, and *A* the area. Figure 2-2 shows how the relative amount of edge, determined through Patton's formula, compares for a circle (always a value of 1.00), a square, and a rectangle for the same-size area.

An even simpler method for measuring edge uses two diagonal transect lines through square sample plots of 0.8 square kilometer (Baxter and Wolfe, 1972). The number of changes between cover types along both lines furnishes an index of edge. Both Patton's method and that of Baxter and Wolfe seem to work equally well; a highly significant correlation between them was found when they were used on the same area (Taylor, 1977).

Indicator Species Indicator species allow assessment of habitat conditions in relation to a herbivore population. Preferred or staple foods that occur at low to moderate frequency can serve as indicator species. Since

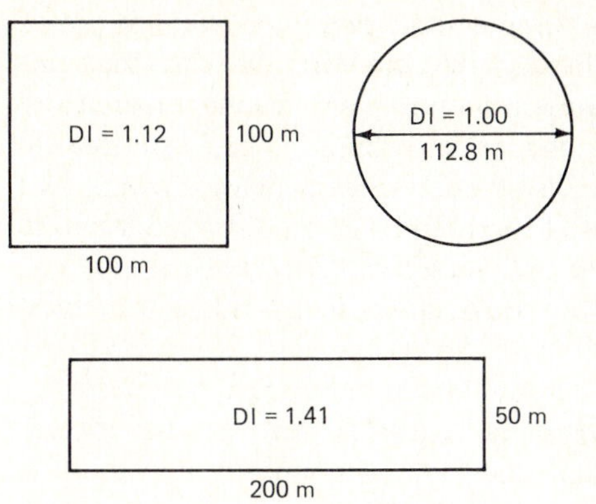

FIGURE 2-2
How the shape of areas affect the amount of edge. Patton's (1975) formula is $DI = \frac{TP}{2\sqrt{A \cdot \pi}}$ *or* $DI = TP/2\sqrt{A \cdot \pi}$ where DI (diversity index) is equal to the TP (total perimeter) divided by 2 × the square root of area × π. For a circle this value is always 1.00, and for a square it is 1.12. This particular rectangle has a DI of 1.41, 41 percent greater than that for a circle of the same area (by permission, The Wildlife Society).

many herbivores have fairly consistent food preferences, changes in the conditions of indicator species can reflect changes in the herbivore population. Sometimes exclosures are constructed around sample plots to exclude herbivores. The indicator species protected within the exclosure may, after a given period of time, be compared with those outside the exclosure. The difference represents a measure of grazing or browsing intensity. Exclosures should not be used for extended periods because some important food species become senescent after several years of protection from herbivores (Tueller and Tower, 1979).

Young balsam fir (*Abies balsamea*) and white birch (*Betula papyrifera*) serve as indicator species for moose. Moose populations in central Newfoundland in excess of 9.7 per square kilometer (12 per square mile) completely halted growth and regeneration in birch and inflicted "extremely severe" damage on balsam fir, particularly on areas of more fertile soil (Bergerud and Manuel, 1968). These researchers estimated that these same tree species could persist without serious damage in areas of moose populations at about 2.35 per square kilometer (6 per square mile).

In practice, though, indicator species should be used cautiously. The time of year for sampling must be selected to show the most pronounced use and then standardized in subsequent years. Managers should not rely upon a single indicator species, but should instead use several. Some species may be very important in the diet of a herbivore, yet they are unsuited for use as an indicator species because of their high abundance. Inkberry (*Ilex glabra*), for example, is an abundant shrub along the coastal plain of the southeastern United States, important in the diet of the deer. Yet it shows little evidence of being browsed because of its great abundance in some areas (Harlow, 1979). Preferred species that occur at low densities may be browsed heavily regardless of the herbivore population size and are thus unable to reflect changes in herbivore numbers.

Condition of Individual Animals Some of the more important means of habitat evaluation are indirect, through the condition of individual animals. When wildlife biologists capture live animals for various research and management purposes or when they operate hunter check stations to examine freshly killed game animals, they can take measurements that reliably indicate the animals' physical condition. The physical condition, in turn, tells something about quantity and quality of food, disease frequency, parasite levels, and various types of stress, all of which relate to habitat quality.

Weights of animals vary with habitat conditions. In the deer family (Cervidae), antler size, including diameter and circumference of the beam, relates directly to quality and quantity of food. Amount of fat deposited, particularly kidney fat, is a good indicator of nutritional condition, which, when seasonally adjusted, serves as a sign of habitat quality.

Another physical characteristic useful for habitat assessment is the condition of femur marrow. The marrow consists of fat, water, and nonfat residue. When an animal is under nutritional stress, the marrow fat is depleted and the relative amount of water increases. Biologists examine the marrow fat by breaking open the femur of a dead animal. The marrow of a healthy animal has a whitish appearance and a waxy feel (low water content), whereas that of an animal under severe nutritional stress is pink or reddish and more watery. More precise marrow fat determinations are made by a compression method and three drying methods. After reviewing all available techniques, Hunt (1979) concluded that oven-drying afforded the most useful combination of simplicity and precision.

Poor nutrition, brought on by poor habitat quality, can lead to increases in parasitism and diseases. Thus parasite levels furnish another index of habitat condition. Eve and Kellogg (1977) counted abomasal parasites in deer and found that deer from populations below carrying capacity had low parasite loads, while those from populations exceeding carrying capacity had heavy loads.

Analysis of blood samples can reveal a great deal about habitat conditions in some species. Controlled experiments have shown that assays of hematology, blood chemistries, and hormones relate to protein and energy intake in deer and pronghorns (Seal, 1977). Additional studies of free-living deer in Minnesota where habitat conditions were determined independently, showed a correlation between blood assays and habitat quality (Seal et al., 1978).

Observed Rate of Increase

There is probably no more sensitive indicator of habitat quality than a wildlife population's observed rate of increase. This rate is the outcome of the struggle between reproduction and immigration on the one hand and mortality and emigration on the other. Populations living in high-quality habitats tend to have higher rates of increase than populations of the same species living in poorer habitats. Methods for estimating rate of increase appear in Chapter 4.

Evaluation of Natural Communities

Wildlife management developed as an art and science in response to the needs of economically important species, mainly game animals. Although successful in many cases, game management was practiced piecemeal, typically addressing the needs of one species at a time. A broader segment of the public became interested in the welfare of wildlife in the mid 1960s, and sportsmen too began to advocate more conservation programs directed at so-called nongame species. Then came the National Environmental Policy Act in 1969, requiring environmental impact statements for federally funded projects likely to affect the environment. Environmental assessments of this sort require thorough examinations of entire biotic communities, not just the few game species. So between the increasing public concern for nongame wildlife and the requirements of environmental impact statements, there emerged needs for rapid and accurate techniques to evaluate natural communities. The first group of methods was developed by ecologists wrestling with theoretical questions on the diversity of animal life.

Species Diversity The measurement of species diversity is important in assessing the biological value, natural richness, and uniqueness of an area. The most basic and objective measure of species diversity is simply the number of species within a particular group (birds, for example) found per sample (Peet, 1974). But used alone, this measurement, called species richness, can be misleading. Table 2-5 contains diversity data from two hypothetical study sites. Both contain 100 (50 pairs) of breeding birds of the same 5 species. Area A has 46 pairs of species 1, and 1 pair for each species 2, 3, 4, and 5. Area B contains 10 pairs of each of the 5 species. Both areas have the same species richness, yet extremely different distributions of the species and, presumably, differences in community diversity. In other words, the species evenness differs sharply between the two areas. Any serious diversity measurement needs to account for both species richness and species evenness, so that the greater the number of species and the more even the distribution, the higher the diversity value. The Shannon-Wiener formula, derived from information theory, is one of the most widely used diversity measures

$$H = -\sum p_i \log p_i$$

where H is the diversity (or heterogeneity) value and p_i is the percentage importance for each species (Shannon and Weaver, 1949).

The diversity values (H) are computed in Table 2-5. Area 1 has a value of 0.3896731, while area B's value is 1.609438. The diversity value for area B is 4.13 times that of area A. Natural logs were used in this example; common logs work equally well so long as all calculations are made with the same type of log. It is not necessary for each area under comparison to contain an equal number of species. Most will not.

Diversity indexes, however, are not without their drawbacks. Some mathematical ecologists have questioned their precision and their validity under most field conditions. But a far worse danger may lie in the almost unequivocal acceptance of diversity as a management objective by many wildlife agencies. Since diversity indexes are fairly easy to obtain, they may be used even to the exclusion of other management objectives.

The Shannon-Weaver and other related indexes measure only alpha diversity, or diversity within specific communities or habitat types (Samson and Knopf, 1982). In practice, this often means that habitats with light to moderate disturbance will score higher diversity values. These higher

TABLE 2-5
A hypothetical comparison of diversity between two study sites
Species richness and overall abundance are equal but species evenness differs sharply

Survey data—area A			
Species	p_i	$\log p_i$	$p_i \log p_i$
1	.92	−0.0833816	−0.0767111
2	.02	−3.912023	−0.0782405
3	.02	−3.912023	−0.0782405
4	.02	−3.912023	−0.0782405
5	.02	−3.912023	−0.0782405

$-\Sigma\, p_i \log p_i = -0.3896731$; $\Sigma\, p_i \log p_i = 0.3896731$; $H = 0.3896731$

Survey data—area B			
Species	p_i	$\log p_i$	$p_i \log p_i$
1	.20	−1.6094379	−0.3218876
2	.20	−1.6094379	−0.3218876
3	.20	−1.6094379	−0.3218876
4	.20	−1.6094379	−0.3218876
5	.20	−1.6094379	−0.3218876

$-\Sigma\, p_i \log p_i = 1.609438$; $\Sigma\, p_i \log p_i = 1.609438$; $H = 1.609438$

values, in turn, usually result from the infusion of earlier succession species which flourish in the wake of human disturbance. The trouble is that although the total diversity scores are higher, they result from bringing in such ubiquitous species as starlings, mockingbirds (*Mimus polyglottos*), and robins (*Turdus migratorius*) while excluding such late-successional species as spotted owls (*Strix occidentalis*) or red-cockaded woodpeckers (*Picoides borealis*). Starlings and robins will fare well without active management on their behalf. It makes little sense to favor robins at the expense of red-cockaded woodpeckers. Perhaps all species are ultimately equal, but some clearly need more help than others.

Samson and Knopf (1982) recommended two other levels of diversity that they thought more important than alpha diversity. These were diversity between habitat types (beta diversity) and diversity over large regions (gamma diversity). They warned against the current emphasis on alpha diversity and suggested that gamma and beta diversity are more important considerations.

Standard Evaluation Procedures

The primary means through which wild lands can be protected from development is acquisition by a conservation organization. But private groups such as the Nature Conservancy and public agencies such as the U.S. Fish and Wildlife Service have limited funds with which to acquire such areas for refuges, sanctuaries, or reserves. If all wild lands cannot be saved, conservationists must devote their funds toward serving the most valuable of these areas. Plant ecologists and wildlife specialists have devised standard evaluation procedures for determining habitat quality in the broadest sense of the term. Through methods such as those described below, several wild areas can be compared, the one ranking highest receives highest priority for acquisition.

All habitat evaluation procedures rank certain characteristics, including physical and biological properties, for each proposed area. Rankings are usually made on a scale of 1 to 10. The different procedures vary according to the characteristics measured and the relative importance given each.

One such evaluation method was developed in conjunction with the Nature Conservancy in Britain (Goldsmith, 1975) and uses four main categories:

1 *Extent* (*E*)—area or, for linear features, length

2 *Rarity* (*R*)—recorded for each habitat type and calculated from R = 100 − % area per candidate area

3 *Plant species richness* (*S*)—the number of flowering plant species in a 20 by 20–meter sample plot

4 *Animal species richness* (V)—assumed to relate to vertical stratification on a 1 to 4 (grassland to woodland) scale

For each habitat type, an index of ecological value (IEV) is determined by

$$\text{IEV} = \Sigma\,(E \times R \times S \times V)$$

The summed products for all habitat types within each proposed area are then used to rank the areas.

The U.S. Fish and Wildlife Service has developed its own official habitat evaluation procedure (HEP). Three-person teams are used to determine, through field reconnaissance, literature reviews, and established ranking methods, the habitat suitability values for a number of different wildlife species. These suitability values are combined to yield an index of habitat quality. For proposed areas, the habitat value is obtained by multiplying the size of the area times the habitat quality index (Flood et al., 1977; Schamberger and Farmer, 1978).

Birds are conspicuous, easily identified, and ubiquitous. The diversity of bird species should itself be a good, reliable measure of overall community diversity, and several evaluations of their usefulness have been made. One such method (Graber and Graber, 1976) combined a habitat evaluation index (HEI) with a faunal (avian in this case) index (FI). The HEI measured overall composition of the habitat under study, and the FI provided the best measure of habitat quality.

Lines and Perry (1978) tested their habitat evaluation methods against avian diversity. Although they obtained poor correlations between selected population estimates and habitat indexes, they found significant correlations after combining population estimates for all species. Such consistent patterns between basic habitat evaluation methods and the actual occurrences of avian diversity have awarded more confidence to habitat evaluation methods.

These and other evaluation methods are important conservation tools and are undergoing continual refinements. Drawbacks exist, perhaps the most serious of which is the subjectivity used in applying rank values. This subjectivity means that different workers may assign substantially different values to the same area. This problem should diminish as refined methods become more objective.

Remote Sensing

Technological advances that so often seem to threaten wildlife can also provide useful conservation tools. One new set of tools important in habitat evaluation is remote sensing. The older form of remote sensing is high-altitude photography, often done with different film types, particularly infrared. Differences in reflectivity among various vegetation types are detectable through infrared photographs. High-altitude photography can also determine water conditions, including extent of certain forms of pollution.

In July, 1972, the first Landsat remote sensing satellite began orbiting the earth, transmitting data of considerable importance to natural resource management. Two and a half years later a second Landsat entered service. These satellites maintain orbits of about 800 kilometers (500 miles), circling the earth every 9 days. Landsat data are projected as images, each representing an area of the earth's surface of 185 by 178 kilometers. The images consist of 7.5 million digital picture elements (pixels) with each pixel representing an area on earth of 79 by 79 meters. Landsat images show some constructed features including towns, larger roads, and farms. Color composites (see Figure 2-3) show clear water as black, natural vegetation as bright red, and agricultural fields as pink. More detailed information such as weather, moisture conditions, and chemical measurements can be obtained from remote study areas in which special data collection platforms are placed (Heaslip, 1976).

Landsat data can be used to monitor large-scale trends in land use such as forest clearings or conversion of land for agricultural uses. Other uses include detection of thermal water pollution and location of areas of seasonal forage production. Landsat may also aid in location of potential study sites for more detailed field investigations (Anderson, et al., 1980).

All detailed remote sensing data must be verified through ground truths, special trips by field teams into the study areas. The remotely sensed data are then calibrated according to measurements taken during the ground truths.

HABITAT MANAGEMENT

Once the habitat within a given area has been evaluated, the manager must decide what, if any, manipulations are needed to fulfill the management objectives. If the area still exists in an undisturbed state and the objective is to maintain pristine conditions, the manager may decide to forego any manipulations and concentrate instead upon protection from disturbance. This same prescription often applies to the habitat of a threatened or endangered species. Should the management objective be to maximize wildlife diversity, then managers may create limited disturbances through use of fire or timber cutting, in the pattern most likely to achieve maximum diversity within that particular region. Game species usually reach their highest numbers under early-to-mid successional stages, so game managers commonly manipulate habitat to maintain those favorable stages. Other

FIGURE 2-3
A Landsat color composite, showing the Wichita Wildlife Refuge and Fort Sill Military Reservation. The pink area near the center is the city of Lawton, Oklahoma. Across the lower portion appears the Red River.

useful methods for habitat management include planting and manipulation of edge.

Adjustment of Seral Stages

The most common form of habitat management is adjustment of the seral or successional stage so that the management area becomes better suited to the desired species.

Setting Back Succession *Fire* The least expensive, most natural and universal way of setting back succession in terrestrial habitats is fire. Although fires are often popularly regarded as destructive and harmful to wildlife, ecologists have long recognized the importance of fire in the perpetuation of many wild plant and animal species, particularly those best adapted to early- to mid-successional stages.

But as natural and important as fire is, its use in habitat management does have definite drawbacks. Prescribed fires can and do get out of control. Their use is subject to safety regulations and to air pollution laws. Moreover, fire is not always beneficial to wildlife. Burns that are too hot from combustion of too much fuel or fires occurring at the wrong time of year can have serious consequences for local wildlife populations.

Wild fires have long been part of the environment of North American grasslands or prairies. Consequently, the plant and animal life of those regions are well adapted to the occurrence of fire, and many species fare better in the months and years following a burn. Researchers at the Woodworth Station of the Northern Prairie Wildlife Research Center in North Dakota, for example, have compared use and reproductive success on recently burned and unburned grassland (Kirsch and Kruse, 1972). They found that 52 percent of the duck nests on burned grassland were successful compared with only 33 percent on unburned grassland. Sharp-tailed grouse produced more than twice as many broods per unit area on burned as upon unburned grassland. Production of upland plover (*Bartramia longicauda*) was 67 percent higher on burned than unburned prairie.

Fires also changed the vegetation at the Woodworth Station. There was a 30 percent increase in the number of plant species on burned over unburned plots. Flowering, seed production, and height also increased among the plants on burned areas.

Many of the gallinaceous birds of North America fare better in habitats that are periodically burned. Prairie chickens need nesting and brood-rearing habitat that can best be provided by prescribed burning at three- to five-year intervals (Kirsch, 1974). Small, freshly burned areas are strong attractants for male prairie chickens as booming grounds (Cannon and Knopf, 1979). Bobwhite quail on the Welder Wildlife Refuge in Texas show consistently higher use of experimentally burned areas than they do for unburned control areas (Wilson and Crawford, 1979).

Deer populations often increase in the years following fire. Burned areas in a pinyon-juniper woodland produced about three times the herbaceous forage per unit area than unburned areas, and mule deer (*Odocoileus hemionus*) made greater use of the burns (McCulloch, 1969).

On the other hand, fires reduce wildlife populations of species best adapted to late-successional conditions. Intense fires in a boreal forest region in Alaska resulted in a 60 percent reduction in the breeding population of Alaskan spruce grouse (*Canachites canadensis*), and numbers on unburned areas remained the same (Ellison, 1975).

In a comprehensive review of the effects of fire, Bendell (1974) found such inconsistency that he cautioned against making firm generalizations. Bendell found that bird and mammal diversity showed only a slight increase in grassland and shrub regions following fires, and a slight decrease in forested regions.

Grazing Livestock grazing sets back succession. National wildlife refuges along the Gulf Coast serve as important wintering areas for waterfowl. Some cattle are kept on these refuges under grazing leases because they help retard succession and maintain better feeding areas for waterfowl. Livestock grazing must be carefully regulated, however, as overgrazing not only destroys wildlife habitat but also may permanently decrease productivity for both wildlife and livestock (see Chapter 10).

Logging The cutting of forests sets back succession and may thus favor early- to midsuccessional species. White-tailed deer are usually favored by limited amounts of logging in small cutting units, so long as native vegetation is allowed to grow in the cutover areas. As a general rule, cutover forests produce considerably more browse than do mature forests (Halls and Alcaniz, 1968; Swift, 1960).

Mechanical Treatments Much more expensive than fire, many mechanical treatments are safer and the outcome more predictable. In some of the western rangelands into which brush has invaded, succession can be set back through bulldozing. Two bulldozers can treat large areas by dragging a heavy cable or chain between them, ripping out brush and small trees by the roots (Yoakum, 1971; Yoakum and Dasmann, 1971).

Disking is commonly employed on agricultural lands to set back succession. It is particularly useful for breaking up dense stands of sod-forming perennial grasses to favor the food-rich annuals (Burger, 1973).

Mowing is a good way to create edge and it helps stimulate production of annuals as well. Some stands of brush, including honeysuckle (*Lonicera* spp.) and sumac (*Rhus* spp.) eventually become senescent. Mowing them restores vigor (Burger, 1973).

Herbicides Chemical herbicides offer a relatively inexpensive and potentially highly selective means of manipulating succession. Herbicides can be sprayed, injected, or even painted onto undesirable plants. Although unquestionably effective, herbicides also pose environmental hazards and many are potentially carcinogenic or mutagenic. Until their long-term effects are better understood and more refined methods of application developed, most authorities (Burger, 1973; Yoakum et al., 1980) warn against their widespread use in wildlife habitat management.

Advancing Succession Succession will, of course, advance naturally, so long as disturbances do not intervene. So the principal means of advancing succession is the exclusion of disturbances such as fire, grazing, logging, or mechanical or chemical disturbances.

The main problem with advancing succession for management objectives is time. As shown in Table 2-2, decades or even centuries may pass before mature communities are established. This time problem is the reason why conservationists battle so hard to preserve representative samples of old growth forests wherever they still occur. Few have the patience to wait several centuries for the mature forest to reestablish itself.

Planting This approach to habitat management is, of course, far more expensive than fire and most other methods used to set back succession. But planting also leads to faster and more predictable results (Burger, 1973).

Woody perennials are typically planted to develop thermal and escape cover, though many species such as blackberry (*Rubus* spp.) and wild plum (*Prunus* spp.) also provide food. All woody perennials aid in stabilizing soils, thus reducing erosion. The planting of woody perennials also helps advance succession.

Perennials, being long-lived, put much of their yearly production into vegetative growth. Annuals, on the other hand, can perpetuate themselves only through seed production and thus they typically provide more food than do perennials. It is no coincidence that our most popular and productive agricultural crops including wheat and corn are annuals.

Annuals, then, are planted in wildlife management mainly to produce foods. This can easily be done on farms merely by leaving grain unharvested in odd corners, along fences, and other places where modern farm machinery is hard to maneuver. Under other conditions annuals may be planted exclusively for wildlife. Burger (1973) recommended several mixtures of annual seeds for application in various regions of the United States.

Whenever planting is used, it must be planned carefully. A clump of shrubs or a stand of annuals in the middle of a large, barren field will prove virtually useless. Instead, cover should be planted where it is in close proximity to food and vice versa.

Construction of Cover This approach can aid many species, including rabbits and quail. Artificial nest boxes like those constructed for the wood duck (*Aix sponsa*) and nesting platforms built for mallards (*Anas platyrhynchos*) provide important nesting cover. Nesting platforms can be useful for species other than game. They have been successfully placed on topped trees for ospreys and on telephone poles for double-crested cormorants (*Phalacrocorax auritus*) (Yoakum et al., 1980).

Brush piles act as important sources of cover for protection from both predators and severe weather. They have been successfully used for the benefit of rabbits, wild turkeys (*Meleagris gallopavo*), collared peccary (*Tayassu tajacu*), and several species of quail and sparrow (Yoakum et al., 1980).

Manipulation of Edge

Another common type of habitat management involves the manipulation of edge. Since so many game and nongame species seem to be favored by edge, management is often directed at increasing the amount of edge per unit area. This is often done by setting back (or in a few cases, by advancing) succession in small, irregular patches that, whenever possible, follow natural topographic features. If planting is used, the planted areas are shaped to maximize edge. Either way, the manipulated habitat becomes a mosaic of smaller patches providing a larger variety of habitat types per unit area.

Inherent edge is that which naturally occurs because of abrupt changes in soil type, topography, or microclimate (Thomas et al., 1979). It is a relatively permanent feature of the habitat, so habitat managers devote their efforts toward creating induced edge. Mowing in irregular patches is a quick and dependable way to create induced edge. Forest clearcuts can be designed so as to maximize induced edge. Plantings too can be done with edge in mind.

Because some species are not favored by edge, occasionally it becomes necessary to manage habitat so as to minimize edge. This seems especially true of many tropical species that evolved in large tracts of contiguous, mature rain forests. The best management strategy under such conditions is preservation of large tracts of mature forest.

Mitigation

Mitigation is a special and potentially very important form of habitat management. Basically, mitigation is a procedure for compensating for habitat losses caused by development. This compensation is done by securing and improving similar habitats elsewhere, so that the wildlife population losses in the developed area will be offset by population increases in the mitigated area. In the United States, federal agencies such as the Army Corps of Engineers and the Bureau of Reclamation are required under the Fish and Wildlife Coordination Act of 1958 to mitigate losses imposed by development projects. The cost of mitigation is supposed to be borne by the developing agency as a part of the project costs. Selection of mitigation sites and recommendation for specific habitat improvements are the responsibility of the U.S. Fish and Wildlife Service. The service selects mitigation sites using the same habitat evaluation procedures (HEP) described earlier.

Like many other methods in wildlife management, mitigation is easier to describe in theory than it is to implement in practice. Rappaport et al. (1977) cited as problems inadequate initial assessment of project impacts, poor interagency cooperation, and biological ineffectiveness of the little mitigation actually carried out. Suggestions for improvement included a more firmly worded amendment to the 1958 act, more rigorous enforcement of existing laws, or both. Parenteau (1977) suggested a stronger role for the U.S. Fish and Wildlife Service (along with increased funding) and advocated mitigation criteria that incorporate nonconsumptive as well as consumptive values of wildlife. While there remains room for improvement, mitigation represents a very important means of providing wildlife habitat.

SUMMARY

The basic components of wildlife habitat include food, cover, water, and space. Species that are the same number of steps from the primary producers or green plants are at the same trophic level. Most of the energy transferred between trophic levels is lost as dissipated heat, and the losses are greatest when they involve homeothermic species. The role played by a species in a community is called its niche.

Though variously defined, carrying capacity is the number of animals of a particular species that can be maintained on a given area. Carrying capacity is not a fixed value but rather one subject to change. One of the most common causes of changing carrying capacity is biotic succession, the predictable change in the plant community through time. Managers often use this principle to manipulate habitat toward the stage that best suits the desired species.

Another common form of habitat manipulation employs the edge effect. For species with limited mobility and varied habitat requirements, the greater the edge, the higher the carrying capacity. Species diversity also tends to increase with increasing edge.

Habitat evaluation can assess suitability for a single species, for an array of featured species, or for an entire natural community. Food production, availability of cover, and edge are commonly measured habitat components. For herbivores, certain plants may be used as indicator species to show the size of a population in relation to carrying capacity. Indirect, but thoroughly practical methods of evaluating habitat involve determining the condition of individual animals and measuring the rate of population increase. These methods work because condition of individual animals and reproductive success reflect the overall quality of the habitat.

Evaluation of natural communities is a newer practice and one that is becoming more important in wildlife management. Species diversity, measuring both the total number of species within a particular area and their percentage occurrence is a major tool for this type of evaluation. The U.S. Fish and Wildlife Service and other agencies have developed standard evaluation procedures which require rapid assessments of sets of species. Such evaluations are usually done to select the richest and most diverse areas for protection.

The most common form of habitat management is adjustment of the successional stage to favor the desired species, whether by setting succession back or by aiding its advance. Planting is a more expensive but generally faster and more reliable means of habitat management. Woody perennials are planted to furnish cover, and annuals are seeded to provide food. Another common habitat management practice, often used in conjunction with manipulation of succession, is manipulation of edge. Increases in edge usually favor game species and boost overall species diversity. But there are edge-sensitive species for which the best management may be to minimize the amount of edge.

An important form of habitat management is mitigation, or the practice of compensating for habitat lost in one area by improving habitat elsewhere. Whenever habitat is lost through federal development projects, the developing agency is by law required to secure and improve habitat elsewhere. The U.S. Fish and Wildlife Service evaluates potential mitigation sites and makes recommendations.

REFERENCES

Anderson, W., W. Wentz, & B. Treadwell. 1980. A guide to remote sensing information for wildlife biologists. In S. D. Schemnitz (Ed.), Wildlife management techniques manual (4th ed.). Washington, D.C.: The Wildlife Society.

Baxter, W., & C. Wolfe. 1972. The interspersion index as a technique for evaluation of bobwhite quail habitat. In J. Morrison and J. Lewis (Eds.), *Proc. 1st National Bobwhite Quail Symp.* Stillwater, Okla.: Oklahoma State University.

Bendell, J. F. 1974. Effects of fire on birds and mammals. In T. Kozlowski and C. Ahlgren (Eds.), *Fire and Ecosystems.* New York: Academic Press.

Bergerud, A., & F. Manuel. 1968. Moose damage to balsam fir-white birch forests in central Newfoundland. Newfoundland. *J. Wildl. Manage.* 32: 729–746.

Blair, R. M. 1959. Weight techniques for sampling browse production on deer ranges. In *U.S. Forest Service, Techniques and methods of measuring understory vegetation.* Asheville, N.C.: Southeastern and Southern Forest Experiment Stations.

Burger, G. 1973. *Practical wildlife management.* New York: Winchester Press.

Cannon, R., & F. Knopf. 1979. Lesser prairie chicken responses to range fires at the booming ground. *Wildl. Soc. Bull.* 71: 44–46.

DeVos, A., & H. Mosby. 1971. Habitat analysis and evaluation. In R. Giles (Ed.), *Wildlife management techniques.* Washington, D.C.: The Wildlife Society.

Ellison, L. N. 1975. Density of Alaskan spruce grouse before and after fire. *J. Wildl. Manage.* 39: 468–471.

Eve, J., & F. Kellogg. 1977. Management implications of abomassal parasites in southeastern white-tailed deer. *J. Wildl. Manage.* 41: 169–177.

Flood, B., M. Sangster, R. Sparrowe, & T. Baskett. 1977. A handbook for habitat evaluation procedures. *Resour. Publ. 132.* Washington, D.C.: U.S. Fish and Wildlife Service.

Fretwell, S. 1972. *Populations in a seasonal environment.* Princeton, N.J.: Princeton University Press.

Goldsmith, F. 1975. The evaluation of ecological resources in the countryside for conservation purposes. *Biol. Conserv.* 8: 89–96.

Gullion, G. W. 1970. Factors influencing ruffed grouse populations. *Trans. N. Am. Wildl. and Natur. Resour. Conf.* 35: 93–105.

Graber, Jean W., & R. K. Graber. 1976. Environmental evaluations using birds and their habitats. *Biol. Note No. 97. Illinois Nat. Hist. Surv.*

Grange, W. B. 1948. Habitat chronology of the town of Remington, Wisconsin. In *Wisconsin grouse problems.* Pub. 328, Madison, Wisc.: Wisconsin Department of Conservation.

Gysel, L., & L. Lyon. 1980. Habitat analysis and evaluation. In S. Schemnitz (Ed.), *Wildlife management techniques manual* (4th ed.). Washington, D.C.: The Wildlife Society.

Halls, L., & R. Alcaniz. 1968. Browse plants yield best in forest openings. *J. Wildl. Manage.* 32: 185–186.

Harlow, R. F. 1977. A technique for surveying deer forage in the southeast. *Wildl. Soc. Bull.* 7: 185–191.

Harlow, R. F. 1979. In defense of inkberry—dangers of ranking deer forage. *Wild. Soc. Bull.* 7: 21–24.

Heaslip, G. B. 1976. Satellite viewing our world: the NASA Landsat and the NOAA SMS/GOES. *Environ. Manage.* 1(1): 15–29.

Hunt, H. M. 1979. Comparison of dry-weight methods for estimating elk femur marrow fat. *J. Wildl. Manage.* 43: 560–562.

Jordan, P., D. Botkin, & M. Wolfe. 1971. Biomass dynamics in a moose population. *Ecology* 52: 147–152.

Kirsch, L. M. 1974. Habitat management considerations for prairie chickens. *Wildl. Soc. Bull.* 2: 124–129.

Kirsch, L., & A. Kruse. 1972. Prairie fires and wildlife. *Proc. Ann. Tall Timbers Fire Ecology Conference.*

Leopold, Aldo. 1933. *Game management.* New York: Charles Scribner's Sons.

Lines, I., Jr., & C. Perry. 1978. A numerical wildlife habitat evaluation procedure. *Trans. N. Am. Wildl. and Natur. Resour. Conf.* 43: 284–301.

Martin, A., H. Zim, & A. Nelson. 1951. American wildlife and plants: a guide to wildlife food habits. New York: McGraw-Hill.

McCulloch, C. Y. 1969. Some effects of wildfire on deer habitat in pinyon-juniper woodland. *J. Wildl. Manage.* 33: 778–784.

Mosby, H. S. 1980. Reconnaissance mapping and map use. In S. Schemnitz (Ed.), *Wildlife management techniques manual* (4th ed.). Washington, D.C.: Wildlife Society.

Nudds, T. D. 1977. Quantifying the vegetative structure of wildlife cover. *Wildl. Soc. Bull.* 5: 113–117.

Parenteau, P. A. 1977. Unfulfilled mitigation requirements of the Fish and Wildlife Coordination Act. *Trans. N. Am. Wildl. and Natur. Resour. Conf.* 42: 179–184.

Patton, D. R. 1975. A diversity index for quantifying habitat "edge." *Wildl. Soc. Bull.* 3: 171–173.

Peet, R. K. 1974. The measurement of species diversity. *Ann. Rev. Ecol. and Syst.* 5: 285–307.

Rappaport, Ann, J. Mitchell, & J. Nagy. 1977. Mitigating the impacts to wildlife from socioeconomic developments. *Trans. N. Am. Wildl. and Natur. Resour. Conf.* 43: 274–283.

Samson, F., & F. Knopf. 1982. In search of a diversity ethic for wildlife management. *Trans. N. Am. Wildl. and Natur. Resour. Conf.* 47: 421–431.

Schamberger, M., & A. Farmer. 1978. The habitat evaluation procedures: their application in project planning. *Trans. N. Am. Wildl. and Natur. Resour. Conf.* 43: 274–283.

Seal, U. S. 1977. Assessment of habitat condition by measurement of biochemical and endocrine indicators of the nutritional, reproductive, and disease status of free-living animal populations. In *Classification, inventory, and analysis of fish and wildlife habitat.* Proceedings of a national symposium. Office of Biological Service, Washington, D.C.: U.S. Fish and Wildlife Service.

Seal, U. S., M. Nelson, L. Mech, & R. Hoskinson. 1978. Metabolic indicators of habitat differences in four Minnesota deer populations. *J. Wildl. Manage.* 42: 746–754.

Shannon, C., & W. Weaver. 1949. The mathematical theory of communication. Urbana, Ill.: University of Illinois Press.

Strelke, W., & J. Dickson. 1980. Effect of forest clear-cut edge on breeding birds in east Texas. *J. Wildl. Manage.* 44: 559–567.

Swift, L. W. 1960. Wildlife management and protection in the forests of the United States. *Proc. 5th World Forestry Congress.* Seattle, Washington. 5: 1794–1800.

Taylor, M. W. 1977. A comparison of three edge indexes. *Wildl. Soc. Bull.* 5: 192–193.

Thomas, J., C. Maser, & J. Rodiek. 1979. Edges. In J. W. Thomas (Ed.), *Wildlife habitats in managed forests—the Blue Mountains of Oregon and Washington.* USDA Forest Service Agriculture Handbook No. 553.

Tueller, P., & J. Tower. 1979. Vegetation stagnation in three-phase big game exclosures. *J. Range Manage.* 32: 258–263.

Wilson, Marcia M., & J. A. Crawford. 1979. Response of bobwhites to controlled burning in south Texas. *Wildl. Soc. Bull.* 7: 53–56.

Yoakum, J. 1971. Habitat improvement. In R. D. Teague (Ed.), *A manual of wildlife conservation.* Washington, D.C.: The Wildlife Society.

Yoakum, J., & W. P. Dasmann. 1971. Habitat manipulation practices. In R. D. Giles (Ed.), *Wildlife management techniques.* Washington, D.C.: The Wildlife Society.

Yoakum, J., W. Dasmann, H. Sanderson, C. Nixon, & H. Crawford. 1980. Habitat improvement techniques. In S. Schemnitz (Ed.), *Wildlife management techniques manual.* Washington, D.C.: Tbe Wildlife Society.

WILDLIFE POPULATIONS

Although the means of managing wildlife are generally carried out through manipulation of habitat, the objectives are usually defined in relation to one or more wildlife populations. Wildlife managers may want to increase a species population, as they would when managing a rare or endangered species. Or they may want to increase the harvest of game species, a task which is *not* the same as increasing numbers (see Chapter 7, Management for Harvest). In some cases, the objective may be to stabilize a population at a given level. When populations are too high, are causing economic damage, or both, the management objective may be to reduce the population through various types of population control. Some of these objectives can be met through habitat manipulation, but others require direct and systematic manipulation of the population itself.

Habitat conditions determine carrying capacity for a given species. When populations are low relative to carrying capacity, birth rates tend to be relatively high. Populations at carrying capacity have relatively low birth rates and higher death rates. Both birth and death rates thus change in relation to population density. This density-dependence is a fundamental concept in the management of wildlife populations.

POPULATION DYNAMICS

A population is comprised of all members of a particular species within a definable area. The basic components of any population include a birth rate, a death rate, a sex ratio, and an age structure (Cole, 1957). When populations change, they do so through shifts in these four components. This section also includes discussion of dispersal, a critical population phenomenon still poorly understood.

Birth Rate

The actual birth rate, usually called fecundity, is measured as the number of live births per female over a given period of time, usually one year. In egg-laying species, birth rate is measured in terms of egg production per female. For many calculations used in population analysis, the birth rate is defined as the number of female births per female.

A number of factors can affect a population's birth rate. A decline in the quality or quantity of food per member of a population may cause a corresponding decrease in birth rate. The frequency of births (birth intervals) affects a population's birth rate, though most temperate zone and arctic species reproduce once yearly in spring and early summer. Among mammals, birth rates tend to be higher when populations are low relative to carrying capacity (K) and gradually diminish as K is approached. The effects of fecundity rates upon population growth are shown in Table 3-1. Note that when the fecundity rate doubles, the potential rate of growth for populations more than doubles.

Death Rate

A population's growth shows as it nears K and, once K is reached, the population grows no more. As shown above, part of the reason for the decline in growth is a reduction in birth rates. But in most cases, the majority of the decrease stems directly from an increase in the population's

TABLE 3-1
EFFECT OF FECUNDITY RATE ON POPULATION GROWTH
Assume 50:50 sex ratio and sexual maturity by the end of the first year, 1 litter or clutch per female per year and no mortality

	Population size					
	Year					
Mean litter or clutch size	**0**	**1**	**2**	**3**	**4**	**5**
1	2	3	4	7	10	15
2	2	4	8	16	32	64
5	2	7	25	86	300	1,052
10	2	12	72	432	2,594	15, 571
20	2	22	242	2,663	29,294	322,271

death rate. Causes of death that increase along with population size are termed density-dependent forms of mortality and may include starvation, diseases, parasitism, predation, or combinations thereof. These forms of mortality include those most important to terrestrial populations; thus there is a fairly consistent positive relationship between death rate and population size.

Death rates are rarely constant between different age classes. Among mammals, mortality rates are generally higher among the juvenile and very old age classes. Animals in their middle or prime years experience the lowest mortality rates. Mortality rates among birds are less well known, in part because birds are more mobile, often smaller, and hence carcasses more difficult to recover. But the main problem in determining death rates in avian populations is aging the birds. In most avian species, it is impossible to age them beyond their first year or two. This difficulty in obtaining reliable data on age structure and death rates has led ornithologists to assume that postfledgling mortality rates are constant, irrespective of age class. This is only an assumption and, until adequately tested, should be taken as such. In fact, a comparison of published avian mortality rates with maximum observed longevities led Botkin and Miller (1974) to reject the hypothesis of age-independent mortality. They suggested that the simplest alternative explanation was that mortality increases gradually with age.

Age Composition

For almost all vertebrate populations, both fecundity and mortality rates change with age. Thus, at any given time, a population's age composition has much to do with its population dynamics. The effect of age at first breeding upon population growth is shown in Table 3-2. Age composition has little if any effect on population growth for species that typically reproduce during their first year. But the greater the age at first breeding, the greater the effect that age composition can have upon population growth.

The first problem a biologist may encounter when trying to determine age structure is finding a reasonably precise aging technique. Precise aging for birds is generally possible only when samples of the population are permanently banded as fledglings. Mammals are a bit easier to age; such characteristics as tooth wear, tooth eruption, cementum annuli rings, eye lens weights, epiphyseal fusion, and growth rings in horns are used as indicators of age (see Taber, 1971, for a detailed review of aging techniques).

Aside from techniques that permanently mark animals during their first year, no aging method is 100 percent accurate, so errors are inevitable.

TABLE 3-2
EFFECT OF AGE AT FIRST BREEDING UPON POPULATION GROWTH
Assume 50:50 sex ratio, a fecundity rate of 2.0 per adult female per year and no mortality.

	Population size					
	Year					
Age at 1st breeding (years)	**0**	**1**	**2**	**3**	**5**	**10**
1	2	4	8	16	64	2,048
2	2	4	6	10	26	288
3	2	4	6	8	18	120
4	2	4	6	8	14	72
5	2	4	6	8	12	52

Some authorities merely accept these inevitabilities, assuming that the errors will "average out" in the analysis. However, as Caughley (1977) has shown, even errors on the order of 10 percent can substantially affect analysis and the conclusions drawn from it. Errors in aging would only "average out" if all age classes were equally represented in the population, a highly unlikely condition. Moreover, assuming that there is an equal chance of aging animals too young or too old, the magnitude of errors resulting from erroneous aging will depend on the difference in the size of the age classes. For a specific example of this effect, plus a discussion on the uses and abuses of age ratios, see Chapter 4.

Sex Ratio

A sex ratio of about 50:50 at birth is the general rule among most species of vertebrates (Caughley, 1977; Emmel, 1976). Departures from this ratio (principally through differential mortality) influence population dynamics in various ways, depending on a species' mating habits. Population growth in monogamous species (those that form a pair bond) will decline with a departure from an even ratio, regardless of whether males or females incur the higher mortality rate. In polygynous species (those in which each male mates with more than one female), changes in sex ratio can have major effects on population growth. Reproductive rates within a polygynous species are, within limits, a function of the number of breeding-age females. Thus, assuming all females breed, a population with a ratio of 1 male to 4 females can produce up to 1.6 times as many young as could the same-sized population with a 50:50 sex ratio. Conversely, the same polygynous species with a ratio of 4 males to 1 female could produce only about 40 percent as many young as could the same population with a 50:50 sex ratio (Table 3-3).

Dispersal

The movement an animal makes from its birth or release site to the place where it reproduces or would have reproduced had it survived is termed "dispersal" (Howard, 1960). Dispersal is essential for long-term survival of populations. Local habitat conditions change through time, so that sooner or later they become unsuited for a population of a particular species. Populations would become extinct if it were not for successful dispersal of some of its members (usually subadults) into areas that furnish better habitat.

Yet despite its importance in both population ecology and wildlife management, dispersal remains poorly understood and, in Caughley's view (1977), the most difficult of all population processes to investigate. Rather than study dispersal directly, biologists often must content themselves with estimates of rate of spread by documenting the rate at which a species expands its geographic range. Such range expansion furnishes at least a rough index of dispersal. Humphrey (1974) studied range expansion in the nine-banded armadillo and estimated its rate of spread to be about 10 kilometers per year. Surprisingly, eight species of big-game animals released in New Zealand expanded their ranges at rates of from 0.6 to 8.7 kilometers per year (Caughley, 1963), all considerably less than the armadillo.

TABLE 3-3
THE EFFECTS OF SEX RATIO UPON POPULATION GROWTH
Assume that each population is composed of 1,000 animals and that each female bears 2.0 young

Sex ratio (males:females)	Number of females	Number of young produced
1:5	833	1,666
1:4	800	1,600
1:3	750	1,500
1:2	667	1,334
1:1	500	1,000
2:1	333	666
3:1	250	500
4:1	200	400
5:1	167	334

The Logistic Equation

A mathematically simple way of looking at the relationship between the size of a population and the rate at which it grows is through the logistic equation. To understand how this equation works, first consider the simplest representation of population change, the percent increase (or decrease). If a population contains 100 individuals one year and 105 the next year, it grew 5 percent. Should the population continue to grow at that rate, the following year it would contain $105 \times .05 + 105 = 110$ (the fraction would be rounded off to the nearest whole animal) and the year after it would have $110 \times .05 + 110 = 116$. The calculation can be simplified by merely multiplying the population of each year times 1.05. In this case the 1.05 is called lambda (λ).

Populations tend to grow exponentially and because of this biologists have found it convenient to substitute for lambda an exponential value to depict populations growing at a constant rate. They use the base of natural logarithms (2.71828) with the value r representing the power to which the base is raised. A positive value for r shows population growth, a negative r means a decreasing population, and an r of 0 indicates a stable population. Table 3-4 compares percent growth, lambda, and r. Note that r is the natural log of lambda.

Biotic potential and intrinsic rate of increase are other terms used interchangeably with r, to indicate the maximum rate at which a population

TABLE 3-4
RATES OF INCREASE OR DECREASE AS PERCENT, LAMBDA, AND *r*

% Increase or decrease	Lambda	*r*
−100	0.00	
−75	0.25	−1.386
−50	0.50	−0.693
−25	0.75	−0.288
0	1.00	0.000
+10	1.10	+0.095
+25	1.25	+0.223
+50	1.50	+0.406
+75	1.75	+0.560
+100	2.00	+0.693
+200	3.00	+1.099
+500	6.00	+1.792

can grow with a stable age distribution when no resource is limiting (i.e., a theoretically perfect environment). Thus, in its simplest form the exponential growth equation is

Change in population size/Change in time interval (usually 1 year) $= rN$

where r is the intrinsic rate of increase and N is the population size at the start of the time interval.

An example of a population growing at or very near its biotic potential is that of the white-tailed deer in Michigan's George Reserve. Early in 1928 two bucks and four does (the latter assumed to be pregnant) were released in a fenced 458-hectare (1,146-acre) enclosure from which the original deer had long been extirpated. Nearly 6 years later, in December of 1933, biologists conducted a drive count and tallied 160 deer. Two deer carcasses were found, bringing the total to at least 162. Drive counts, though, rarely can account for all animals in a population, and McCullough's (1979) reconstruction of the population data indicated that the real total was probably about 181. For the population to increase from 6 to 181 within 6 years, it had to grow at an average rate of 76.5 percent. If, for the sake of simplicity, one is bold enough to assume that the rate of increase was constant for those 6 years, the lambda was 1.765, hence r was 0.588. The resulting growth curve, shown in Figure 3-1, is essentially the J-shaped curve. The carrying capacity (K) of the habitat invariably slows and then halts population growth entirely. This biological inevitability can be approximated mathematically by adding the expression $(K-N)/K$ to the exponential equation so that

Change in population size/Change in time interval $= rN[(K-N)/K]$

Note that the closer the population is to K, the greater the effect of the added expression. For example, if K is 300 and the population is 10, then

$(300-10)/300 = 0.967$

But if K is 300 and the population is 275, then

$(300-275)/300 = 0.083$

which, when multiplied times rN, will diminish its value considerably.

The addition of K changes the shape of the resulting population growth curves from the J shape to the sigmoid or S shape and converts the exponential equation to the logistic equation (see Figure 3-2). The number of animals annually added to the population is highest at the inflection point, midway along the sigmoid curve at $K/2$. Beyond the inflection point, growth slows steadily and then stops when K is reached at the asymptote.

The basic equation and the sigmoid growth curve illustrate how a population will grow from a very small number to a much greater one in a finite environment. Most wildlife managers, however, never actually deal with populations that move completely through the range of values

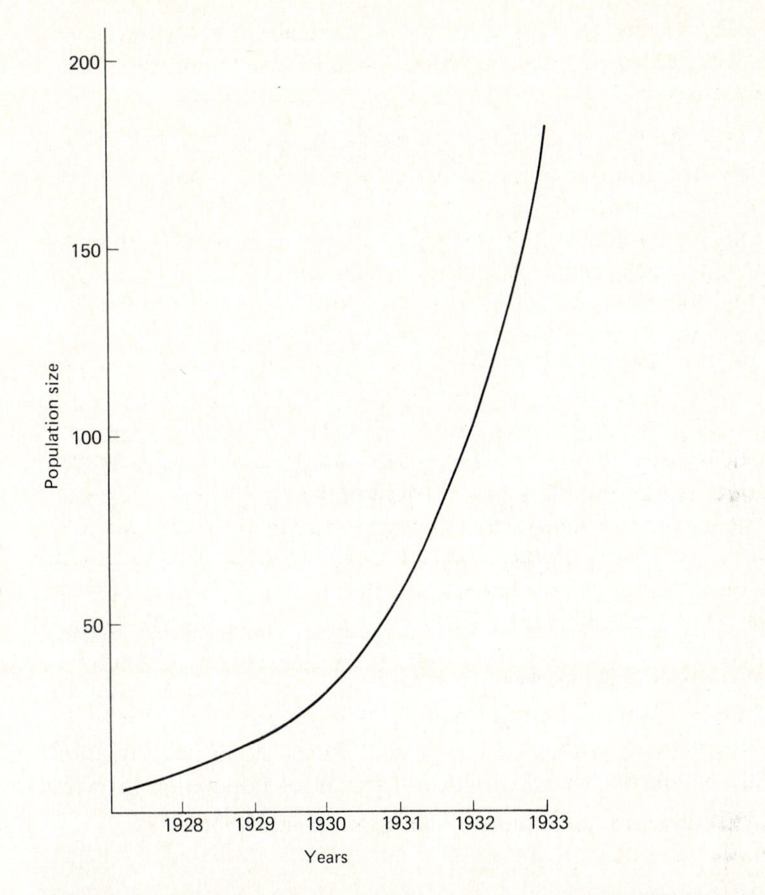

FIGURE 3-1 Population growth of the George Reserve deer herd. The initial and final population estimates are from McCullough (1979). Points in between are based on an assumption of a constant rate of growth of 76.5 percent annually, the average required for the observed increase (copyright © 1979, The University of Michigan Press, by permission).

depicted by the sigmoid curve. Endangered species specialists may work on populations confined to the lower half of the curve, and game managers, more often than not, deal with populations that fall somewhere between the inflection point and K.

Moreover, as time is irreversible, the sigmoid growth curve implies that populations invariably continue to grow until K is reached. In reality, populations change continually, usually in response to changes in habitat conditions. Although K is a fixed value within the logistic equation, it may in the real world fluctuate as succession advances or is set back. And while r is generally fixed within the equation as a theoretical maximum, the actual rate of increase can change in response to shifts in age structure,

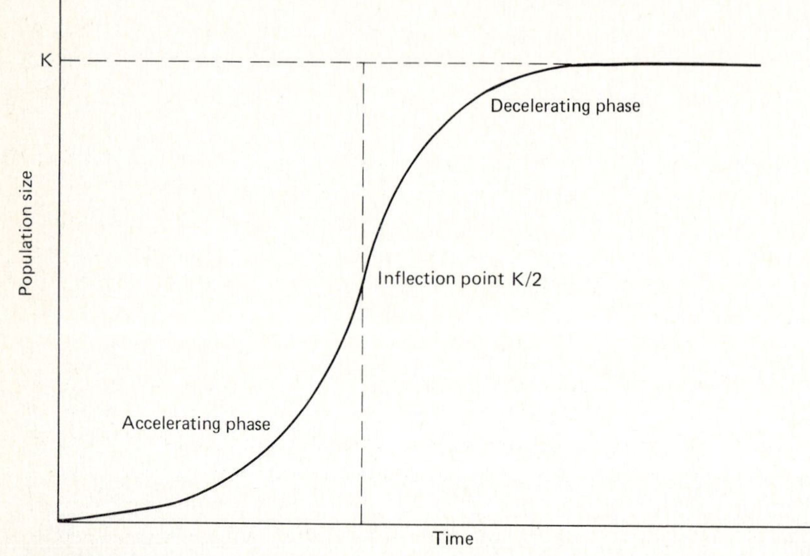

FIGURE 3-2
The S-shaped or sigmoid growth curve generated by the logistic growth equation. Midway through the curve is the inflection point, $K2$, the point of maximum rate of increase and productivity (see Chapter 7, Management for Harvest).

fecundity rates, mortality rates, and sex ratios. Thus, the logistic equation is useful only in that it provides a general model of population growth. More precise analysis requires more specific and complex models.

Major Population Fluctuations

While all populations fluctuate to some extent, some of them do so very markedly. Major population fluctuations are of two main types, cycles and irruptions. Cycles are major population changes that occur at regular

FIGURE 3-3
A continuum between strongly *r*-selected species and strongly *K*-selected species. These traits should always be viewed as relative measures, since each species' life history must compromise to some extent between *r*- and *K*-selection.

Pests	Game species		Endangered species
Norway rat	Cottontail rabbit	White-tailed deer	Golden eagle
House mouse	Mallard	Elk	Whooping crane
Rock dove (pigeon)	Bobwhite quail	Moose	African elephant
		Black bear	California condor

r-selected ——————————————————— K-selected

intervals, usually either the 3- to 4-year cycle characteristics of microtine rodents in arctic regions, or the 9- to 10-year cycle found in populations of snowshoe hare, lynx (*Lynx canadensis*), and ruffed grouse. The longer cycles typically occur in boreal forest.

The 9- to 10-year cycles were first described by the British ecologist Charles Elton after he reviewed a century or so of fur-trapping records of the Hudson's Bay Company's Northern Department (Elton, 1924; Elton and Nicholson, 1942). Since then most ecologists have accepted cycles as valid population phenomena that occur under certain conditions (Keith, 1963).

Although the existence of cycles has generally become accepted, the causes of these cycles remain controversial. Two types of causes have been postulated: extrinsic, those caused by forces outside the populations themselves, and intrinsic, those forces from inside the populations. Extrinsic factors include such varied phenomena as sunspots, long-term climatic cycles, infectious diseases, declines in quality and quantity of foods, and predation. Intrinsic factors consist of physiological stresses brought on by crowding, changes in social systems, and, most recently, genetic changes within the populations (Krebs et al., 1973).

Certain extrinsic phenomena correlate nicely with cycles, but actual causal mechanisms have not been found. The 9- to 10-year cycle correlates fairly closely with the 18.6-year modal cycle of the moon, so that two population peaks occur in each full cycle (Archibald, 1977). Archibald could only speculate on possible causal mechanisms, suggesting that changes in moonlight might have affected behavior, weather, or photoperiod.

Irruption, the other type of major population fluctuation, is of more concern to the wildlife manager. Irruptions occur at irregular intervals and seem to occur in more temperate and perhaps even tropical regions than do cycles. Furthermore, these severe population increases can have serious economic consequences.

An example of an economically serious irruption is that of house mouse (*Mus musculus*) populations of South Australia. At irregular intervals these populations explode, resulting in heavy crop losses both in the fields and in storage facilities. The region has seasonally variable rainfall, with most rains falling during late winter and early spring. Fields become waterlogged and mice cannot be found there. Summers are hot and dry, causing the soils of the fields to crack, providing cool and relatively moist retreats for mice that reinvade from surrounding areas. Should rain fall during the otherwise dry summer, it creates favorable burrowing conditions within and below the cracks. The mice breed rapidly and successfully and the population irrupts (Newsome, 1969). At least in this case, the irruption is triggered by rainfall patterns.

From the 1920s through the 1940s many of the cervid populations in the United States built up sharply, causing many wildlife managers a great deal of concern. At that time, most wildlife authorities (cf. Leopold et al., 1947) suspected that the elimination of natural predators such as the gray wolf and the cougar was responsible for these "irruptions." In more recent years, this view has been challenged (Caughley, 1970, 1977). Caughley believed that increases in the cervid populations were merely the normal oscillations caused by the interaction of these herbivores with their food supply. The natural regulation of cervid populations remains a controversial topic and is treated in greater detail in Chapter 6.

Population Patterns

The patterns of population growth seen in wild animals were shaped through natural selection as means of adapting to environmental conditions. A useful distinction between the two main kinds of population strategies emerged from a theoretical study of island biogeography MacArthur and Wilson, 1967). According to this theory, the first animals reaching uncolonized islands encounter abundant resources and their populations expand rapidly. Natural selection during this phase favors rapid reproduction to take advantage of newly discovered resources, an evolutionary strategy known as "*r*-selection." With time though, the island becomes crowded, resources per animal decline, and the rapid growth of the *r*-selection phase gives way to a more conservative strategy of population growth better suited for increased competition and dwindling resources. This more conservative strategy is called "*K*-selection."

The concept of *r*- and *K*-selection has been applied to species that dwell on continents as well as those on islands. The first lesson that wildlife managers can learn from *r*- and *K*-selection is that certain population characteristics are generally linked together (see Table 3-5). For example, *r*-selected species reproduce rapidly when environmental conditions are favorable, and they are therefore capable of withstanding very high rates of mortality. Population turnover is rapid and *r*-selected species are usually best adapted to early- to midsuccessional stages. *K*-selected species, on the other hand, have relatively low reproduction rates, higher ages at first breeding, and can sustain lower levels of mortality. They are generally larger animals with lower metabolic rates, have longer life spans, and invest more time and effort in the care of a smaller number of young. The distinction between *r*- and *K*-selected species is not absolute but should instead be regarded as different ends of a continuous scale. Most species fall somewhere between the two extremes. Strongly *r*-selected species include the mallard and the cottontail rabbit. The whooping crane (*Grus americanus*) and the African elephant are clear examples of *K*-selected species.

TABLE 3-5
A COMPARISON OF *r*- AND *K*-SELECTED POPULATION CHARACTERISTICS
Note that *r*- and *K*-selected traits are always compared in relative, rather than absolute terms (modified from Pianka, 1970, by permission, University of Chicago Press)

Characteristic	*r*-selected	*K*-selected
Intrinsic rate of increase (*r* max)	High	Low
Variance in *r* (observed rate of increase)	High	Low
Juvenile mortality	High	Low
Population turnover	Rapid	Slow
Age at first breeding	Younger	Older
Body size	Small	Large
Life span	Short	Long
Dispersal rate	Rapid	Slow
Population stability	Low	High
Successional stage	Early to mid	Mid to late

The second lesson that wildlife managers can learn from *r*- and *K*-selection is that practices of managing one do not necessarily apply to the other. Extreme *r*-selected species can, like the house mice in South Australia, build up rapidly enough to inflict severe damage on crops and property. Management must try to control damage, either through habitat manipulation or through direct population reduction. Extreme *K*-selected species are often endangered. They rarely exist in very high numbers and generally lack the capacity to increase rapidly under improved habitat conditions. Moreover, *K*-selected species are adapted to stable, late-successional habitats, those least able to withstand human disturbance. Waterfowl species, many of which are *r*-selected, increase quickly with improved habitat conditions. In contrast, despite the vast time, money, and effort the U.S. Fish and Wildlife Service has invested in an admirable effort to protect and preserve the whooping crane, its population recovers very slowly.

POPULATION GENETICS

Most wildlife management is a numbers game, the success of which is measured by population changes in relation to management objectives. Only recently have long-term genetic changes begun to receive attention from wildlife managers. One reason for this neglect was that changes in numbers are often obvious and relatively easy to measure, whereas genetic changes, being more subtle, escaped notice. Science has recently given wildlife managers the tools to measure genetic changes in populations. At the same time, managers have begun to encounter problems that can be solved only through use of population genetics.

Populations are composed of individuals that live, reproduce, and die. Numbers of individuals within populations ebb and flow with changing habitat conditions. If wildlife managers are to assure perpetuation of wild populations far into the future, they must think not in terms of numbers per unit area, but in terms of preservation of gene pools. A gene pool is the sum total of the genetic diversity contained within a population or even within an entire species. Unlike populations, gene pools can increase only through outbreeding or through mutation. Mutations are extremely rare, estimated at frequencies of between 10^{-4} and 10^{-6} for mammals (Smith et al., 1977) and hence cannot be counted on for rapid replenishment should the gene pool shrink. Conservation can restore numbers but it cannot restore losses to the gene pool.

Problems in Wildlife Management

The gene pool of wild species can get into trouble in two opposite ways. The more common problem, usually associated with endangered species, is a severe reduction in the size of the gene pool. A second type of problem occurs when one wild species interbreeds with another. This interspecies hybridization contaminates gene pools of one or both of the parent species. When one parent species is far more numerous than the other, the less numerous species can be threatened with extinction. Human activities are likely to increase the frequencies of both these problems.

Severe Reduction of the Gene Pool The most common problem in population genetics is severe reduction in a species' gene pool. This essentially irreversible process usually results from reductions in numbers, combined with regional exterminations that contract the species' geographic range. As the range diminishes, the unique genes carried by locally adapted variants (ecotypes) are lost. When this process is allowed to continue, the species eventually becomes confined to small, isolated patches of its former range. The isolation accelerates genetic losses further by preventing genetic exchange or gene flow between populations inhabiting patches of remaining range. The smaller the population is in each patch, the greater the rate of loss of genetic diversity through chance recombinations, a process known as "genetic drift."

Dr. Tom Lovejoy of the World Wildlife Fund considers the genetics of dwindling populations to be one of the two most important scientific questions in conservation. (The other, that of minimum critical size of ecosystems, is addressed in Chapter 8.) He cited as a case in point an isolated band of bighorn sheep (*Ovis canadensis*) studied by Steve Berwick in Montana. There were 12 sheep in the band, but the social structure limited the effective breeding population to 6. At the time of Berwick's study, the population had gone through at least two population bottlenecks and was showing the reduced size, vigor, and other morphological effects of inbreeding. Later, some 35 more sheep were introduced from elsewhere and by 1977 the herd had grown to about 75 larger, more healthy animals (Lovejoy, 1978).

The most serious short-term effect of inbreeding and drift is a decline in fecundity, survival of young, or both, a problem called "inbreeding depression." For example, 15 of 16 species of ungulates showed significant signs of inbreeding depression when inbred lines from captive populations were compared with other captive populations that were not inbred (Ralls et al., 1979).

Longer-term effects of genetic drift include random changes in the phenotypes and depletion of genetic variance (Franklin, 1980). In a small population, the chance effects of drift will eventually fix certain alleles, losing others. Changes in the phenotype characters in the population then come about not as a result of natural selective forces, but instead from this random process that actually excludes part of an already depleted gene pool.

As drift continues to deplete the gene pool of a population, it weakens the population's ability to recolonize former portions of its geographic range, even if the numbers of animals within the population substantially increase. The genetic variance within populations, once thought to be due to hybrid vigor or heterosis, appears instead to reflect genetic adaptations to local habitat conditions (Powell and Taylor, 1979). Over very long periods of time, species with depleted gene pools will be less likely to survive in changing environments. Eventually, speciation as an evolutionary process may completely cease.

Genetic Swamping: Contamination of the Gene Pool Sometimes wildlife managers must contend with the opposite sort of genetic problem, that of genetic swamping through hybridization with a closely related species. In the southeastern United States, from Florida to Texas, there occurred a wild canid called the red wolf (*Canis rufus*). Intermediate in size between the larger gray wolf and the smaller coyote, the red wolf was severely reduced through various forms of predator control. But it was not this reduction alone that brought this species to the brink of extinction; rather it was hybridization with the coyote.

No one knows for certain just why the coyote has greatly expanded its geographic range since the turn of the century, but various theories credit the expansion to human disturbances to habitats, populations, or both.

Whatever the cause, coyotes moved eastward, coming into contact with a much smaller number of red wolves. Intermediate or hybrid animals began showing up where only pure red wolves had once lived. Now the red wolf is critically endangered and, according to the U.S. Fish and Wildlife Service, the genetic swamping continues. Perhaps the species will be saved only in captivity, thanks to a captive breeding program established at the Point Defiance Zoo in Tacoma, Washington (Shaw and Jordan, 1977).

Theories of hybridization between previously distinct species usually cite habitat disturbance as a cause, since disruptions on a large scale can bring previously distinct species into contact. Furthermore, the "new" habitats resulting from the disturbance can aid survival of hybrid offspring that would ordinarily be genetically unfit in undisturbed habitats of either parental species. If hybrids survive without a loss of fertility, they may then backcross with either parent species, threatening them with genetic contamination.

Thus, as human activities disrupt more and more of the earth's natural habitats, more closely related species will be brought into contact than ever before. Whether or not fertile hybrids will result depends in each case upon the genetic distance between the two species. If hybrids are produced, both parental species can be permanently changed genetically, with the less numerous parental species altered more drastically than the more numerous one.

Restocking During the first half of the twentieth century, wildlife managers commonly restocked species into areas where they had once been numerous. Most reintroduced animals were not as well adapted to local conditions as were the original inhabitants, but some became established and multiplied anyway. Some of these restockings were clearly beneficial in areas where the original populations had been completely eliminated. But when animals were restocked simply to boost local numbers, problems arose when the two populations interbred.

The bison originally inhabiting what is now Yellowstone National Park were mountain or wood bison (*Bison bison athabasca*), a subspecies regarded by most authorities as morphologically and behaviorally distinct from plains bison (*B. bison bison*) (Karsten, 1981). Perhaps two dozen wood bison were surviving in Yellowstone by 1902. That same year in a well-intentioned effort to increase the population, authorities introduced 18 cows and 3 bulls from two herds of plains bison (Cahalane, 1945). The bison that inhabit Yellowstone today are a mixed breed and any genetic distinctiveness between the two subspecies has been permanently lost.

As a general rule, then, a species should not be restocked into an area that still contains natives of the same species. If larger populations are desired, they can be achieved simply through improving local habitat conditions, a practice necessary to sustain larger populations anyway.

Allowing local native populations to increase is a less expensive, more efficient procedure and will prevent potentially troublesome problems from the infusion of new genes into the population.

However, in cases in which no native population of a species is present, restocking may be the only way to restore a population into former areas of its geographic range. Population genetics can help in these cases by selecting individual animals most variable genetically, in other words, those that appear to be the least related to one another. In this way the founders of the new population will represent a larger gene pool and will thus be more likely to survive and prosper in the new environment.

Application of Population Genetics to Wildlife Management

There are a number of ways to measure genetic variability. Some phenotypic characters such as unusual colors or markings on plumage or hair indicate genetic differences. Differences in blood types, mitochondrial DNA, and in serological reactions show genetic distinctions. But the primary method used to measure genetic variability is electrophoresis, a technique that separates components of enzymes in an electrical field. Enzymes separate in the electrical field according to differences in molecular structure. These differences within an enzyme for a particular species indicate that the enzyme is polymorphic, that is, controlled by more than one allele. The frequency of polymorphism and the distribution of polymorphic forms furnish an index of overall genetic variability or heterozygosity. Notice the variations in the frequencies of polymorphisms for various vertebrates (Table 3-6). The northern elephant seal (*Mirounga*

TABLE 3-6
MEASURES OF THE PROPORTION OF LOCI POLYMORPHIC PER SPECIES OF POPULATION
The higher the polymorphism, the greater the genetic variability (after Smith et al., 1976, by permission, The Wildlife Management Institute)

Taxon	Proportion of loci showing polymorphism (%)
Sunfish	14–36
Sparrow	33
House mouse	30
Field mouse	20
White-tailed deer	36
Elephant seal	0
Humans	28

augustirostris) displays no polymorphism, presumably because the species suffered a close brush with extinction when its numbers dropped to perhaps 20 in the early 1890s (Bonnell and Sealander, 1974).

Population genetics can thus be used in wildlife management to deal with problems of severe reduction in gene pools, hybridization between species, and selection of genetically more variable individuals for restocking. More specific applications, including estimation of minimum effective population sizes and the effect of population size on genetic diversity, are reviewed in Chapter 8. In addition, refined methods of population genetics may prove useful in comparing population characteristics with genetic ones, and even for defining management area boundaries.

SUMMARY

The primary population parameters include birth rate, death rate, age composition, and sex ratio. All four influence one another and all four determine the populations status and rate of growth. Another important population parameter is dispersal, or the movement individual animals make from their birthplaces to the places they reproduce. Only through dispersal can wild animals naturally replenish populations that have become locally extinct.

A general pattern for population growth can be derived from the logistic equation. Real populations may depart substantially from the pattern generated by the logistic, but it nonetheless remains the best general approximation.

Some populations undergo major fluctuations. If these fluctuations occur at regular intervals, usually either 3 to 4 years or 9 to 10 years, they are called cycles. (Examples of cyclic species are the microtine rodents of the arctic and the lynx and snowshoe hare of the boreal regions of Canada.) Severe population increases at irregular intervals are called irruptions. Irruptions take place in Australian house mice and in many deer populations, among others.

Populations characteristically tend to form distinct patterns that can be thought of as opposite ends of a scale. At one end are species that play the percentages through rapid reproduction, high turnover rates, and short life spans. They typically are best adapted to the inherently unstable conditions of early- to midsuccessional stages. Such species are known as *r*-strategists. The other end of the scale features species which are slow to mature, reproduce slowly, and provide their offspring with considerable parental care. These are called *K*-strategists and are best adapted for stable, late-successional stages. Extreme *r*-strategists are likely to become pests, while extreme *K*-strategists are likely candidates for endangered species lists.

Recently wildlife managers have begun addressing problems in population genetics. These can include either severe reduction of gene pools, typically problems for rare or endangered species, or contamination of a gene pool through interspecies hybridization. Both these problems increase due to human activity, and managers can expect to see them at greater frequencies in the future. Another, though potentially less serious, genetic problem can arise when members of one subspecies are introduced into the geographic range of another subspecies, resulting in interbreeding and offspring less suited for local habitat conditions than the original subspecies. Fortunately, more techniques are being developed to help wildlife managers measure and assess genetic variability, a consideration that over the long term may prove more important than mere numbers alone.

REFERENCES

Archibald, H. L. 1977. Is the 10-year cycle induced by a lunar cycle? *Wildl. Soc. Bull.* 5: 126–129.

Bonnell, M. L., & R. K. Sealander. 1974. Elephant seals: genetic variation and near extinction. *Science* 184: 908–909.

Botkin, D. B., & R. S. Miller. 1974. Mortality rates and survival of birds. *Am. Nat.* 108: 181–192.

Cahalane, V. H. 1944. Restoration of wild bison. *Trans. N. Am. Wildl. Conf.* 9: 135–143.

Caughley, G. 1963. Dispersal rates of several ungulates introduced into New Zealand. *Nature* 200: 280–281.

Caughley, G. 1970. Eruption of ungulate populations, with emphasis on Himalayan Thar in New Zealand. *Ecology* 51: 53–72.

Caughley, G. 1977. *Analysis of vertebrate populations.* New York: John Wiley & Sons.

Cole, L. C. 1957. Sketches of general and comparative demography. *Cold Springs Harb. Symp., Quant. Biol.* 22:1–15.

Elton, C. S. 1924. Periodic fluctuations in the numbers of animals: their causes and effects. *British J. Exp. Biol.* 2: 119–163.

Elton, C. S., & M. Nicholson. 1942. The ten-year cycle in numbers of the lynx in Canada. *J. Anim. Ecol.* 11: 215–244.

Emmel, T. C. 1976. *Population biology.* New York: Harper & Row.

Franklin, I. R. 1980. Evolutionary change in small populations. In M. E. Soule and B. A. Wilcox (Eds.), *Conservation biology: an evolutionary-ecological perspective.* Sunderland, Mass.: Sinauer Assoc. Inc.

Howard, W. E. 1960. Innate and environmental dispersal of individual vertebrates. *Am. Midland Nat.* 63: 152–161.

Humphrey, S. R. 1974. Zoogeography of the nine-banded armadillo (*Dasypus novemcinctus*) in the United States. *Bioscience* 24: 457–462.

Karsten, P. 1981. Studbook of the wood bison (*Bison bison athabascae,* Rhoads). Calgary, Alberta: *Calgary Zool. Soc.*

Keith, L. B. 1963. *Wildlife's ten-year cycle.* Madison, Wisc.: University of Wisconsin Press.

Krebs, C. J., M. S. Gaines, B. L. Keller, Judith H. Myers, & R. H. Tamarin. 1973. Population cycles in small rodents. *Science* 179: 35–41.

Leopold, A., L. K. Sowls, & D. L. Spencer. 1947. A survey of over-populated deer ranges in the United States. *J. Wildl. Manage.* 11: 162–177.

Lovejoy, T. E. 1978. Genetic aspects of dwindling populations: a review. In S. A. Temple (Ed.), *Endangered birds: management techniques for preserving threatened species.* Madison, Wisc.: University of Wisconson Press.

MacArthur, R. H., & E. O. Wilson. 1967. *The theory of island biogeography.* Princeton, N.J.: Princeton University Press.

McCullough, D. R. 1979. The George Reserve deer herd. Ann Arbor, Mich.: University of Michigan Press.

Newsome, A. E. 1969. A population study of house mice temporarily inhabiting a South Australian wheatfield. *J. Anim. Ecol.* 38: 341–359.

Powell, J. R., & C. E. Taylor. 1979. Genetic variation in ecologically diverse environments. *Am. Scient.* 67: 590–596.

Ralls, K., K. Brugger, & J. Ballou. 1979. Inbreeding and juvenile mortality in small populations of ungulates. *Science* 206: 1101–1103.

Shaw, J. H., & P. Jordan. 1977. The wolf that lost its genes. *Nat. His.* 86(10): 80–88.

Smith, M. H., H. O. Hillestad, M. N. Manlove, & R. L. Marchinton. 1976. Use of population genetics for management of fish and wildlife populations. *Trans. N. Am. Wildl. and Natur. Resour. Conf.* 41: 119–133.

Smith, M. H., M. N. Manlove, & James Joule. 1977. Genetic organization in space and time. In *Populations of small mammals under laboratory conditions. Spec. Pub. Ser.* vol. 5, Pymatuning Laboratory of Ecology, Pittsburgh, Pa.: University of Pittsburgh.

Taber, R. D. 1971. Criteria of sex and age. In R. H. Giles (Ed.), *Wildlife management techniques.* Washington, D.C.: The Wildlife Society.

POPULATION ESTIMATION AND ANALYSIS

"Abundance" is the general term referring to actual numbers, trends, or both, of wildlife populations. On very rare occasions, wildlife managers, like their counterparts in range management, are able to count an entire population. Such a complete census offers the obvious advantage of furnishing a full inventory of the population. Whooping cranes, for example, are large, conspicuous, and gregarious, characteristics that make possible complete censuses every year. Bison have similar traits and likewise can often be completely counted. Wild animals, though, are rarely as cooperative as livestock and their habitats are never so open and homogeneous as an improved pasture. In the vast majority of cases managers must forego the complete census and instead make estimates of abundance.

ESTIMATION OF ANIMAL ABUNDANCE

Wildlife managers need reliable estimates of abundance for various reasons. Estimates are needed to establish hunting regulations for game. Abundance itself is often a key consideration in determining management objectives and then in measuring progress toward meeting them.

Estimating abundance can be tricky, however. Most techniques require sampling, and sampling errors can be serious. In addition, most techniques rely upon assumptions which, if not met, offer very inaccurate and unreliable estimates. Prudent managers then should select their methods carefully and interpret results cautiously.

Despite these drawbacks, wildlife managers and researchers continue to use and refine techniques to estimate abundance. If done correctly they usually do afford useful and reliable results. Moreover, computers have greatly aided use of these methods, first by providing means of testing the assumptions and second by allowing more rigorous analysis of large data sets impossible in earlier times.

This chapter introduces only a small set of the wide array of techniques currently used to estimate abundance. More detailed and complete accounts are available elsewhere (Davis, 1982; Davis and Winstead, 1980; Sen, 1982).

The Sample Census

When complete censuses are not practical, managers who want estimates of population size must rely upon sample censuses. Sample censuses are usually done by selecting sample areas randomly for censusing. The results of such sample censuses are then extrapolated for the entire area over which the population estimate is desired. If sample areas are correctly selected and the method carefully applied, sample censuses can provide estimates of actual numbers; these estimates, divided by the area under study, can furnish estimates of population density.

The Index

What if managers want to know something about the abundance of a particular species over an entire county, state, or region? Alternatively, suppose that a management objective can be met merely through investigating population trends? In either case, an index or estimation of population trends will fill the need. Indexes estimate relative abundance rather than actual numbers. They have one tremendous advantage over sample censuses; they can be applied quickly over large areas at relatively low cost. For this reason, most state and provincial wildlife agencies use indexes for estimating relative abundances of game species. Remember, though, that indexes never determine actual size of a population, only its relative abundance. Their application is often at least as much art as science.

Selecting a Method

Before choosing a method, a wildlife biologist must consider carefully the management objectives, the characteristics of the species, the nature of the habitat and terrain, and the time and personnel available.

Biological Considerations When managers need to estimate abundance of a particular species, they must first consider its biological characteristics. Is the species diurnal? How homogeneous is the habitat in which it occurs? Table 4-1 presents a comparison of characteristics that make a population relatively "easy" or "difficult" to estimate. A population with a relatively high density is easier than one with a low density because data may be obtained more efficiently and allow calculation of statistical confidence limits, a useful if elusive addition to the abundance estimate.

Administrative Considerations Management objectives can greatly influence selection of a method. Is the population to be harvested at or near the point of maximum sustained yield? Is the species endangered? Other practical matters must be considered. How much time and personnel are available to estimate abundance and over how large an area? Can special equipment be issued or purchased? How much travel, by land, water, or air, can be afforded? These are just a few basic administrative considerations that must be taken into account in choosing a method.

Methods

The first four methods described here are normally used for sample censuses. The last three normally provide indexes. Many more methods for both sample censuses and indexes exist, and the reader can find more of them described in Caughley (1977) and Eberhardt (1971, 1978).

Mark-and-Recapture One common technique for estimating abundance is that of mark-and-recapture. A sample of the population is captured, marked, and released at capture sites. Then a second sample is

TABLE 4-1
CHARACTERISTICS OF SPECIES AND HABITAT CONDITIONS THAT MAKE CENSUSING EASY OR DIFFICULT

Easy	Difficult
Conspicuous by sight or sound	Cryptic, silent
Homogeneous habitat	Habitat highly varied
Open land habitat	Densely forested or brushy habitat
Easy access by road	Accessible only by foot or horseback*
Normally occurs at high density	Normally occurs at low density
Diurnal	Nocturnal

*Aerial survey methods can now make censusing of conspicuous animals much easier even in remote regions.

captured a brief time later and the population estimated by the ratio of marked to unmarked animals. The simplest form of mark-and-recapture is known variously as the Petersen index, the Lincoln index, and the Petersen estimate. The first two names are misleading since the technique yields an estimate of actual numbers and is therefore a sample census and not an index.

An example of the Petersen estimate is shown on Table 4-2. These hypothetical data could have been collected for fox squirrels (*Sciurus niger*) in the southeastern United States. The biologist set out a 20 by 20 grid of 400 stations or potential trap sites. From these 400 stations 200 were selected at random and live traps were placed there. Fifty-two squirrels were caught, marked, and released. One week later the experiment was repeated, again with 200 randomly chosen of the 400 stations. Forty-nine squirrels were caught, 24 of which were marked from the previous session. By using the ratio of marked to unmarked squirrels against the number known to be marked, one can make a simple calculation that estimates population size. The area sampled is that covered by the grid plus a border equivalent to one-half the distance between stations. Thus if the stations were 50 meters apart, a 20 by 20 grid plus border would sample an area of 100 hectares. The population density would then be 1.06 per hectare.

In theory, the Petersen estimate is elegantly simple and clear. In practice, it can be applied to almost any catchable vertebrate, from fish through frogs, from falcons to felids. But the technique is labor-intensive, time-consuming, and therefore expensive. It cannot be applied over large areas. And it requires at least four assumptions:

TABLE 4-2
AN EXAMPLE OF THE PETERSEN ESTIMATE OR LINCOLN INDEX
(See text for explanation)

	Number	**Symbol**
Number captured, marked, and released during capture session	52	M
Total number of captures during 2d trapping session	49	n
Number of marked animals recaptured during 2d trapping session	24	m
Population size	?	N

Use the formula: $$\frac{M}{N} = \frac{m}{n}$$

$$\frac{52}{N} = \frac{24}{49}$$

$$N = 106$$

1 Each individual has an equal chance of being captured.

2 There are no births nor immigrations into the study area between the first and second trapping sessions.

3 There is no differential mortality or emigration between marked and unmarked members of the population.

4 No marks are lost.

If any of these assumptions prove false, serious errors in population estimation can result. The first assumption, that of equal catchability, is the most troublesome. Should marked animals become trap-prone, an underestimation will result. If the marked animals become trap-shy, the population will be overestimated. Should marked animals die at a greater rate than unmarked ones, the population will be overestimated, the same problem that will occur if marks are lost.

The assumption of equal catchability can be strengthened a bit by randomizing trap placement between the two capture sessions, as stated in the example above. Otherwise, marked animals will likely be captured in numbers disproportionately high, since their home ranges fall within the trapping sites. Some biologists recommend using a different capture method for the second trapping session to deal with the assumption of equal catchability. The second and third assumptions can be dealt with simply by avoiding the season of births and by minimizing the time between the first and second capturing sessions. Dual marks (e.g., two leg bands or two ear tags) will ease problems associated with the fourth assumption, as will minimizing the time lag between capture sessions.

Otis et al. (1978) used computer simulation to evaluate the assumption of equal probabilities of capture. They concluded that many published studies had been done with inadequate capture efforts and sample sizes. As minimum guidelines, they suggested live trapping on at least a 12 by 12 grid, spaced so that at least four traps fall within an average estimated home range. The shortest trapping period recommended was 8 to 9 days.

Transect Surveys Many conspicuous species such as large mammals and migratory birds can be counted directly on sample areas. The survey routes or transects are randomly or systematically placed throughout the study area. An observer then walks or flies in a light aircraft along each transect route at the same time of day, season, and weather conditions, and counts animals seen within a constant distance of the survey line. The length and width of the transect lines provide estimates of the area sampled and thus allow an estimate of density.

In practice, observers rarely, if ever, manage to count all animals (Caughley, 1974a; Caughley and Goddard, 1972; Caughley et al., 1976), so populations tend to be underestimated. Furthermore, conditions of terrain and vegetation often render the technique useless. Nonetheless, where conditions permit, transect surveys are a rapid, reliable, and relatively inexpensive means to obtain estimates of population density.

Aerial transect surveys are most commonly used to count large mammals in grasslands, savannas, or other relatively open habitat types. High-wing aircraft are usually used and streamers are attached to the wing struts so that when the plane is flown at a constant altitude, the streamers delimit a known area, (say, 100 meters) to the observer (Caughley, 1977).

Biologists also use aerial transect counts to estimate numbers of waterfowl on wetland habitats where visibility is no problem. With practice, observers can reliably estimate the number of hundreds of birds or large mammals at a glance. However, most observers photograph large flocks or herds to verify their preliminary counts.

Sometimes transect surveys are conducted on foot. Deer and upland game birds have been counted in this manner through use of the Hahn or King censuses. Either type is similar in route selection to aerial surveys. Routes are randomly selected and the observer walks the transect lines at a certain time of day during a particular season of the year. Since strip or transect widths are known, the area can be calculated to furnish a density estimate. Table 4-3 gives an example of a King census.

These ground surveys can be carried out with a transect of fixed width or indefinite width (Eberhardt, 1968, 1978). The former is simpler to use and is recommended (Caughley, 1977) for populations that have a fairly high density. At lower densities, however, transects of indefinite width allow inclusion of all animals sighted, in turn permitting greater precision. The distance of each animal from the transect line must be reliably estimated. Furthermore, the probability of seeing an animal at varying distances must be determined and the results incorporated into the density calculations (Eberhardt, 1968, 1978; Caughley, 1977).

Change-in-Ratio Method The basic technique was introduced by Kelker (1940, 1944) and is sometimes called Kelker's method. When a population is composed of individuals easily recognized as members of two or more classes such as sex, age, or phenotypic variant, the population size can be estimated when one form is differentially added or removed. The method was developed primarily for use on deer populations in which bucks only were hunted. Similar to the Petersen estimate, Kelker's method requires an initial survey to determine the ratio between classes prior to removal or addition, followed by a second survey after the manipulation. Since the number added or removed is known, it can be combined with the two ratios to yield an estimate of population size. An example of Kelker's method appears in Table 4-4.

Two other basic variations can be used. Members of both (or all) groups

TABLE 4-3
AN EXAMPLE OF A KING CENSUS

Ten kilometers of transect line are randomly placed through a study area. A biologist walks the lines, counting a total of 11 grouse within 25 meters on either side of the transect line. What is the estimated population density?

$$\text{Average area surveyed} = \text{line length} \times \text{twice the strip width}$$
$$= 10 \text{ km} \times 2 \times .025 \text{ km} = 0.5 \text{ km}^2$$

$$\frac{11 \text{ grouse}}{0.5 \text{ km}} = \frac{x}{1.0 \text{ km}^2}$$

$$x = \frac{22 \text{ grouse}}{\text{km}^2}$$

can be added or removed so long as the additions or removals are disproportionate. The nearer the additions or removals are to each other, the less accurate the resulting calculation is to the population size. Another variation involves the use of two or more species as groups, and one species added or removed more than the other (Caughley, 1977; Chapman, 1955; Rupp, 1966). If members of more than one group are added or removed, the formula becomes more complex with a distinction made between C_x representing the changes in the first group and C_y the second. C becomes the difference between C_x and C_y and the formula is

$$N_1 = (C - p_2C)/(p_2 - p_1)$$

TABLE 4-4
AN EXAMPLE OF THE USE OF THE CHANGE-IN-RATIO METHOD, ALSO KNOWN AS KELKER'S METHOD

An initial survey of a deer population reveals a ratio of 60 bucks per 100 does. Twenty-five bucks are then removed by hunting, and a second survey yields a ratio of 44 bucks per 100 does. What was the total population size before the hunt?

Use the formula $$N_1 = \frac{C - p_2C}{p_2 - p_1}$$

where N_1 = population size at first survey
p_1 = proportion of males (bucks) in N_1
p_2 = proportion of males in N_2
C = number of males removed (negative sign) or added (positive sign)

In this case, the calculations are:

$$N_1 = \frac{-25 - (.306 \times -25)}{.306 - .375} = \frac{-17.35}{-.069}$$
$$= 251.4, \text{ about 251 animals prior to hunting}$$

where N_1 = population size at first survey
p_1 = proportion of x individuals in N_1 population
p_2 = proportion of x individuals in N_2 population

Pellet Group Counts Where conditions of terrain and climate permit, biologists can estimate the population densities of some large ungulates by estimating the densities of fecal pellets and the rates at which they are deposited. The usual assumption is that cervids deposit an average of 13 pellet groups each day (see Table 4-5). In practice, biologists have found that the average of 13 is far from precise and can be quite variable. Deposition rate should be determined for species, region, and season before the technique can be employed with confidence. By clearing a series of sample plots of "old" pellet groups and then counting the numbers of pellet groups found on those plots a known number of days later, biologists can convert pellet group density into an estimate of population density.

This technique is obviously ill-suited for areas of dense, tall ground cover, where finding pellets would be difficult and where the clearing of vegetation from plots might seriously bias survey results by attracting or repelling animals from the plots. Likewise, pellet group surveys work poorly in warm, humid climates, where rapid decomposition of pellets could restrict the time interval between surveys.

Roadside Counts A very common method of estimating abundance is the roadside count. The biologist records the number of animals seen per kilometer of road driven at a fixed speed at a constant time of day (usually early or late) at a constant time of year. This technique is relatively inexpensive and rapid, making it suitable for extensive surveys over large areas. However, roadside counts, unlike transect surveys, can be used only as indexes of abundance and not for density estimates. For one thing, roads are rarely if ever laid out either randomly or systematically. For another, the presence of the road itself, plus gravel and atypical flora, render roads unsuitable as representative habitat types. Nevertheless, road counts, done in the same manner over several years or between different regions in the same year, can provide useful indexes.

Road Kills Wildlife biologists have often been interested in the frequencies of road-killed animals as indexes to population trends (Overton, 1971). One problem until recently has been lack of standardization. However, interstate highways in the United States provide fairly uniform road conditions, on which Case (1978) evaluated the use of road kills. He noted that 7 years of road-kill reports from Nebraska's interstate highways showed peaks during May and October, presumably relating to periods of reproduction and dispersal. Surprisingly, road-kill rates did not correlate with daily traffic volume but did correlate with average vehicle speed. If

TABLE 4-5
AN EXAMPLE OF A PELLET GROUP SURVEY

Two hundred sample plots of 10 m^2 each are randomly placed within a study area of 1 km^2. A biologist clears the "old" pellet groups from the plots and returns 60 days later, finding a total of 33 pellet groups on the 200 plots. What is the estimated density of deer?

$$\text{Pellet group density } (x) - \frac{33}{2{,}000 \text{ m}^2} = \frac{x}{1{,}000{,}000 \text{ m}^2}$$

$$x = 16{,}500 \text{ pellet groups/km}^2$$

$$\text{Pellet groups per deer during the 60 days} = 13 \times 60 = 780$$

$$\text{Deer density } \frac{16{,}500}{780} = 21.2 = 21 \text{ deer/km}^2$$

standardized for vehicle speed, road kills could provide a useful index of population trends.

Scent Station Surveys The U.S. Fish and Wildlife Service developed scent station surveys to estimate trends in coyote populations. A survey line consists of 50 stations placed at 0.48-kilometer (0.3-mile) intervals on alternate sides of a secondary or tertiary dirt road (Linhart and Knowlton, 1975). Each station consists of sifted soil with a diameter of 0.91 meter (3 feet). A synthetic chemical attractant goes in the center of each station. Attracted to the chemical's potent smell, the coyote leaves footprints in the fine, sifted soil. Checking stations the morning after they are set out, the biologist tallies the visitation rate as "no. of stations visited / total scent station nights." This technique is suitable for use over extensive areas and records visits (though probably not with equal effectiveness) of other carnivore species besides the coyote.

POPULATION ANALYSIS

Wildlife biologists and managers often need to know more about populations than mere abundance. They may want to know something about the rate of population increase, for example. Mortality and survival rates are important population characteristics too, particularly in species harvested for sport or furs. These kinds of information are obtained through population analysis.

Estimating Rate of Increase

Wildlife managers interested in sustained harvest or game populations (Chapter 7 and in control of predator populations (Chapter 9) can benefit by knowing something about their species' rate of increase. In a general sense the maximum intrinsic rate of increase (r_{max}) can be crudely estimated by multiplying maximum known clutch or litter size times the maximum proportion of females likely to breed under ideal conditions. But these ideal conditions rarely occur in the wild. Managers are more likely to concern themselves with more subtle changes in their population's observed rate of increase ($\bar{r}$).

Age Ratios A popular means of estimating a population's rate of increase is through use of age ratios and their graphical depiction, age pyramids. The popularity of the method is due more to ease of data collection and treatment than to inherent precision or accuracy. All one needs to do is to age accurately a sample of a population and then to compute the ratio for each age class. Often, rate of increase is estimated by only three age classes: juvenile, subadult (fully grown but not yet breeding), and adult (capable of breeding). The basic assumption is that the higher a population's rate of increase, the higher the proportion of juveniles, subadults, or both, in the population. Thus, age ratios are used to compare changes between years for a single population or within the same year for different populations. State wildlife departments have commonly used age ratios to estimate rate of increase in deer and other big game.

Yet for all its convenience and simplicity, the age ratio method has largely lost its credibility, a casuality to the age of computer simulation. In retrospect it is remarkable that use of the method persisted for so long. A moment's reflection about what age ratios really mean should raise serious doubts in the minds of wildlife managers. If the ratio of young animals to adults increases, there has been (1) an increase in the number of young brought about by an increase in fecundity, (2) a decline in the number of adults caused by an increase in mortality, or (3) some combination of (1) and (2). With age ratios used alone, there is no way to tell which change occurred. The converse is true if the ratio of young to adults decreases. Should changes in mortality rates affect all age classes more or less equally, then there would be no change detected through age ratios. In short, a change in age ratio indicates that there has been a change in fecundity, survival, or both. Consistent age ratios imply that there has been no change in age-specific fecundity or survival. That is very little information on which to base a management decision.

Caughley (1974b) confirmed the preceding suspicions about age ratios through computer simulation. He simulated a series of populations, half of which were rapidly increasing ($r = 0.2$) and half of which were plunging toward extinction ($r = -0.2$). These conditions exceed the normal rates of change in most wild populations. Caughley then examined the changes in age ratios resulting from these two radically different population trends.

He concluded that the age ratios sampled from the two radically different populations were indistinguishable. Caughley's example demonstrated that the changes in age ratios did not and could not affect actual changes in the populations and added that "Age ratios unsupported by other information seem to be statistics in search of an application" (Caughley, 1974b:557).

Observed Population Growth A far more reliable means of estimating $\bar{r}$ is through two or more estimates of population size taken at different times (usually 1 year apart). By computing lambda as N_{t+1}/N_t the manager can estimate r for that time interval by $\log_e \lambda$ (Chapter 3). This, of course, assumes that accurate and consistent means are used to estimate N each time.

When N is very low relative to K, resources available per animal are quite high, so that subsequent population growth approaches $\bar{r}_{max}$. During such circumstances, $\bar{r}$ is about r_{max} as it likely was for the George Reserve deer population.

Most managed populations never become so low in numbers as to permit an estimate of r_{max} from $\bar{r}$. Caughley and Birch (1971) have noted that, if a population has been fluctuating around a mean presumed to be K and is artificially reduced below that number, r_{max} can be approximated by the rate at which the population recovers. This is done by extrapolating backward along the logistic growth curve, using a linear regression of $\log_e (K - N)/N$ on time.

Estimating Mortality and Survival

The pattern of mortality and survival within a population tells a great deal about the population's strategy for survival. Furthermore, mortality patterns may sometimes be partitioned into that due to hunting and that due to other causes, a useful tool for game managers. More recent refinements have made it possible to test whether or not hunting mortality actually adds to a population's overall mortality, helping answer a controversial question in wildlife management (see Chapter 7).

Life Tables As shown in Chapter 3, mortality rates generally change with age. These patterns of mortality and survivorship for each age class are tallied in life tables. The earliest life tables were constructed for human populations and date back at least to the third century A.D. (Hutchinson, 1978). In more recent times actuaries have employed life tables to calculate human mortality patterns for insurance companies. It wasn't until the late 1940s that a wildlife biologist patiently collected a data base with which one of the first life tables for a wild population would be constructed.

Adolph Murie wanted to determine how wolf predation affects the population of Dall sheep (*Ovis dalli*) in Mt. McKinley National Park. He collected several hundred sheep skulls and determined each animal's age at death by counting the annual growth rings on the horns. Most of these deaths were presumed to be from wolf predation. Murie (1944) found very few skulls from sheep that had died in middle age (years 2 to 10) and concluded that the wolves preyed primarily on the young and the aged, thus having little real effect on the population.

In a classic paper, Edward Deevey (1947) used Murie's data to calculate a life table. The approach won rapid acceptance among wildlife specialists, who have used various types of life tables ever since.

Life tables come in several forms. The most basic is the dynamic or age-specific life table, in which a cohort of animals (often adjusted to 1,000 for convenience in calculation) born at the same time is followed until the last member has died. The distribution of ages at death forms the basis of the table. Animals alive at the beginning of the age interval are noted as l_x, those dying within the interval as d_x, and the mortality rate (d_x/l_x) as q_x. Survival rate is $l - q_x$.

The dynamic life table is far easier to understand in theory than it is to implement in practice. The practical difficulties associated with locating a large number of young animals born at the same time and then determining age at death for each prevent widespread use of dynamic life tables. Biologists find it far more convenient to use time-specific life tables, in which a sample is taken from a population over a brief period of less than one year and the members sampled are aged. Mortality rates are calculated from the age distribution, assuming that the age distribution of the population is stable.

Studies done by capturing, marking, and releasing animals over a period of several years can be used to compile composite life tables. In banding young waterfowl, biologists rarely capture enough in a single year to construct a life table, so they pool data from young birds captured over several years and treat them as a single, large cohort. Table 4-6 is a composite, age-specific life table from data compiled for African or Cape Buffalo in the Serengeti. The pattern of mortality is typical for most large mammals (Caughley, 1966) with lower mortality rates in the middle years. Note that this life table, like most others, is for one sex.

Despite their widespread use, life tables have disadvantages. If they are calculated through use of marked animals, there can be no differential mortality due to the marks and no loss of marks. If hunters return or report kills of marked animals, the rate of hunter return (compliance) must be estimated reliably. Life tables based upon shot samples of unmarked animals require reliable aging techniques and the assumption that the age distribution of the shot sample is representative of the entire population. Should analyses be based upon aged remains of found skulls (as were those

TABLE 4-6
LIFE TABLE FOR FEMALE AFRICAN BUFFALO (*SYNCERUS CAFFER*) IN THE SERENGETI
(Modified from Sinclair, 1977, copyright © The University of Chicago Press, used by permission)

Age in years	1_x	d_x	q_x
0	1000	330	.33
1	670	94	.14
2	576	11	.02
3	565	12	.02
4	553	21	.04
5	532	16	.03
6	516	29	.06
7	487	20	.04
8	467	35	.07
9	432	52	.12
10	380	44	.12
11	336	73	.22
12	263	67	.25
13	196	56	.29
14	140	49	.35
15	91	43	.47
16	48	26	.54
17	22	15	.68
18	7	7	1.00

of Murie), the aging techniques must be accurate and there should be no differential destruction by age class either by decomposition or scavenging. Many analyses, particularly those of time-specific tables, assume a stable population ($r = 0$) and a stable age distribution. Corrections can be made (cf. Caughley, 1977; Sinclair, 1977) if the rate of increase does not equal zero, but they require an accurate approximation of the real rate. With all these conditions and assumptions, it is not surprising to discover serious errors. Caughley (1966), after reviewing life tables for mammals published before 1965, concluded that only two of these, Deevey's analysis of Murie's data and Caughley's own analysis of Himalayan thar (*Hemitragus jemlahicu*), were faultless. Since Caughley's evaluation, Murphy and

Whitten (1976) have reanalyzed Murie's data, adding their own plus more data from the National Park Service. They concluded that the Dall sheep population studied by Murie did not have a stable age structure, thereby violating the assumption necessary for Deevey's (1947) analysis.

The potential values of life tables are great but so are the pitfalls. Besides dealing with the difficulties in meeting assumptions, the would-be users must also invest a great deal of time and effort, in other words expense, in gathering the data. They should consult carefully more detailed evaluations such as Caughley's (1966, 1977), Eberhardt's (1971), and Seber's (1973).

Survivorship Curves If the number of survivors in each age class of a life table are plotted against time, the resulting figure is called a "survivorship curve." The *y* axis may be either arithmetic or logarithmic, with the latter more commonly used. The graphical form is easier to understand and furnishes a convenient basis through which data from different populations can be compared.

Survivorship curves come in three basic forms, based upon mortality patterns. Most mammals have moderate juvenile mortality, very little middle-aged mortality, and higher mortality in older age classes. The resulting survivorship curve is a convex one, type I (see Figure 4-1). Many birds and reptiles experience fairly constant rates of mortality independent of age class. The survivorship curve for such mortality patterns, known as type II, plotted logarithmically, forms a straight diagonal line. Type III, a deep concave curve, reveals extremely high mortality rates in very young age classes, followed by fairly high survivorship thereafter. This pattern of survivorship is typical for many insects and fishes.

The Chapman-Robson Method If a stable population experiences constant mortality rates for several years, the age classes become distributed geometrically. Thus, if a biologist suspects rather constant mortality rates irrespective of age classes, survival rates may be estimated by the Chapman-Robson method. Developed for fisheries population studies (Chapman and Robson, 1960; Robson and Chapman, 1961), this technique can be applied to terrestrial wildlife (Paulik, 1963). It has an advantage over the life table in that the principal assumption, in this case constant, age-independent mortality, can be tested from the age distribution data by the chi-square test (see Eberhardt, 1971, and Seber, 1973, for testing procedures). Variance can also be calculated, making possible direct comparisons between populations.

The basic Chapman-Robson formula is

$$S = T/(n + T - 1)$$

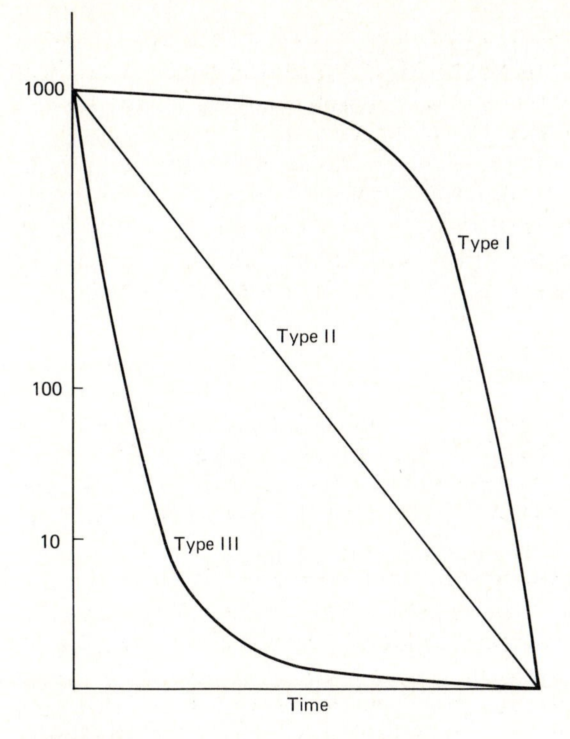

FIGURE 4-1
The three basic forms of survivorship curves.

where S = survival rate
n = sample size
$T = \Sigma$ (age class$_x$)(number in age class)

Table 4-7 presents a sample problem along with calculations for S. Note that the first age class is always coded as 0.

Often a population may generally have age-independent mortality for all except the first or last age classes. Seber (1973) explains how such age classes may be pooled or truncated.

Stochastic Methods During the 1970s, mortality studies, particularly those relating to studies of banded game birds, underwent revolutionary changes. Computer technology expedited complex statistical procedures, resulting in far more sensitive, rigorous, and realistic models than were ever possible before.

Bird banding studies before the 1970s relied upon life table approaches (Table 4-8). The use of life tables requires assumptions that normally cannot be tested. Furthermore, models derived from life tables are

TABLE 4-7
AN EXAMPLE OF THE CHAPMAN-ROBSON METHOD

Age class (x)	Number in sample n_x
0	548
1	251
2	114
3	50
4	19
5	13
6	5
	$n = 1{,}000$

$$T = 0(548) + 1(251) + 2(114) + 3(50) + 4(19) + 5(13) + 6(5) = 800$$

$$S = \frac{T}{n + T - 1} = \frac{800}{1000 + 800 - 1} = 0.445$$

The survival rate is 0.445 and the mortality rate is $1 - S$ or 0.555

TABLE 4-8
A SIMPLE EXAMPLE OF THE USE OF TAG (BAND) RETURN DATA TO ESTIMATE MORTALITY
Suppose that 1,000 young birds of the year are banded and released shortly before the hunting season; bands from the cohort are returned each season over the next few years; assume constant hunting pressure each season and a 100 percent band return rate (Modified from Caughley, 1977, by permission John Wiley & Sons, Ltd.)

	YEARS						
	0	1	2	3	4	5	6
Recoveries m_x	322	74	20	7	3	2	1
Initial no. M 1,000							
Total mortality:* q_x	.77	.73	.65	.77	.57	§	§
Hunt mortality (constant):† $\bar{u}_x$	.322	.322	.322	.322	.322	§	§
Mortality from other causes:‡ W_x	.448	.408	.328	.448	.248	§	§

*To compute total mortality for each year, use the formula $q_x = 1 - (m_{x+1}/m_x)$.
Weighed mean annual mortality is calculated as $\bar{q} = mo_x/\Sigma m_x = 322/426 = .75$.
†Hunting mortality ($\bar{u}_x$) can be computed only for the first year's return in this model and $\bar{u}_x = 322/1{,}000 = .322$.
‡Mortality from other causes may be estimated by subtracting $\bar{u}_x$ from q_x to give W_x.
§Too few data for meaningful calculation.

"deterministic"; that is, they deal with observed rates and fixed variables. Improved methods are "stochastic," in which variables are allowed to change at random, providing probability estimates.

Improvements began with Seber (1970, 1972), who developed a method for analysis of band recovery data that allowed for annual variations in survival rates, hunting pressure, and recovery rates. Assumptions inherent in the technique could be tested statistically using chi-square or Z statistics. The principal drawback of Seber's method was that it excluded juvenile birds as they usually incur mortality rates different from those of adult birds.

Johnson (1974) designed a model similar to Seber's, but that included a formula to estimate differential survival of juvenile birds. Like Seber's, Johnson's technique allows annual variations in key parameters and permits calculation of variance estimates.

Brownie et al. (1978) extended Seber's method further. They recognized that all band recovery models are based upon two important parameters, the annual survival rate (S) and the annual band recovery rate (f). (Figure 4-2 shows the critical fates that can befall any banded game bird.) Then they developed a detailed series of 13 stochastic models. These models test whether or not the values for S and f vary with such factors as age, sex, capture and handling history, and calendar year.

Data for these stochastic models are tabulated much like those in Table 4-8. The rows and columns are totaled and the distribution of the data tested against the distributions expected, given certain assumptions regarding S and f.

FIGURE 4-2
Three critical fates of a banded game bird alive at the start of the year (after Brownie et al., 1978).

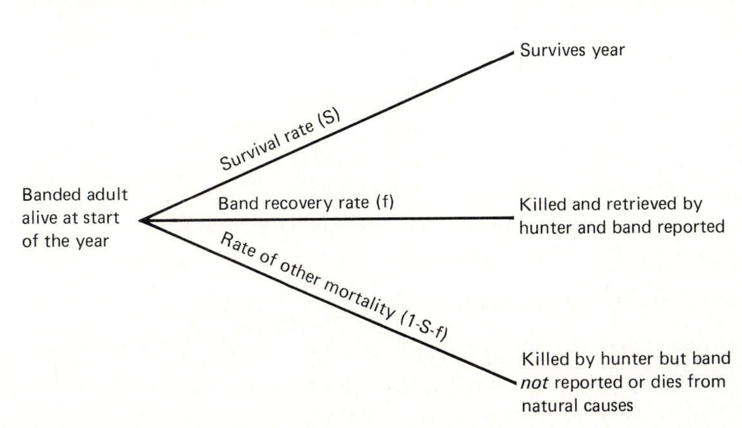

Actual computations of stochastic models lie well beyond the scope intended here. However, anyone designing studies based upon band recovery data should consult carefully Seber (1973) and Brownie et al. (1978) before embarking on such an expensive and laborious task.

SUMMARY

Animal abundance can be estimated in three general ways. For conspicuous animals in open terrain within small discrete areas it may be possible to obtain a complete census, thereby determining the actual size of a population. More often, though, biologists must rely upon sample censuses, where censuses are done on sample areas selected randomly or systematically. When abundance must be estimated for very large areas, the most practical approach may be the use of an index. Indexes measure population trends instead of actual numbers, but they do have the advantage of being quickly and easily applied over large areas.

Before selecting a specific method, the biologist or manager should decide which of the three general approaches is most realistic and which will satisfy the management objectives. He or she must then consider the characteristics of both the species and the habitat, plus the limits imposed by time and personnel.

Several methods are reviewed in this chapter. The mark-and-recapture method is an intensive type of sample census and is normally used only in research. Transect surveys, both from the air and from the ground can be used more extensively and are well suited for waterfowl, big-game, and other conspicuous species to furnish estimates of actual numbers. For species with two easily distinguishable sexes or age classes, the selective addition or removal of one of them can afford an estimate of numbers through the change-in-ratio method. Hooved mammals can be censused under some conditions by the pellet group count.

Roadside counts normally allow estimates only of population trends and are therefore suitable only for use as indexes. But they are fast, inexpensive, and easily standardized. The frequency of road kills can prove useful in estimating population trends, provided that road conditions are constant, such as those afforded by interstate highways. A more specialized index is the scent station survey, developed by the U.S. Fish and Wildlife Service to monitor trends in coyote populations. It also has promise as an index for other carnivore populations.

A rate of population increase is a good measure of the overall condition of a population in relation to its habitat. Age ratios have long been used to estimate rates of increase, the assumption being that the greater the proportion of younger age classes the greater the rate of increase. However, this technique, although easily applied, yields ambiguous results. A far more reliable estimate of rate of increase can be obtained through use of census data for several different years, providing an observed rate of increase.

Life tables have provided the most popular basis for estimating mortality and survival. Most life tables, though, depend upon troublesome and unrealistic assumptions, so they should be used very carefully. Plotted data from a life table form a survivorship curve.

The Chapman-Robson survival estimator can be a useful alternative to the life table. The method's primary assumption, that of a geometric age distribution, can be tested statistically and, as variance can also be computed, comparisons can be made between populations.

During the 1970s biologists working on waterfowl survival rates developed a series of alternatives to the life tables. These methods are quite complex and rely upon powerful computers to process the rather copious data. But these stochastic methods allow statistical testing of all major assumptions and are therefore more reliable than traditional life table analyses.

REFERENCES

Brownie, C., R. D. Anderson, K. P. Burnham, & D. S. Robson. 1978. Statistical inference from band recovery data—a handbook, U.S.D.I. Fish and Wildlife Service. *Res. Publ. No. 131.*

Case, R. M. 1978. Interstate highway road killed animals: a data source for biologists. *Wildl. Soc. Bull.* 6: 8–13.

Caughley, G. 1966. Mortality patterns in mammals. *Ecology* 47: 906–918.

Caughley, G. 1974a. Bias in aerial survey. *J. Wildl. Manage.* 38: 921–933.

Caughley, G. 1974b. Interpretation of age ratios. *J. Wildl. Manage.* 38: 557–562.

Caughley, G. 1977. Analysis of vertebrate populations. New York: John Wiley & Sons.

Caughley, G., & L. C. Birch. 1971. Rate of increase. *J. Wildl. Manage.* 35: 568–663.

Caughley, G., & J. Goddard. 1972. Improving the estimates from inaccurate censuses. *J. Wildl. Manage.* 36: 135–140.

Caughley, G., R. Sinclair, & D. Scott-Kemmis. 1976. Experiments in aerial survey. *J. Wildl. Manage.* 40: 290–300.

Chapman, D. G. 1955. Population estimation based on change of composition caused by a selective removal. *Biometrika* 42: 279–290.

Chapman, D. G., & D. S. Robson. 1960. The analysis of a catch-curve. *Biometrics* 16: 354–368.

Davis, D. E. (Ed.). 1982. CRC handbook of census methods for terrestrial vertebrates. Boca Raton, Fla.: CRC Press, Inc.

Davis, D., & R. Winstead. 1980. Estimating the numbers of wildlife populations. In S. D. Schemnitz (Ed.), Wildlife management techniques manual (4th ed.). Washington, D.C.: The Wildlife Society.

Deevy, E. S., Jr. 1947. Life tables for natural populations of animals. *Quart. Rev. Biol.* 22: 283–314.

Eberhardt, L. 1968. A preliminary appraisal of line transects. *J. Wildl. Manage.* 32: 82–88.

Eberhardt, L. 1971. Population analysis. In R. H. Giles (Ed.), Wildlife management techniques (3d ed., rev.). Washington, D.C.: The Wildlife Society.

Eberhardt, L. 1978. Transect methods for population studies. *J. Wildl. Manage.* 42: 1–31.

Hutchinson, G. E. 1978. An introduction to population ecology. New Haven, Conn.: Yale University Press.

Johnson, D. H. 1974. Estimating survival rates from banding of adult and juvenile birds. *J. Wildl. Manage.* 38: 290–297.

Kelker, G. H. 1940. Estimating deer population by differential hunting loss in the sexes. *Proc. Utah Acad. Sci., Arts and Letters* 17: 65–69.

Kelker, G. H. 1944. Sex ratio equations and formulas for determining wildlife populations. *Proc. Utah Acad. Sci., Arts and Letters* 20: 189–198.

Linhart, S. B., & F. F. Knowlton. 1975. Determining the relative abundance of coyotes by scent station lines. *Wildl. Soc. Bull.* 3: 119–124.

Murie, A. 1944. The wolves of Mt. McKinley. *Fauna of the National Parks of the U.S. Fauna Ser. No. 5* Washington, D.C.: U.S. Department of the Interior, National Park Service.

Murphy, E. C., & K. R. Whitten. 1976. Dall sheep demography in McKinley Park and a reevaluation of Murie's data. *J. Wildl. Manage.* 40: 597–609.

Otis, D. L., K. P. Burham, G. C. White, & D. R. Anderson. 1978. Statistical inference from capture data on closed animal populations. *Wild. Monogr. No. 62,* Washington, D.C.

Overton, W. S. 1971. Estimating the numbers of animals in wildlife populations. In R. H. Giles (Ed.), Wildlife management techniques (3d. ed., rev.). Washington, D.C.: The Wildlife Society.

Paulik, G. J. 1963. Estimates of mortality rates from tag recoveries. *Biometrics* 19: 28–57.

Robson, D. S., & D. G. Chapman. 1961. Catch curves and mortality rates. *Trans. Am. Fisheries Soc.* 90: 181–189.

Rupp, R. S. 1966. Generalized equation for the ratio method of estimating population abundance. *J. Wildl. Manage.* 30: 523–526.

Seber, G. A. F. 1970. Estimating time-specific survival and reporting rates for adult birds from band returns. *Biometrika* 57: 313–318.

Seber, G. A. F. 1972. Estimating survival rates from bird band returns. *J. Wildl. Manage.* 36: 405–412.

Seber, G. A. F. 1973. The estimation of animal abundance and related parameters. New York: Hafner Press.

Sen, A. R. 1982. A review of some important techniques in sampling wildlife. *Occas. Paper No. 49.* Canadian Wildlife Service.

Sinclair, A. R. E. 1977. *The African Buffalo: a study of resource limitations of populations.* Chicago: University of Chicago Press.

CHAPTER 5

BEHAVIOR IN WILDLIFE MANAGEMENT

Wildlife managers employ a wide array of techniques in measuring habitat requirements, conducting habitat evaluations, and estimating abundance, fecundity, and mortality in wildlife populations. The human dimension studies reviewed in Chapter 1 have added a great deal to the effectiveness of management programs by providing tools to measure public attitudes and to estimate economic and other values of wildlife. Much less attention, however, has been given to the application of animal behavior in wildlife management.

In part this lack of attention stems from the fact that humans are strongly visually dependent and rely less upon other sensory mechanisms such as smell. We often cannot perceive the same stimuli as other animals can and thus cannot interpret many behaviors. But more importantly, behavior is often difficult to measure, particularly in the field. Behavioral traits, unlike those pertaining to habitats or populations, tend to be: (1) continuous, rather than discrete, (2) made up of both heritable and environmental components whose relative importance remains generally unknown, and (3) fluid or changeable (Thompson 1968). Small wonder, then, that wildlife professionals often see behavior as more interesting from an anecdotal standpoint than from a scientific one.

Animal behavior is a dynamic field, filled with fresh ideas, new tools, and improved methods of data collection and analysis (cf. Lehner, 1979, for a comprehensive review). Wildlife managers can now take advantage of these new developments to address important management problems. The same computer technology that aids other areas of wildlife management is making possible more thorough and quantitative studies of animal behavior. Meanwhile, more and more detailed field investigations integrate behavior with ecology, in effect taking behavior out of the laboratory and putting it into the subject's natural environment.

This chapter introduces animal behavior in relation to several management problems. Other management problems related to behavior are covered in Chapters 7, 8, and 9.

BEHAVIOR AND POPULATION REGULATION

One of the principal ways in which behavior affects wildlife management is the manner in which it influences population size. Any discussion of this phenomenon needs to begin with an explanation of how animals use space.

Home Range Sizes

The area that an individual animal uses for obtaining food, mates, and caring for its young is called its "home range" (Burt, 1943). Biologists determine home range sizes by locating animals repeatedly through retrapping, repeated direct observations of individual animals, or radiolocations of animals fitted with radio transmitters. Once a biologist has plotted the locations on a map, the home range size can be computed.

Studies of animal movements show that, as a general rule, animals remain within particular areas; that is, they maintain home range fidelity. Although home range sizes between animals within the same species vary according to local habitat conditions and population density, they are nonetheless generally predictable.

Kleiber (1961) demonstrated that an animal's basal metabolic rate (an index of its total energy requirements) increases with its weight raised to the ¾th power. Thus, a 10-kilogram animal does not have twice the basal metabolic rate as one weighing 5 kilograms, but instead $10^{3/4}/5^{3/4}$, or 1.68 times as high a rate. Reasoning that basal metabolism was at least a crude index of total energy requirements, McNab (1963) compared home range sizes in a large number of mammalian species with Kleiber's findings on metabolic body size. He found strong and statistically significant correlations (see (Figure 5-1). Herbivores had much smaller home ranges than did "hunters," a group including granivores, fructivores, insectivores, and carnivores. This ecological distinction is consistent with the second law of thermodynamics. Home range sizes seemed to be largely tied to an animal's energy requirement, which was in turn dictated by its body size and trophic level.

Schoener (1968) tested body weight in birds against sizes of feeding territories (the best approximation of home ranges available). He discov-

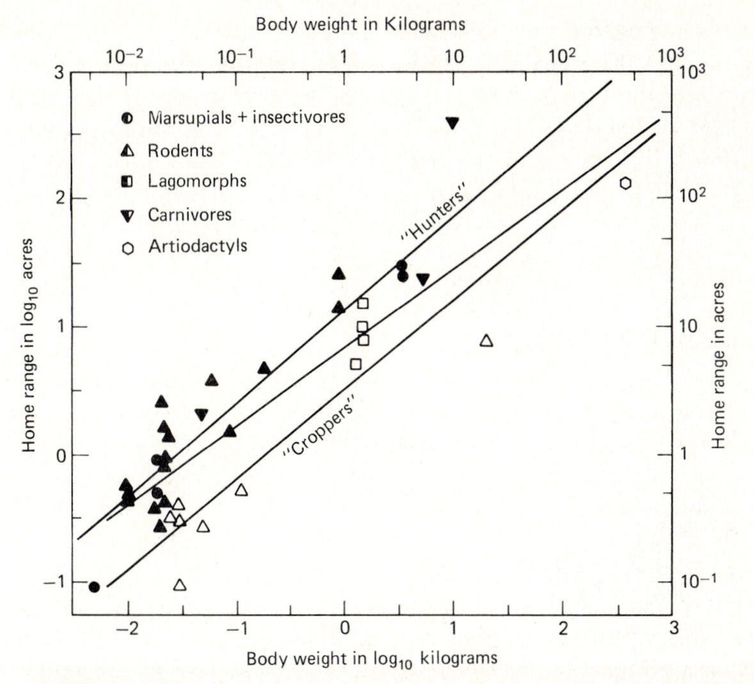

FIGURE 5-1 The relationship between home range size and body weight in some mammals. The center line indicates the fitted curve for pooled data from the "croppers" and the "hunters." (After McNab, 1963, © The University of Chicago Press. Reprinted by permission.)

ered a rather consistent correlation (see Figure 5-2) much like the one McNab found for mammals. Likewise, he learned that predatory or carnivorous birds had much larger feeding territories than did omnivorous or herbivorous ones.

Gregarious species tend to have larger home ranges than do solitary ones since the groups must range over enough area to provide food for all their members. Nonetheless, home range sizes are still tied to body sizes, which are in turn related to energy requirements, to determine how large an area is needed. The area an animal requires is thus ultimately tied to the habitat in which it lives through the habitat's ability to provide food and cover.

Territoriality

All motile animals have home ranges. Some species are also territorial. Most definitions of territoriality refer to it as a behavioral trait through which an individual, a mated pair, or a social group of animals maintains

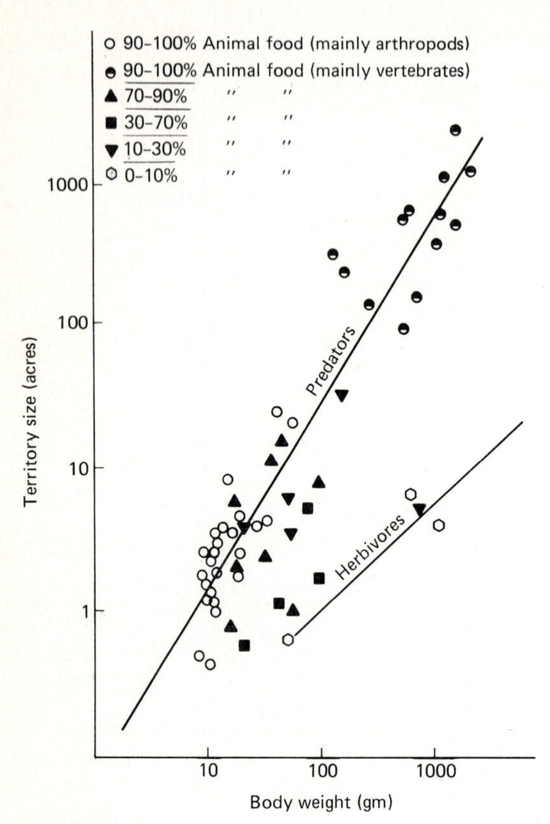

FIGURE 5-2
The relationship between size of feeding territories and body weight for various species of birds. (From "Sizes of feeding territories among birds; by T. W. Schoener, *Ecology,* 19: 123–141, © Ecological Society of America. Reprinted by permission.)

exclusive use of an area through active defense against other members of the same species (conspecifics). Realizing that active defense may be observed only rarely, some authorities have adopted more functional definitions of territoriality. Schoener (1968) simply defined territories as "exclusive areas." Such exclusive areas may be spatial, as they are for many birds, or temporal as they are for many mammalian carnivores. Davies (1978) described a statistically testable criterion by recognizing territoriality "whenever individuals or groups of conspecifics are spaced out more than would be expected from a random occupation of suitable habitats" (see Figure 5-3).

Whether members of a particular population live clumped together in groups or spaced out in individual territories depends largely upon habitat.

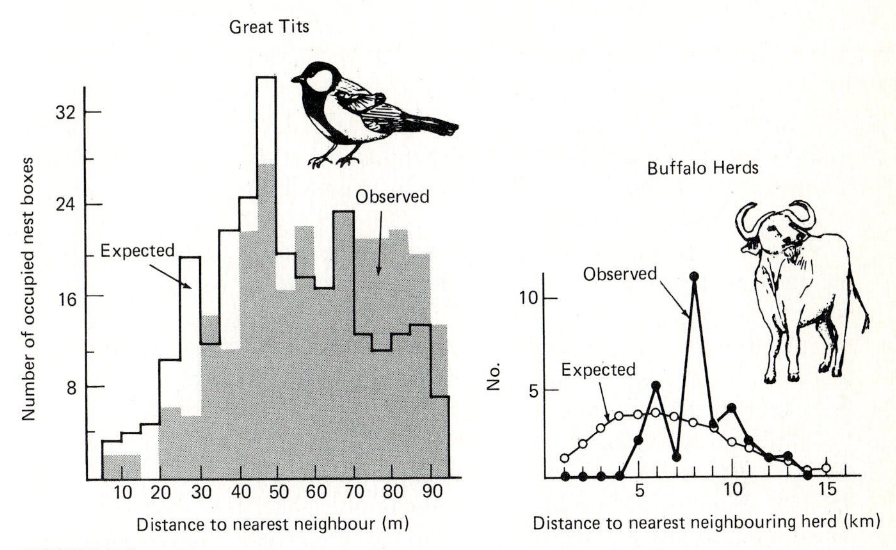

FIGURE 5-3
One practical test for territoriality is "spacing out," or the difference between the observed spatial distribution and that expected by chance alone. (From "Territory and breeding density in the Great Tit [*Parus major*]," by J. R. Krebs, *Ecology,* 52: 2–22, © Ecological Society of America, and *The African Buffalo* by A.R.E. Sinclair, © University of Chicago Press. Reprinted by permission.)

When food is abundant and fairly evenly distributed, the animals are likely to be spaced out more or less uniformly. In contrast, populations that must rely on unpredictable and patchily distributed food are likely to live in larger, more mobile groups. Group living makes location and exploitation of patchily distributed food easier, whereas living solitarily or in pairs allows more efficient use of locally abundant and predictable food (Davies, 1978).

Types of Territories The classification of territories presented here is based on the system developed by Nice (1941) and refined by Wilson (1975).

Type A. A type A territory is a large, defended area in which requirements for mating, rearing young, and gathering at least most of the food are met. It is most prevalent among bottom-dwelling fishes, arboreal lizards, insectivorous birds, and some small mammals.

Type B. A type B territory is one in which all breeding activity occurs but not most of the food gathering. Wilson (1975) cites the nightjar (*Caprimulgus europaeus*) and the reed warbler (*Acrocephalus scirpaceus*) as examples.

Type C. One of the smallest territories, the type C includes only a small defended area around the nest of colonial nesting birds such as gulls, herons, or pelicans.

Type D. Type D territories, also called leks, include only the pairing and mating areas, defended by polygynous males. Examples include the Uganda kob (*Kobus kob*) and some gallinaceous birds such as the prairie chicken or pinnated grouse.

Type E. Like type C territories, type E territories are quite small. They include only defended roosting positions maintained by bats and birds that roost in large groups.

Since birds are generally conspicuous, diurnal, and vocal, territoriality among them is rather obvious and easily confirmed. Avian territories range from type A through type E.

Territoriality among mammals is usually more subtle than it is among birds. Not only are mammals less conspicuous and more nocturnal, they are also less mobile in relation to their home range sizes than are birds. Thus, territoriality is generally more difficult to detect. Exceptions exist, of course, such as in the conspicuous territoriality of the howler monkey (*Alouatta palliata*), whose loud and frequent vocalizations warn away intruders. Most mammalian territories are more like those found in various European mustelids. Weasels (*Mustela nivalis*), stoats (*Mustela erminea*), and otters (*Lutra lutra*) maintain strict territories within each of the two sexes, marking boundaries with glandular scent regularly applied to conspicuous "posts." Females typically maintain smaller exclusive territories within the larger ones of the males (Erlinge, 1968; Lockie, 1966).

Group Territories Among mammals, especially the carnivores and the primates, social groups have exclusive group territories as do, for example, various species of baboons.

Group territories are especially common among the carnivores that practice cooperative hunting. Gray wolves maintain group territories, scent marking the peripheries and attacking intruders (Mech et al., 1971; Mech, 1972). Spotted hyenas (*Crocuta crocuta*) form groups of 10 to 100 individuals, often sharing a communal den site (Kruuk, 1966, 1972a). These groups, known as "clans," are quite territorial, and large-scale battles along the boundaries of their group territories are common. Prides of African lions are not as cohesive as are wolf packs or hyena clans. They do, however, keep group territories located where prey is sufficiently abundant to support the pride year-round (Schaller, 1972).

Dominance Hierarchies

Like territories, dominance hierarchies are maintained through intraspecific aggression. In fact, the two phenomena are so closely related that some authorities (cf. Leyhausen, 1971; Noble, 1939) view them as different ends of a continuum. Domestic cats (*Felis domesticus*) form what Wilson (1975) terms "relative dominance hierarchies" under very crowded conditions. Even the highest-ranking cat yields to the most subordinate one when at or near the subordinate's sleeping place, the rough equivalent to a type E territory. The wild ancestors of domestic cats were most likely solitary and had exclusive territories. Placed artificially close together by people, groups of domestic cats bridge the gap between territoriality and dominance hierarchies.

The simplest and most primitive form of dominance is called a "despotism." One animal dominates all others in the group with no rank distinctions below that of the dominant. Most dominance, however, is hierarchical, comprised of several layers, forming a "peck order," a term resulting from the fact that the earliest recognized dominance hierarchies were described for domestic chickens.

Dominance hierarchies require that individuals be able to recognize each other, a condition that implies at least some reliance on memory. Dominance hierarchies occur among invertebrates, fishes, and amphibians, as well as mammals and birds (Wilson, 1975). Apparently some form of dominance hierarchy is likely to emerge whenever circumstances require inherently aggressive animals to live close together. The advantage is obvious. Once dominance is firmly established, fighting among group members is minimized, so much so, in fact, that a biologist can observe a group of animals closely over quite a long time without realizing that a hierarchy exists. The result of reduced fighting is decreased energy expenditure and less chance of injury.

The dominant, or alpha, individual in the group has every advantage when it comes to feeding or mating. Subordinate individuals, too, gain security from constant conflicts along with the possibility of rising in rank with age, experience, and the inevitable attrition among older, more dominant members.

Species that maintain group territories virtually always have dominance hierarchies as well. Separate male and female hierarchies exist in packs of gray wolves (Rabb et al., 1967) and in clans of spotted hyenas (Kruuk, 1972a). Cape hunting dogs (*Lycaon pictus*) have such well developed and subtle hierarchies that in a 10-month field study of their hunting behavior, Estes and Goddard (1967) saw no sign of dominance conflicts.

Territoriality, Dominance Hierarchies, and Population Regulation

Many wildlife management objectives involve population manipulations. Managers may ask whether or not territoriality or dominance hierarchies can limit population densities, in essence dictating carrying capacities. Like many other questions in ecology, this one is neither simple to answer nor absolute.

The late Paul Errington devoted most of his professional life to the study of population regulation. Investigations of the muskrat (*Ondatra zibethicus*), a territorial aquatic rodent, convinced Errington that, at least among territorial species, intraspecific competition (territoriality) was the principal force regulating population densities. The immediate cause of death might be predation, disease, or starvation; however, territoriality, through limiting the number of animals that an area could maintain, was the ultimate limiting factor (Errington, 1956).

Wynne-Edwards (1962) brought the role of social behavior into sharp controversy with the publication of his book, *Animal Dispersion in Relation to Social Behavior.* Reduced to its simplest elements, the Wynne-Edwards hypothesis stated that territoriality and dominance evolves primarily for the good of the group rather than the individual. The benefit to the group is self-regulation of population density, thereby preventing overexploitation of food supplies. According to this hypothesis, territoriality and dominance hierarchies evolve through a process called group selection. Wynne-Edwards proposed that certain groups of animals possessing genes for territoriality and related traits are favored by natural selection over other groups within the same species that lack such genes. The latter groups are more likely to become extinct through a boom-and-bust system of population growth.

The Wynne-Edwards hypothesis flew in the face of conventional evolutionary theory because it maintained that selection operates on the group level rather than the individual level. This heretical view and its implications for the social regulation of animal populations became quite controversial in the 1960s.

The main weakness of the Wynne-Edwards hypothesis was that territoriality and dominance can both be explained through individual selection, a much simpler proposition. Territorial behavior clearly benefits the territory holders by ensuring sufficient food and cover and ultimately by affording improved reproductive success. Dominance confers considerable advantage on dominant animals, which are more likely to mate and pass on their genes. Subordinates, too, can be favored by an arrangement that promotes group harmony in the short run along with the chance to rise in rank and mate successfully in the long run.

Another problem for the Wynne-Edwards hypothesis is that few, if any, populations of wild animals seem to regulate themselves. Nonetheless, the hypothesis periodically manages to creep back into the wildlife literature. The inherent efficiency and orderliness of strict, self-regulating, homeostatic populations appeal to many people, perhaps as some ultimate example of nature's wisdom. A brief and speculative essay by Peterle (1975) suggested that a Wynne-Edwards type of mechanism may have regulated white-tailed deer populations in pristine North America. The essay drew strong criticism from Smith (1976), who pointed out that territoriality, the only practical means through which deer could regulate their populations, had never been reported for white-tails. Coblentz (1977) added to the criticism, noting the complete lack of evidence that deer populations ever regulate themselves.

Territoriality would be more likely to regulate populations if territory sizes were fixed. Territory sizes, though, often change with shifts in available resources, a flexible feature of territoriality that has been recognized for a long time. Huxley (1934) compared territories with elastic disks, noting that they expand and contract with changes in pressures, though only within certain limits. Kendeigh (1941) reported that territory size in house wrens (*Troglodytes aedon*) is compressible when more birds enter the local population.

Some of the best evidence that territoriality regulates population size comes from deliberate removal experiments. One of the first removal experiments resulted from attempts to measure the effects of avian predators on spruce budworms (*Choristoneura fumiferana*). The investigators tried to shoot all territorial male birds (36 species) within a 16-hectare (40 acre) tract of woods, intending to measure the response of the local spruce budworm population. In their census conducted prior to shooting, they tallied 148 territorial males. Once the shooting began, however, the researchers were surprised to find that "new" males took over territories almost as soon as the "old" residents had been shot. More than 300 males, the original territory holders and their replacements, were shot during the experiment (Stewart and Aldrich, 1951). The obvious conclusion was that dozens of male birds lived in the area as "floaters" without territories. Territoriality, then, was apparently limiting the number of male birds able to breed. Unfortunately, this and most other removal experiments on birds have concentrated mainly upon removing males, generally the more conspicuous and territorially more active of the sexes. Such experiments may fail to answer the more important question of whether or not territoriality limits the number of breeding females (Brown, 1969).

Compressible territories on the one hand and the confirmed existence of floaters on the other make generalizations about the effects of territoriality upon population sizes a risky business. Furthermore, within any given territorial species, the effects of territoriality may change drastically with population size, exerting virtually no effect at lower densities, but having significant influence at high densities. Figure 5-4, based on Brown's

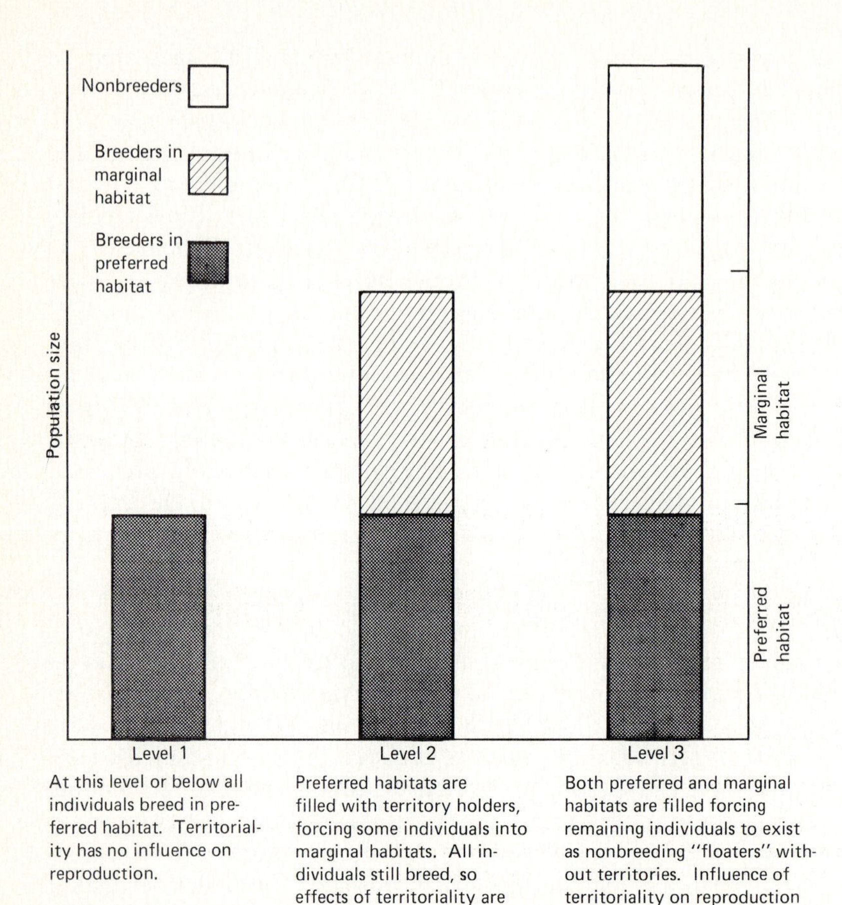

FIGURE 5-4
How the effects of territoriality might likely increase with increasing population density. (Modified from Brown, 1969, © The Wilson Ornithological Society. Used by permission.)

(1969) three-level classification, illustrates how the effects of territoriality can change with density. Obviously, the size of the area under study, along with the availability of adjacent, unoccupied habitat, should be considered in evaluating the effects of territoriality on any particular population.

Dominance hierarchies have the same basic potential for influencing population size as does territoriality. For most social species, group sizes tend to increase along with population densities (Caughley, 1977). Should food or some other critical resource increase in availability, populations can grow and more members can be supported within each group. When conditions take a turn for the worse, the reverse happens. Less food or other resource is available to lower ranking members, forcing them to disperse or starve. Either way, the dominance hierarchy can reduce density under declining habitat conditions.

Thus territoriality and dominance, though offering benefits to the individual, may at times affect or even limit some populations. The effects are sharply density-dependent. Territoriality probably limits local populations in the most favored habitat types more frequently than it regulates populations over larger areas (cf. Krebs, 1971). Perhaps the safest generalization about territoriality and dominance hierarchies is that both commonly act as buffers (level 2 in Figure 5-4) against the more extreme fluctuations that would occur in their absence. Wildlife managers therefore cannot count on behavior to regulate local populations.

BEHAVIOR AND MANAGEMENT PROBLEMS

Wildlife managers can put animal behavior to work for them. A knowledge of social behavior, for example, may allow more effective applications of either control measures or methods aimed at population increases. Other management problems arise when animals either become extremely wary of humans or else lose all fear of them.

Population Control in the African Elephant

Management of the African elephant is difficult for two interrelated reasons. The first is habitat reduction, as more and more of the species' range gives way to croplands. This range reduction causes elephants to concentrate in protected areas such as national parks, leading to the second problem. Herbivores the size of elephants can radically alter their environment, changing forests to woodlands, woodlands to savannas, and savannas to grasslands. Not only will dense concentrations of elephants alter the natural habitat within protected areas, they will also attempt to disperse into neighboring farmlands, causing serious damage to crops and sometimes threatening human safety. As population densities increase, local reductions sometimes become necessary.

Effective control measures must take into account the distinctive traits of elephant social behavior. The primary social group is the matriarchal unit, led by an old female and averaging 10 to 12 members. Sometimes the concentrations of these groups become quite high, with population densities estimated at 1.33 per square kilometer (0.51 per square mile) in North Bunyoro National Park (Laws, 1974) and 5 per square kilometer (1.9 per square mile) in Lake Manyara National Park (Douglas-Hamilton, 1973). African elephants are not territorial.

Early control programs tried to reduce populations through selective culling from the matriarchal units. This practice directed at older animals often removed the matriarchs, causing larger social groups to form around the dwindling number of surviving matriarchs (Laws, 1974). Finally, since older adult females are much less fecund than are younger adults, removal of older elephants does little if anything to reduce the rate of population increase.

In view of these problems, authorities such as R. M. Laws advocate nonselective culling whenever control is necessary. This nonselective control takes advantage of the behavior of the matriarchal group. When alarmed, members of a unit tend to bunch around the matriarch, making her easy to spot. Control officials can then shoot the matriarch first, knowing that the other unit members can easily be dispatched in the confusion that results.

Even though such nonselective control uses elephant behavior to the managers' advantage, it too has drawbacks. As the elephant's range declines, so too does its total population size. Deliberate population reductions will become more difficult to justify as the total numbers of elephants decline. Also, because matriarchal group members are usually closely related, removal of entire groups could hasten the decline of the gene pool (see Chapters 3 and 8). Poaching of elephants increases along with the price of ivory on the world market. If this trend continues, the elephant in Africa may soon become officially recognized as an endangered species (see Chapters 8 and 11).

"Surplus" Killing by Carnivores

Carnivorous mammals occasionally kill far more prey animals than they could possibly need for food. The fox or weasel in the hen house is a familiar example of "surplus" killing directed at domesticated animals, yet this same puzzling behavior occurs at times with wild prey in natural habitats.

Two separate incidents of such surplus killing of wild prey led Kruuk (1972b) to evaluate this inefficient and seemingly aberrant phenomenon. Kruuk studied red fox (*Vulpes vulpes*) predation upon a breeding colony of black-headed gulls (*Larus ridibundus*). Fewer than 3 percent of the gulls killed were actually eaten. Kruuk found a statistically significant correlation between the darkness of nights and the frequency of kills. As Kruuk himself could capture nesting gulls on the darkest of nights by hand, he concluded that the birds were especially vulnerable at such times, particularly since the colony was located in exposed terrain.

Later, in the Serengeti, Kruuk discovered that spotted hyenas killed in one night 82 Thompson's gazelle (*Gazella thompsonii*) and severely injured 27 others within 10 square kilometers (3.9 square miles) of open grassland. This unusually high rate of predation took place during heavy rains and storms on an extremely dark night.

After comparing his observations with those of other field investigators, Kruuk developed a behavioral explanation for surplus killing. He concluded that mere satiation of hunger has little if any effect upon catching and killing prey. Satiation does, however, appear to affect searching and hunting, which, under normal circumstances, would prevent excess catching and killing from occurring. Surplus killing, then, takes place only under extenuating circumstances such as very dark nights in open terrain or the unnatural confines of a pen or hen house, where searching and hunting are unnecessary for overcoming the prey's antipredator defenses. Under natural conditions, these circumstances are too rare to have significant impacts upon wild prey populations.

The kinds of surplus killing reviewed by Kruuk can have important management implications. Predators can and do kill excessive numbers even of their natural prey. But the fact that such heavy losses occur occasionally should not be taken as conclusive proof that predators are significantly affecting their prey. These incidents happen under unusual circumstances. Moreover, where such incidents are likely to occur, their effects could be reduced by management devoted to increasing cover for the prey.

Reactions to Human Disturbance

Many people believe that wild animals are naturally shy and elusive toward humans. Some species, especially the larger mammals, are deemed incompatible with people, invariably retreating to more remote regions after human disturbance. The truth is that wild animals generally can coexist with people, provided that their habitat requirements are met and that their presence is tolerated. Even the gray wolf, that legendary enemy of people, does not voluntarily retreat to more distant realms at the first whiff of human scent. Instead, these carnivores are killed soon after people and their livestock move into a region. Conflicts arise, old prejudices are rekindled, and the wolf population is destroyed (Mech, 1974). The effects of human disturbance upon wildlife are usually difficult to evaluate because they tend to occur simultaneously with habitat modifications and direct population reductions. Small wonder then, that local disappearances have been so readily attributed to "natural shyness" of the animals themselves.

What is known about the responses of animals to severe disturbance has been summarized by Geist (1971). Following the disturbance, an animal

1 Becomes excited if it senses an object, odor, or sound associated with the disturbance

2 Avoids the locality where it experienced the disturbance

3 Generalizes to similar objects, odors, or sounds and becomes excited upon sensing them.

This process is called "sensitization" and can impose physiological stress and cause local range shifts, with potentially serious consequences for a local population of animals. Excitation is itself costly because it elevates metabolism, resulting in an increased energy use. Prolonged excitation can induce weight loss, reduction in fecundity, and increased mortality in young animals. Shifts in local habitat use can reduce carrying capacity. Disturbance and the stress resulting from it may even increase a population's susceptibility to disease.

Habituation versus Sensitization Although the behavioral stress imposed by human disturbance can have severe consequences, most behavior itself is flexible and adaptable. A stimulus initially regarded as threatening may, if repeated without negative reinforcement, eventually become ignored. This most basic form of behavioral adaptation is called "habituation."

Geist (1971) reported two quite remarkable cases of habituation. A protected herd of mule deer inhabits a small town in Waterton Lakes National Park. The animals seem to be indifferent toward the town's human residents, who in turn tolerate the deer, paying no mind as the animals use buildings as shelter or feed and rest quietly on residential lawns.

Geist's own studies on bighorn sheep in Banff National Park were facilitated by many years of work by a former park warden. The warden wanted to mark some of the bighorns to find out if any were being shot outside park boundaries. First, he habituated adult sheep by teaching them to lick salt from his hand. Eventually he habituated them so completely that he could clamp permanent tags into their ears without upsetting them.

How fast habituation occurs or whether it occurs at all depends upon several factors apart from the characteristics of the particular species. An animal will have more difficulty habituating to an intense stimulus, such as a very loud noise, than to a milder one. Constancy is important because irregular variations in the stimulus may confuse the animal, adding to its stress and its reluctance to habituate. Obviously, the greater the frequency of the stimulus, the more rapidly habituation can proceed.

Disturbance and habituation are reviewed in two cases described below. Habituation was fairly complete in the first, and not at all in the second.

Deer and Snowmobiles The growing popularity of snowmobiles in many parts of the United States and Canada has led wildlife managers to wonder what effect so many of these fast, noisy machines might have upon deer. Researchers in Minnesota (Dorrance et al., 1975) compared home range sizes, movements, and distances from trails maintained by radio-collared white-tailed deer in two study areas. The St. Croix State Park, an area of 12,600 hectares (31,500 acres), received an average use of 195 snowmobiles daily on winter weekends, and 10 snowmobiles a day on winter weekdays. In contrast, snowmobiles were banned except by project personnel on the Mille Lacs Wildlife Management Area, 15,400 hectares (38,500 acres) of similar habitat.

Effects of the snowmobiles were subtle at St. Croix. The traffic displaced deer from areas immediately adjacent to snowmobile trails but otherwise had no detectable effects. The deer had habituated to snowmobiles and probably would have remained unaffected except perhaps during the severest of winters.

Deer at the Mille Lacs Wildlife Management Area increased their movements and shifted or expanded their home ranges in response to disturbance. They were not habituated to snowmobiles.

Bald Eagles in Winter Suspecting that disturbance might be contributing to the decline of the bald eagle (*Haliaeetus leucocephalus*), Stalmaster and Newman (1978) set about measuring the effects of this phenomenon. They studied eagles along the Nooksack River in Washington where the birds were subjected to increased disturbance through logging, mining, recreation, and construction of housing developments. They found a significant negative correlation between levels of human activity and the distribution of wintering eagles. Adult eagles were more sensitive to disturbance than were younger birds, suggesting that the eagles were not habituating to human activity. The average flight or flushing distance of eagles from approaching humans was nearly twice as far for adult (196 meters) eagles as it was for immature ones (99 meters). Flushing distances were significantly shorter whenever eagles were approached from the forest, implying that forests make good buffer zones against human disturbance. Stalmaster and Newman recommended establishment of buffer zones at least 250 meters around bald eagle wintering grounds to guard against any ill effects from disturbance due to human activity.

Grizzly Bears and Human Safety

Persecuted as a threat to both livestock and humans, grizzly bears survive today only where adequately protected. Thousands of grizzlies still live in Alaska and western Canada. But there are fewer than 1,000 remaining in the lower 48 states and most of these occur in and around Yellowstone and Glacier National Parks. Sharp increases in human use of these parks since the 1960s have been met with corresponding rises in human injuries and deaths inflicted by these bears. The grizzly bear is a threatened species within the lower 48, and the widely publicized and sometimes sensationalized accounts of human deaths and injuries have created one of the most controversial issues in wildlife management in the United States.

The basic problem is a behavioral one. The grizzlies seem to have become more dangerous in recent years. Most authorities attribute the increased danger to habituation. In essence they contend that bears in parks encounter unarmed humans repeatedly, learn that the people pose no real threat, and lose whatever fear they may otherwise have had.

But a recent investigation in Glacier National Park raised some doubts about the validity of this traditional view. There can be little doubt that grizzly bears frequently exposed to humans do in fact habituate to them. The question is whether or not habituated bears become more dangerous or less so. Jope (1982) intensively observed grizzlies in portions of the park heavily used by day hikers. She also compiled and analyzed hikers' reports of bear observations. She learned that bears charged people primarily along trails that received less human use. Bears usually charged when startled at close range. Bears were also startled along trails more frequently used but did not charge nearly so often. Moreover, the frequency of both charges and injuries was significantly higher early in the summer than later in the summer, even though more people were on the trails in late summer. This evidence suggests that habituation of grizzly fear responses may actually contribute to a reduction in the rate of human injuries.

There is one point on which all authorities tend to agree. The most dangerous bears are those that have learned to associate people with food. Grizzly bears are omnivorous and highly opportunistic and once they obtain food from people they learn to generalize from the experience and associate food with human presence, scent, structures, or equipment (McCullough, 1982). Jope (1982) reported that two grizzly bears in her study area obtained food from people during her field investigations. Both bears became aggressive toward hikers subsequently encountered. She cited several other cases in which people had been killed by grizzlies that had been accustomed to eating people's food.

Female bears with cubs can be especially dangerous. Of 45 bear attacks prior to 1970 in which the sex of the attacking animal was known, 43 were females, at least 31 of which definitely had cubs (Herrero, 1970). This defensive reaction or maternal aggression is presumably an adaptation to defend cubs against male bears, which can be cannibalistic. Thus far there is no evidence that female grizzlies with cubs become habituated to humans (Jope, 1982), so this form of danger differs from that posed by habituated bears. Yet since 1970 a larger proportion of attacks have been committed by grizzly bears without cubs. Could this suggest that the behavior of these animals is changing?

The question of whether or not habituation increases or decreases the danger posed by grizzlies is an important and complex one, with important management implications. Some biologists who have reviewed the problem (Geist, 1978; McCullough, 1982) have concluded that some means must be found to reverse the habituation process. Perhaps some means of punishing or repelling bears would accomplish this, although no one has thus far discovered an effective method safe to the user.

It might be fruitful to regard the grizzly bear problem as three distinct problems. The first involves the danger posed by females with cubs. In this case habituation has apparently not occurred, and it is an open question as to whether habituation would cause females with cubs to be more dangerous or less so. Park authorities quite rightly regard such bears as especially dangerous, so they close trails where female grizzlies with cubs are known to frequent.

Grizzlies other than females with cubs present dangers to hikers. In this case, the main danger seems to be surprise at close range. Jope (1982) found that only people without bear bells were charged (bear bells are bells worn or carried to produce noise to warn bears of approaching humans) and so concluded that use of such bells is a valid safety precaution.

But the third aspect of the grizzly problem is probably the most dangerous and contributes to the most human fatalities. Grizzlies that have learned to associate humans with food constitute a real threat to people in campgrounds. In such cases, it seems reasonable to assume that habituation increases this danger. The response of a grizzly toward a hiker along a trail in daylight may be quite different from that of the same bear toward a sleeping hiker at midnight.

Several management guidelines have been implemented or suggested (Herrero, 1970; Craighead, 1979; Jope, 1982):

1 Have the Park Service provide backcountry visitors with warnings about the dangerous and unpredictable nature of grizzly bears.

2 Close to hikers and campers those areas known to contain aggressive bears or bears with cubs.

3 Encourage hikers to wear bear bells.

4 Place campgrounds in areas away from choice grizzly bear habitat. (By 1983 at least one camping area in Glacier National Park was protected within a steel enclosure to shield campers from grizzly bears.)

5 Close garbage dumps and other artificial food sources; these are popular sites for people to see wild bears. This suggestion has been implemented.

6 Consider preserving winter-killed elk or other large animal carcasses for distribution on sites away from areas of human use. The carcasses might then attract grizzlies and hold them near such sites.

The future of grizzly bears in our national parks is far from certain. The National Park Service is caught between conflicting demands: on the one hand a responsibility for human safety and on the other a commitment to maintain natural populations of a threatened species. As more and more people use the parks, the opportunities for more injuries and deaths rise. Perhaps the Park Service has already helped through its extensive efforts to educate and warn the public about responsible behavior and reasonable precautions. But neither the service nor anyone else has come up with effective means of modifying grizzly bear behavior to make these animals less dangerous. This failure led McCullough (1982) to conclude that the human–grizzly bear conflict is one of stalemate.

Even in Leopold's time, the fate of the grizzly bear in the lower 48 states was questionable. Some of his contemporaries seemed unconcerned, pointing out that plenty of the great bears still lived in Alaska and Canada. His response: "Relegating grizzlies to Alaska (and Canada) is about like relegating happiness to heaven; one may never get there" (Leopold, 1949:199).

SUMMARY

Although behavioral phenomena are often more difficult to measure than those pertaining to habitat or population biology, they can relate to wildlife management in many important ways. For example, what role, if any, does animal behavior have in influencing population size? Agonistic behaviors such as territoriality and the formation of dominance hierarchies have the potential for limiting population sizes through limiting the number of individuals that breed.

A review of agonistic behaviors, however, reveals that their effects on population size are variable and often rather weak. Moreover, most mechanisms through which such behaviors might influence populations are hard to explain as products of natural selection. The most likely effects of such behaviors are subtle dampenings of population oscillations, particularly at higher densities, rather than any rigorous limitations upon numbers.

Animal behavior can be put to practical uses in wildlife management. In controlling African elephants, for example, control officials have learned to shoot the group matriarch first, then to finish off the entire social unit. Such a practice accomplishes the necessary reductions without the major disruptions imposed by more traditional methods of culling a few members from each social unit.

Surplus killing by carnivores may occur whenever conditions allow these predators to continue capturing and killing prey. Surplus killing occurs even after the predators' hunger is satiated. This behavior has practical consequences whenever carnivores encounter flocks or herds of livestock without means of escape. Wild prey are seldom as vulnerable to surplus killing, though it has been documented on exceptionally dark nights among concentrations of prey without adequate cover.

Other behaviors of interest to wildlife managers involve reactions to human disturbance. These reactions are essentially learned and thus are subject to changes. Changes can include sensitization, which is an increased wariness toward humans resulting from negative reinforcement, or habituation, the reduction of (in this case fear) responses in the absence of any negative reinforcement. Sensitization can cause animals to abandon certain portions of their range or even interfere with reproduction. Habituation can lead to increases in wildlife-related damage, accidents, or even human injuries.

Perhaps the most controversial example of a management problem involving habituation is that of grizzly bears in national parks. Hunting in those parks has long been forbidden, thus removing the primary source of negative reinforcement and allowing the bears to become habituated to humans. While it is not clear whether or not habituated grizzlies are more dangerous to hikers than nonhabituated ones, it seems certain that grizzly bears that have learned to associate people with food are extremely dangerous to campers. Management thus far has consisted of removal of garbage dumps and particularly aggressive bears, along with efforts to regulate human behavior and activity. These measures may need to be supplemented with some means of reversing the habituation process through negative reinforcement.

REFERENCES

Brown, J. L. 1969. Territorial behavior and population regulation in birds: a review and re-evaluation. *Wilson Bull.* 81: 293–329.

Burt, W. H. 1943. Territoriality and home range concepts as applied to mammals. *J. Mammal.* 24: 346–352.

Caughley, G. 1977. *Analysis of vertebrate populations.* New York: John Wiley & Sons.

Coblentz, B. E. 1977. Comments on deer sociobiology. *Wildl. Soc. Bull.* 5: 67.

Craighead, F. C., Jr. 1979. *Track of the grizzly.* San Francisco, Calif.: Sierra Club Books.

Davies, N. B. 1978. Ecological questions about territorial behavior. In J. R. Krebs and N. B. Davies (Eds.), *Behavioral Ecology: an evolutionary approach.* Sunderland, Mass.: Sinauer Associates, Inc.

Dorrance, M. J., P. J. Savage, & D. E. Huff. 1975. Effects of snowmobiles on white-tailed deer. *J. Wildl. Manage.* 39: 563–569.

Douglas-Hamilton, I. 1973. On the ecology and behaviour of the Lake Manyara elephant. *E. Afr. Wildl. J.* 11: 401–403.

Erlinge, S. 1968. Territoriality of the otter, *Lutra lutra. Oikos* 19: 81–98.

Errington, P. L. 1956. Factors limiting higher vertebrate populations. *Science* 124: 304–307.

Estes, R. D., & J. Goddard. 1967. Prey selection and hunting behavior in the African wild dog. *J. Wildl. Manage.* 31: 52–70.

Geist, V. 1971. A behavioral approach to the management of wild ungulates. In E. Duffey and A. Watt (Eds.), The scientific management of animal and plant communities for conservation. *11th Symp. Brit. Ecol. Soc., London.* London: Blackwell Press.

Geist, V. 1978. Behavior. In J. Schmidt and D. Gilbert (Eds.), *Big game of North America.* Harrisburg, Pa.: Stockpole Books.

Herrero, S. M. 1970. Human injury inflicted by grizzly bears. *Science* 170: 593–598.

Huxley, J. S. 1934. A natural experiment on the territorial instinct. *British Birds* 27: 270–277.

Jope, Katherine. 1982. Interaction between grizzly bears and hikers in Glacier National Park, Montana. M.S. thesis. Corvallis, Oregon State University.

Kendeigh, S. C. 1941. Territorial and mating behavior of the House Wren. *Illinois Biol. Monogr.* 10: 1–120.

Kleiber, M. 1961. *The fire of life.* New York: John Wiley & Sons.

Krebs, J. R. 1971. Territory and breeding density in the great tit, *Parus major. L. Ecology* 52: 2–22.

Kruuk, H. 1966. Clan-system and feeding habits of spotted hyenas (*Crocuta crocuta*). *Nature* 209: 1257–1258.

Kruuk, H. 1972a. *The spotted hyena.* Chicago: University of Chicago Press.

Kruuk, H. 1972b. Surplus killing by carnivores. *J. Zool.* 166: 233–244.

Laws, R. M. 1974. Behavior, dynamics, and management of elephant populations. In V. Geist and F. Walther (Eds.), The behaviour of ungulates and its relation to management. *IUCN Publ. New Series No. 24.* Morges, Switzerland.

Lehner, P. 1979. *A handbook of ethological methods.* New York: Garland STPM Press.

Leyhausen, P. 1971. Dominance and territoriality as complemented in mammalian social structure. In A. Esser (Ed.), *Behavior and Environment: the use of space by animals and man.* New York: Plenum Press.

Lockie, J. D. 1966. Territory in small carnivores. *Symp. Zool. Soc., London* 18: 143–165.

McCullough, D. R. 1982. Behavior, bears, and humans. *Wildl. Soc. Bull.* 10: 27–33.

McNab, B. K. 1963. Bioenergetics and the determination of home range size. *Amer. Natur.* 97: 133–140.

Mech, L. D. 1972. Spacing and possible mechanisms of population regulation in wolves. *Amer. Zool.* 12: 642 (abstract).

Mech, L. D. 1974. A new profile for the wolf. *Natur. Hist.* 83(4): 26–31.

Mech, L. D., L. Frenzel, Jr., R. Ream, & J. Winship. 1971. Movement, behavior, and ecology of timber wolves in northeastern Minnesota. In L. D. Mech and L. Frenzel (Eds.), Ecological studies of the timber wolf in northeastern Minnesota. *USDA Forest Service Res. Paper NC-52.*

Nice, Margaret, M. 1941. The role of territory in bird life. *Amer. Midl. Natur.* 26: 441–487.

Noble, G. K. 1939. The role of dominance in the social life of birds. *Auk* 56: 263–273.

Peterle, T. J. 1975. Deer sociobiology. *Wildl. Soc. Bull.* 3: 82–83.

Rabb, G., J. Woolpy, & B. E. Ginsburg. 1967. Social relationships in a group of captive wolves. *Amer. Zool.* 7: 305–311.

Schaller, G. B. 1972. *The Serengeti lion.* Chicago: University of Chicago Press.

Schoener, T. W. 1968. Sizes of feeding territories among birds. *Ecology* 49: 123–141.

Smith, C. A. 1976. Deer sociobiology—some second thoughts. *Wildl. Soc. Bull.* 4: 181–182.

Stalmaster, M., & J. Newman, 1978. Behavioral responses of wintering bald eagles to human activity. *J. Wildl. Manage.* 42: 506–513.

Stewart, R., & J. Aldrich. 1951. Removal and repopulation of breeding birds in a spruce-fir forest community. *Auk* 471–482.

Thompson, W. R. 1968. Genetics and social behavior. In B. C. Glass (Ed.), *Genetics, biology and behavior.* New York: Rockefeller University Press.

Wilson, E. O. 1975. *Sociobiology: the new synthesis.* Cambridge, Mass.: Belknap Press.

Wynne-Edwards, V. C. 1962. *Animal dispersion in relation to social behavior.* Edinburgh: Oliver and Boyd Publishers.

CHAPTER 

HOW POPULATIONS INTERACT

Chapter 3 introduced the ways in which populations of individual species grow and fluctuate in natural environments. But populations of different species affect one another, often in ways that are subtle but significant. This chapter reviews two of the more important and controversial types of population interactions: competition and predation. Although ecologists often debate the finer points of both processes (Schoener, 1982), one generalization applies to both. Predation and competition are beneficial to natural communities. In the short run, these processes promote population stability and enhance species diversity over longer periods.

Predation and competition can best be understood by introducing them in the context of virtually undisturbed environments, free from human activity. With one rather famous exception, the examples and discussion in this chapter focus on just such conditions. Left to themselves, predation and competition would pose few practical problems for wildlife managers. Their effects would be mostly density-dependent and would relax as populations declined, thereby representing no real threat to a population's long-term well-being.

Real problems with population interactions usually result from human disturbances that upset competitive or prey–predator relationships. Major sources of disruptions include modifications of habitat, introduction of domestic livestock, and release of exotic wildlife. Sometimes livestock or exotics can prove to be superior competitors with native wildlife, particularly in disturbed environments. Livestock are often unable to defend themselves adequately against wild predators. Examples of such practical problems appear in Chapters 9 and 10.

COMPETITION

How do competing species exploit the resources within a community? Why can up to 14 species of large herbivores live in the same game reserve in East Africa? How do three species of ptarmigan coexist in the same parts of interior Alaska? Why hasn't one more aggressive species of pocket gopher eliminated three competitors from Colorado? The answer to each of these questions involves the ways in which natural selection deals with competing species.

The competitive exclusion principle, long a cornerstone of ecological theory, states that no two species can occupy the same niche in the same community. According to this principle, whenever two species compete for the same resources, one of them will ultimately prove the superior competitor, pushing the other completely out. The competitive exclusion principle has been criticized, however, because its reasoning is circular or tautological (Cole, 1960; Hardin, 1960; Peters, 1976). It is also too simplistic in that it fails to consider the effects of other factors such as unusually severe weather or habitat disturbance. These additional considerations can introduce elements of chance into the outcome. Furthermore, if two competing species really do manage to coexist in the same habitat, the obvious conclusion is that they surely must occupy different niches. Thus, no one may successfully disprove the competitive exclusion principle, since to do so would require evidence of the coexistence of two species with exactly the same niches.

Perhaps a better way to state the competitive exclusion principle is that the probability that two or more competing species may coexist in the same area is inversely proportional to their niche overlap. This definition eases away from the "same niche" problem and addresses competition in relative rather than absolute terms.

Nineteenth-century Darwinists viewed competition as ruthless, bloody, beastly business with "nature painted red in fang and claw." The truth is less dramatic but far more interesting. Natural selection has favored competitors that are less alike, pushing them into niches with less overlap. This separation reduces the competitive pressure and refines the needs of each species, fine-tuning them to the community in which they live. Along with this niche divergence, the physical traits of competing species, such as the sizes and shapes of the bills of birds, also diverge, a process known as "character displacement" (Brown and Wilson, 1956). Character displacement further reduces competition, allowing species to exploit different resources. This reduction in competition through natural selection leads to niche partitioning. When biologists study competition in the field, they typically discover that competitors coexist through niche partitioning.

There are two broad categories of competition. Interference competition is the direct displacement of competitors through aggression. An example would be a lion appropriating a kill from a leopard (*Panthera pardus*) by driving the smaller cat away. Exploitation competition takes place whenever one species makes more efficient use of a common resource. Feral burros (*Equinus asinus*), for example, outcompete desert bighorn for available forage.

In each of the following examples, natural selection has reduced the intensity of competition. The result is that the species can coexist within the same general areas.

Alaskan Ptarmigans

Three species of ptarmigan, willow ptarmigan (*Lagopus lagopus*), rock ptarmigan (*Lagopus mutus*), and white-tailed ptarmigan (*Lagopus leucurus*) share common winter ranges in parts of interior Alaska. Where the three species overlap, Moss (1974) learned that the willow ptarmigan, true to its name, feeds mainly on willow. The rock ptarmigan relies almost exclusively upon birch buds and catkins, and the white-tailed ptarmigan makes use of alder cones and birch (see Table 6-1). Although the diets overlap slightly, they are sufficiently distinct to prevent intensive competition. Moss (1974) also reported that in Iceland, where the rock ptarmigan occurs without competitors, it prefers willow to birch. Apparently the feeding niches in those two species overlap partially, yet enough distinction is maintained to allow the species to share a common range.

Pocket Gophers

The plains pocket gopher (*Geomys bursarius*), the Mexican pocket gopher (*Cratogeomys castanops*), the valley pocket gopher (*Thomomys bottae*), and the northern pocket gopher (*Thomomys talpoides*) are all territorial and aggressive. Miller (1964) investigated competition among the four species in Colorado. He discovered that the two critical factors were competitive ability (of the interference type) and soil tolerance. In competitive ability they ranked as follows: plains pocket gopher, Mexican pocket gopher, valley pocket gopher, and northern pocket gopher. But in their abilities to dig in soils with more rock and gravel, their ranking was exactly the opposite: northern pocket gopher, valley pocket gopher, Mexican pocket gopher, and plains pocket gopher (see Figure 6-1). The plains pocket gopher could claim the finest textured soils, there out-

TABLE 6-1
WINTER FOOD HABITS IN THREE SPECIES OF ALASKAN PTARMIGAN
(Modified from Moss 1974, © American Ornithologists Union. Reprinted by permission); figures may not total to 100% due to rounding error

	Percent of diet in volume		
Species	**Willow**	**Birch buds & catkins**	**Alder cones**
Willow ptarmigan	95	5	0
Rock ptarmigan	2	89	7
White-tailed ptarmigan	2	41	56

competing the other species. Its limited soil tolerance, however, prevented the plains gopher from displacing the others wherever the soil was coarser. At the other end of the competitive scale, the northern pocket gopher survived not by directly outcompeting any of the other species, but by being able to dig in soils too coarse for them.

East African Ungulates

The vast herds of antelopes and other ungulates that inhabit East and South Africa are impressive not only in their total numbers, but also in the large diversity of potentially competing herbivores sharing the same general area. The ways in which they coexist have received considerable study. Lamprey (1963) compared foraging patterns in 14 species of wild

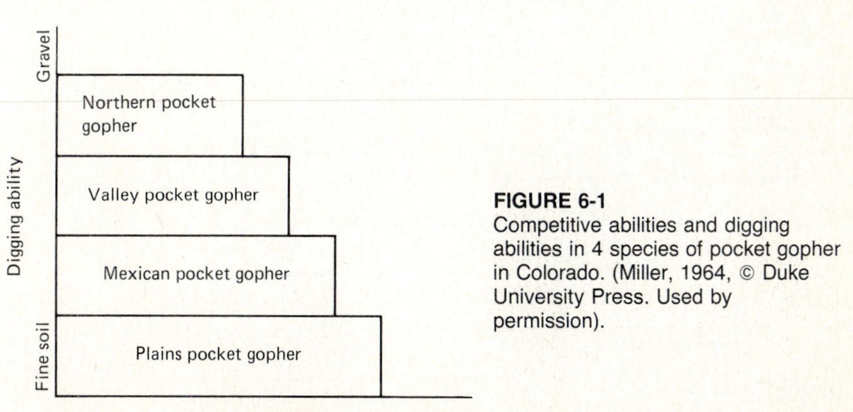

FIGURE 6-1
Competitive abilities and digging abilities in 4 species of pocket gopher in Colorado. (Miller, 1964, © Duke University Press. Used by permission).

ungulates living in a Tanzanian game reserve. He found six major differences in their feeding patterns that allowed efficient use of different parts of the vegetation and the habitat and minimizing direct interspecies competition:

1 Occupation of different habitat and vegetation types (see Figure 6-2)
2 Selection of different types of food within the same habitat type
3 Occupation of different areas during the same season

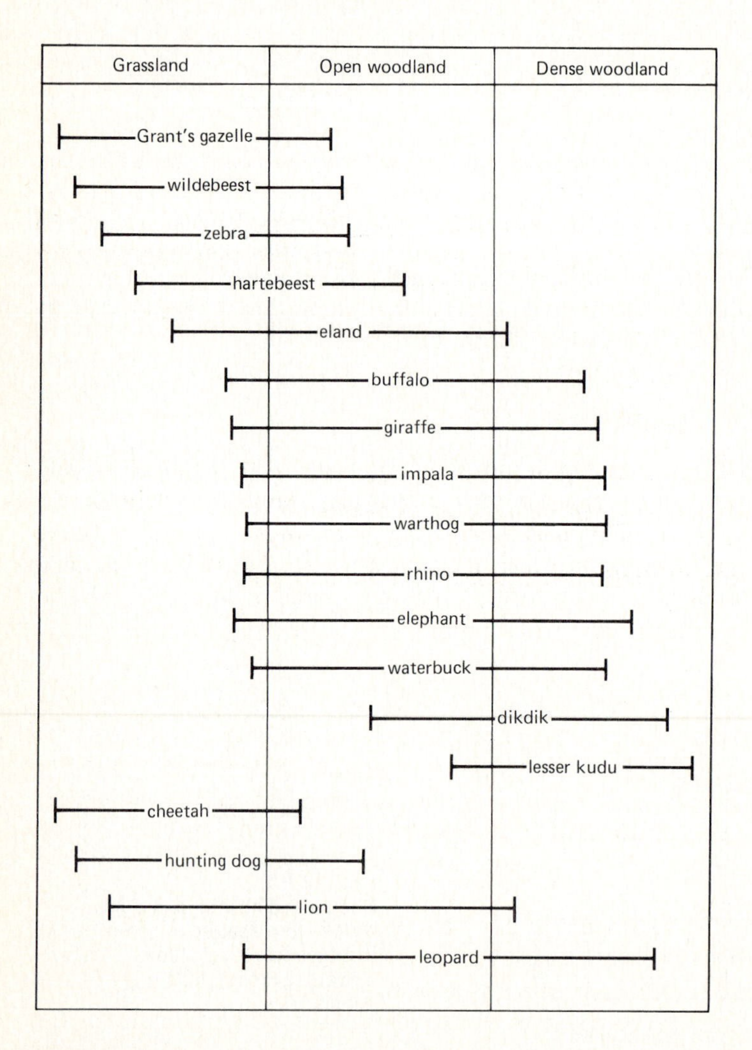

FIGURE 6-2
Habitat use of East African herbivores and carnivores. (Lamprey, 1963, © Blackwell Scientific Publications. Reprinted by permission).

4 Occupation of the same area during different seasons
5 Use of different vertical levels of vegetation
6 Occupation of different dry season refuges when food competition is most critical.

African ungulates also use successional stages in various ways to reduce competition still further. Zebras (*Equus burchelli*) are among the first grazers to take advantage of fresh grasses that emerge following fires in the Serengeti. As they graze, the zebras open up the herb layer through trampling. They also use their powerful teeth to select tougher grass stems over the softer grass leaves. Both the trampling and the selective feeding make grasslands more suitable for foraging by wildebeest (*Connochaetus taurinus*), which follow the zebras in the grazing succession. The additional grazing pressure from the wildebeest favors the growth of shrubs and other dicots. This pattern suggests that numbers of ungulate species in East Africa may be positively rather than negatively correlated (Gwynne and Bell, 1968). In other words, some of the species benefit one another more than they compete.

Large Predators of the Serengeti

The diversity of East African ungulates is reflected in the diversity of predators that prey on them. Part of this diversity among predators is simply due to differences in size, habitat use, and social behavior of the ungulates. Yet all five of the Serengeti large predators—the lion, the leopard, the cheetah (*Acinonyx jubatus*), the wild dog or Cape hunting dog, and the spotted hyena—overlap to a surprising extent, competing for some of the more abundant ungulates such as the Thompson's gazelle.

The hunting methods of the Serengeti predators account to a large extent for the coexistence of the five species (Kruuk and Turner, 1967; Schaller, 1972). The cheetah and the wild dog hunt by daylight, whereas the other three species are predominantly nocturnal. The hyena, the cheetah, and the wild dog attack prey by chasing the animals for some distance. The lion and leopard are both stalkers, though the lion uses a short rush and the leopard often springs from ambush. Lions are year-round residents of the Serengeti woodlands, leopards remain in the thickets and gallery forests, and cheetahs and wild dogs move nomadlike between the woodlands in the dry season and the plains in the wet season (see Table 6-2).

These different hunting patterns result in selection of different species of prey. They also cause different species of predators to select different age classes of animals from within the same species. Schaller (1972) concluded that cursorial predators, such as the cheetah and the wild dog, take larger proportions of the younger and older age classes, since those classes are

TABLE 6-2
BEHAVIORAL CHARACTERISTICS OF SERENGETI PREDATORS THAT RESULT IN DIFFERENTIAL USE OF AVAILABLE PREY
(Modified from Schaller, 1972, © University of Chicago Press. Reprinted by permission)

Behavior types	Lion	Hyena	Leopard	Cheetah	Wild dog
Method of attack	Stalk and short rush	Cursorial	Stalk and ambush	Cursorial	Cursorial
Activity times	Nocturnal	Nocturnal	Nocturnal	Diurnal	Diurnal
Movements	Resident	Seminomadic	Resident	Nomadic	Nomadic
Habitat use	Woodlands	Plains-woodlands ecotone	Thickets river forest	Plains-woodlands ecotone	Woodlands and plains

comprised of individuals least fit to run long distances. Stalkers such as the leopard and the lion tend to take individuals more randomly, not favoring a particular age class. Once again, natural selection has favored different means of obtaining diets sufficiently different to permit coexistence of competing species.

The Effects of Interspecies Competition

Suppose that the early Darwinists had been right in their views on competition. Natural selection would then have favored the most aggressive, relentless competitor. Given time, biological communities might actually decrease in numbers of species, as those less able to compete were forced out.

But natural selection did not invariably select for the aggressive and the relentless. Rather, it favored divergence in feeding strategy and habitat use, resulting in actual reductions in the levels of competition and the extent of niche overlap. Competition served to enhance and even to increase the diversity of species within natural communities.

PREDATION

One of the most controversial subjects in wildlife management is the effect of predation on prey populations. Some people believe that, if unchecked, predators will surely eradicate their natural prey. Others maintain that predation has no real effect because it removes only those individual prey that would have died from other causes. Still others contend that predation benefits prey populations by selectively culling less fit members.

This section introduces some of the biological variables that contribute to the outcome of prey–predator relations, variables that should be considered in evaluating any particular case. Next it reviews some specific examples that have been studied in detail in natural environments. In conclusion, it discusses the role of predation in nature.

Just as with competition, predation poses a problem for wildlife management largely as a result of human activities. Introductions of exotic animals, particularly domesticated species, have in many cases altered the original prey–predator arrangement, resulting in economic and sometimes even ecological problems. Such problems are discussed in Chapter 9. Before assessing predator problems in wildlife management, it is essential to understand how predation functions in a natural environment, with minimal interference from human activity.

Components of Predation

Does predation typically limit a prey population? For this to take place, the predator population must keep the prey population below the level that it would reach if predators were absent. A related question is: Does predation typically regulate a prey population? There is an important distinction between the terms "limit" and "regulate." If food supply, weather, or something else besides predation affects the prey population so that it varies despite the predators, predation might then serve as a partial check on numbers by dampening oscillations that would occur if the predator population were absent. Predation that regulates a prey population is more subtle and less rigid than predation that actually limits a population.

In practice, the effects that predation has on prey populations can vary enormously. Each of the major constituents that affect prey–predator relations deserves some elaboration.

Selection of Different Age Classes One way in which predation's effects can vary is through selection of different age classes of prey. In some cases, predators may simply take members of different age classes randomly, reducing the numbers at least temporarily, but not altering the age structure of the prey population. But predators often take disproportionately high numbers from very young and very old age classes, shifting the population's age structure toward a higher percentage of animals in the middle age groups. This pattern of predation is well-documented for carnivores such as the African lion and the gray wolf, which prey upon large ungulates. When predators select certain age classes over others they change the age structure which in turn alters the reproductive capacities of the prey population. Such change happens only because different age classes have different reproductive values.

Reproductive value is the age-specific expectation to have future offspring (Pianka, 1978). Older age classes clearly have lower reproductive values than do younger ones. Less obvious is the fact that the very youngest age classes usually have lower reproductive values than do slightly older age groups. This is because high mortality rates typically strike the youngest age classes; many of them never reach reproductive age. Those animals surviving this rigorous juvenile period enter age classes that have, on average, the best expectation of producing a large number of offspring.

If predators take a disproportionately high number of prey from age classes with lower reproductive values, leaving a higher percentage of age classes with higher reproductive values, they will do little to lower the overall fecundity of that population. They may even increase fecundity, survival rates (within middle age classes), or both, as they lower population density. The clearest examples of selection of different age classes come from field studies of predators that prey on large ungulates. Under such circumstances, the prey is larger than the predator (see Figure 6-3) as is the case of moose and wolves (Mech, 1966, 1970; Peterson, 1977), and Cape buffalo and lion (Schaller, 1972). Hunting methods also affect selection of different age classes. Schaller (1972) reported that predators such as the cheetah, the wild dog, and the gray wolf take larger proportions of very old or very young prey, as these age classes are comprised of individuals most

FIGURE 6-3
Age distribution of wolf-killed adult moose (modified from Peterson, 1977).

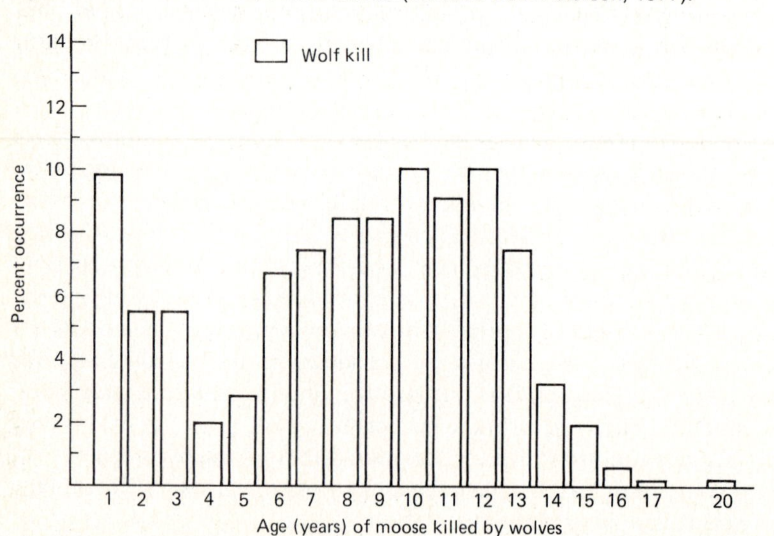

likely to fall behind during a long chase. Stalkers or ambushers, on the other hand, are less selective, their kills more closely resembling the age composition of the population as a whole.

Population Characteristics of the Prey What if prey exhibit unusual population traits such as the periodic irruptions characteristic of microtine rodents? One conclusion is obvious. If the irruption occurs in the presence of predators, then predation at the time of the irruption is not limiting the population. As the irrupting population increases, predation intensifies for two reasons. First, since most predators take more than one species of prey, they are likely to shift their hunting patterns to take advantage of the greater abundance of the irrupting population. This shift is called a functional (Holling, 1959) response to increasing prey numbers, as individual predators each kill more members of a prey population. Should the irruption be fairly localized, additional predators may be attracted to the area, increasing the predation pressure. If the irruption persists for a long enough time, the predator population may grow in response to the greater food supply. Either form of increase for the predator population is known as a numerical response to greater prey numbers (Holling, 1959). Thus, predation pressure generally builds up along with the prey population (Craighead and Craighead, 1956).

Territoriality in the prey population can also influence the effects of predation. Muskrats, for example, are strongly territorial (Errington, 1963, 1967). As shown in Chapter 5, territoriality does not limit a population size so much as it buffers that population from changes that would have taken place if territoriality had been absent. Nevertheless, muskrats live in marshes and marshes are finite. As a marsh fills with muskrat territories, a greater proportion of each year's youth must disperse and move elsewhere, attempting to find suitable habitat. These dispersing young are vulnerable to predation, starvation, exposure, and other forms of mortality. Predation acting under these circumstances would then be largely compensatory to other types of mortality.

Population Characteristics of the Predator If a predator population is affected by factors other than food supply, it may be unable to muster a numerical response to a rise in prey density and will therefore be less likely to affect prey populations. The social behavior of many species of predatory birds and mammals acts to buffer their population growth. Gray wolf packs, for example, maintain group territories, a practice that itself may limit wolf numbers within a given area. But within a pack, the alpha-male and the alpha-female typically deter other pack members from breeding (Rabb, et al., 1967). Male lions often kill young lions sired by mates in neighboring prides, a practice that curbs growth of the lion population (Schaller, 1972). Some biologists have concluded that such social behavior may place an upper limit on the predator's population density before it reaches the limits imposed by its food supply. For several years, the gray wolf population in Algonquin Provincial Park (Ontario), where they hunted deer, and Isle Royale National Park (Michigan), where they killed moose, stabilized at a density of about one wolf per 25 square kilometers (9.7 square miles) (Mech, 1970). The wolf population at Isle Royale, however, has since doubled to at least one per 12.5 square kilometers (4.83 square miles) (Peterson, 1977).

Just as territoriality in a prey population may affect predation patterns, so may territoriality in the predator population. Territoriality seems to be more common among predators than among herbivores. Hornocker (1969, 1970) concluded that the population of territorial cougars he studied in Idaho was incapable of controlling the local mule deer and elk. Other examples of territorial predators include the spotted hyena (Kruuk, 1972), the African lion (Schaller, 1972), and numerous species of predatory birds (Schoener, 1968).

The Availability of Buffer Species Since most predators rely on more than one species of prey, the effects of predation upon any particular prey species is influenced by the availability of other prey acting as "buffer" species. The intensity of predation is generally density-dependent, with predation pressure growing along with the prey population. The buffering effect occurs as predators exert more pressure on more numerous prey and less pressure on scarcer prey.

Snowshoe hare may act as a buffer against increased predation on the ruffed grouse. Rusch et al. (1972) studied great horned owl (*Bubo virginianus*) near Rochester, Alberta, on a 155-square kilometer (60-square miles) study area. During the investigation, the snowshoe hare population increased sevenfold and the ruffed grouse population doubled. The number of owls increased from 10 to 18 and the proportion nesting rose from 20 percent to 100 percent. Yet the percentage of ruffed grouse in the owl's diet actually declined from 23 percent to 0 percent, and the use of snowshoe hare grew from 23 percent to 50 percent. As snowshoe hare and ruffed grouse used essentially the same habitat types, the investigators strongly suspected that the hare population was acting as a buffer against predation on the grouse.

Predation and Species Diversity

The buffering effect of predation helps sustain species diversity among prey populations. Without greater predation pressure on the more numerous prey species, some of them might become so abundant as to outcompete local populations of less numerous prey. Some theoretical ecologists have suggested that over spans of evolutionary time, predation favors an increase in species diversity by permitting greater niche overlap between competing species (Spight, 1967; Roughgarden and Feldman, 1975; Schoener, 1982). Predation may prevent the suppression or even the exclusion of some species by others.

Case Studies Two case studies appear below. The first describes a prey species larger than the predator and the second reports a relationship in which the predator is larger than the prey. Both are based on field studies done in fairly simple ecosystems where confounding variables and buffering effects were minimal. These two case studies are followed by a reevaluation of what many biologists, including Aldo Leopold, have regarded as a classic prey–predator study, the Kaibab deer irruption.

Moose and Wolf at Isle Royale Isle Royale is a 544–square kilometer (210-square miles) island in Lake Superior some 24 kilometers from the nearest mainland. Moose arrived on the island, presumably by swimming from the Ontario mainland, early in the 1900s. By the 1930s the population had grown so large that Murie (1934) termed habitat conditions "most serious" and recommended some form of population reduction. He added, somewhat prophetically:

> Were it known if, and to what extent, our larger predators such as the bear, cougar, or timber (gray) wolf prey on moose, a possible solution to the overpopulation on the island would be to introduce an effective predator. . . . Since one of these predators might possibly do good work in keeping the moose population in check, and since there are few places where large carnivores are tolerated, it would seem desirable to introduce one or more of these on the island.

During brief periods of especially cold weather, ice forms over the surface of Lake Superior from Isle Royale to Ontario. Murie's suggestion was fulfilled in the late 1940s when wolves crossed the ice and arrived on the island. Within a few years, at least one resident wolf pack had become established.

Durward Allen of Purdue University learned about the arrival of the wolves at Isle Royale and took advantage of a unique opportunity to study a prey–predator relationship under pristine conditions. By the time Allen and his associates began field studies in the late 1950s, Isle Royale was a national park, protected from hunting and habitat disturbance. Moreover, the moose was the only species of large herbivore on the island and, with the occasional exception of the beaver and the snowshoe hare, was the only prey for the wolves. The wolf, in turn, was the only species of large predator at Isle Royale. Allen and his students and colleagues have for more than 20 years studied closely this essentially one-prey, one-predator relationship on a protected island.

Each winter at Isle Royale, researchers make aerial surveys to estimate both the moose and the wolf populations. The first survey, done in 1960, indicated that there were some 600 moose, a population density substantially lower than that estimated by Murie before the arrival of the wolves. Presumably the wolves had reduced the moose population. But would the wolves themselves continue to multiply, increasing their pressure on the moose until the prey was eliminated? They did not. Instead, the wolf population stabilized at an average size of about 22 animals for 1959–1973 (see Figure 6-4). At this level, the wolf population posed no threat to the survival and well-being of the moose population.

Yet the moose continued to increase. By 1966 the best estimate was 1,200, and it ranged as high as 1,550 over the next several years (see Figure 6-4). Obviously, wolf predation was not holding the moose population in check.

The wolf population showed no substantial increase until 1974 when the winter count tallied 31. The next year there were 41, and by the winter of 1975–1976, the population reached 44, double the level that it had maintained for 15 years. After the increase in wolf numbers, the moose declined, with the best estimate for 1979 being 750 (Jordan and Wolfe, 1981). Apparently, the two populations were once again seeking an equilibrium, this time at higher levels.

Peterson (1977) attributed the lag time between the moose population increase and that of the wolves to the social organization of these predators. A second pack broke off from the main pack in 1971 and established a group territory of its own. Both packs successfully reproduced in 1973, accounting for the substantial increase in the wolf population observed during the following winter. Probably the second pack would not have become established without the earlier increase in the moose population.

Both populations influenced one another on Isle Royale. Whenever the wolf population increased, it seemed to bring the moose numbers down. But the wolf population itself grew only in response to a larger prey population, and then only after a considerable time lag, imposed by the predator's own social organization. This results in a flexible, dynamic relationship with a built-in feedback to allow one population to adjust to changes in the other's density.

It is impossible to predict just how the moose and wolf numbers will change on Isle Royale in the future. But in the absence of human interference it seems unlikely that the moose population will ever again reach the estimated 2,000 to 3,000 (Mech, 1966) that it did before the

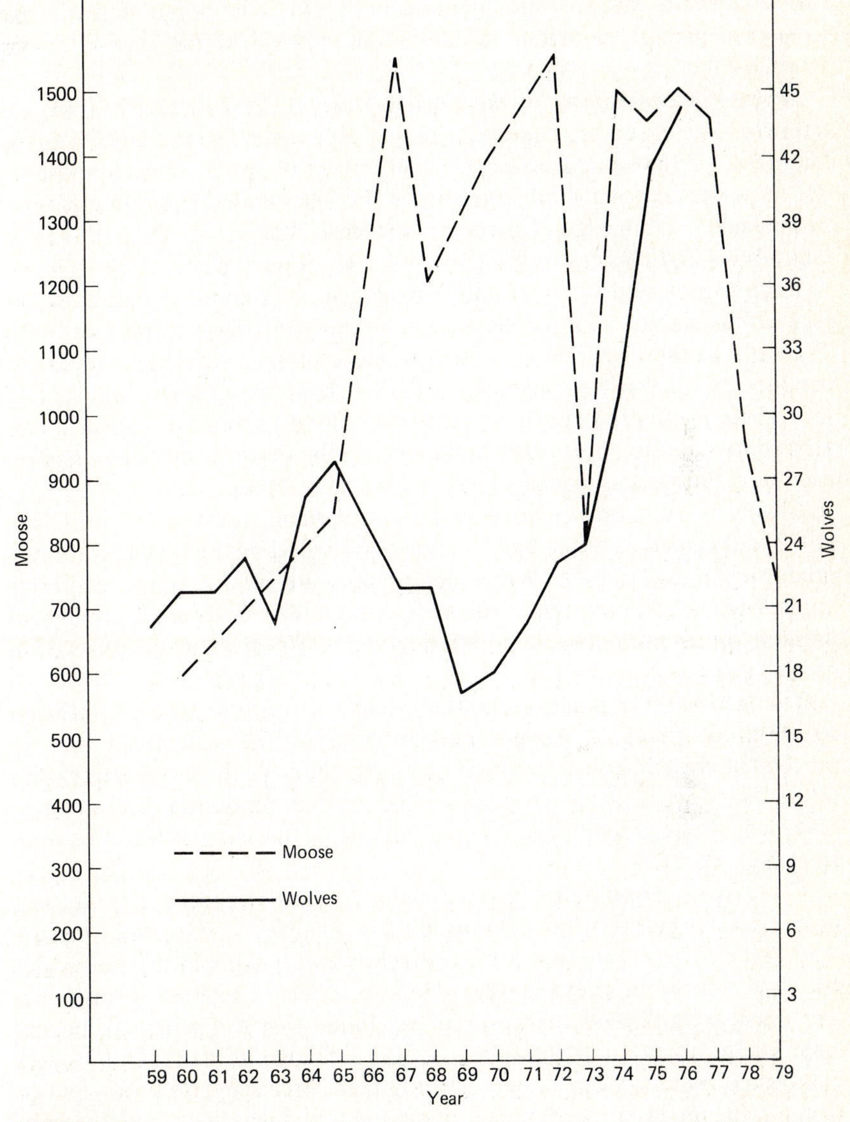

FIGURE 6-4
Population trends of moose and wolf at Isle Royale (data from Mech, 1966; Peterson, 1977; Jordan and Wolfe, 1981).

wolves arrived. And it is equally unlikely that the wolf population will rise to such a high level that it threatens the survival of the moose at Isle Royale.

Jackrabbit and Coyote in the Curlew Valley The coyote is usually an opportunistic predator, taking a wide variety of prey. In the 3,000–square kilometer (1,158-square miles) Curlew Valley along the Utah-Idaho border, however, coyotes feed largely upon the black-tailed jackrabbit (*Lepus californicus*). Utah State University scientists carried out field studies of jackrabbits and coyotes in the Curlew Valley from 1962 to 1970.

Each March and October, field investigators counted the numbers of jackrabbits sighted along a series of randomly distributed transect lines. They also enlisted the aid of a large number of eager wildlife students to conduct drive counts, a technique that yielded reasonable estimates of jackrabbit population densities. As four transect counts were done in and around drive count sites prior to the drives, the drive count data could be used to calibrate the indexes derived from the transect surveys.

Coyote population densities were more difficult to estimate, but Clark (1972) developed a "subjective" estimate bounded by the results from two other survey techniques. While not as accurate as the census methods applied to jackrabbits, this estimate was nonetheless accurate enough to monitor trends in the coyote population and to afford some approximation of its density.

The jackrabbit population declined sharply from 1963 to 1967, reduced by about 66 percent. Wagner and Stoddard (1972) calculated that 69 percent of this change was associated with shifts in the coyote to rabbit ratio. The investigators concluded that coyote predation had at least hastened, if not actually caused, the decline in the jackrabbit population (see Figure 6-5).

The coyote population also declined by an estimated 87 percent, presumably because of the diminished food supply. Clark (1972) discovered that coyote population trends correlated with shifts in the jackrabbit population from the previous year. He also collected a series of reproductive tracts from female coyotes and concluded that the principal mechanism for the coyote population trends was a change in the fecundity rates manifested both in terms of percent of females breeding and average litter size. Some human-induced mortality was evident, including seasonal use of poisoned bait stations (a practice that was later prohibited, see Chapter 9), but Clark concluded that the resulting mortality could not alone account for the population decline.

Like those of the moose and wolves at Isle Royale, the populations of jackrabbits and coyotes in the Curlew Valley were tied to one another. The fluctuations of both predator and prey populations were of much greater magnitude than were those of Isle Royale (see Figure 6-5). This is because of the tendency of jackrabbit populations to fluctuate more widely and to

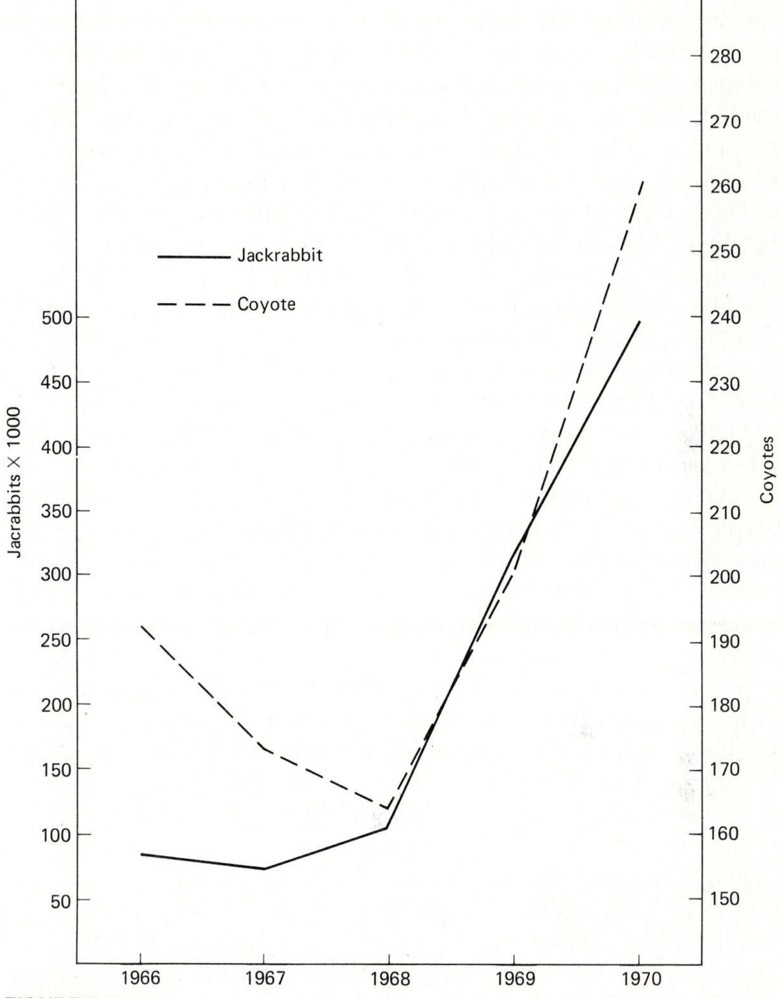

FIGURE 6-5
Population trends of black-tailed jackrabbit and coyote in the Curlew valley (data from Wagner and Stoddard, 1972, © The Wildlife Society. Used by permission).

be less stable than moose populations. Furthermore, coyotes took jackrabbits without selecting any particular age class over another. The March to October mortality for adult jackrabbits averaged 57 percent, while the mortality from birth to October for juveniles averaged 58 percent (Wagner and Stoddard, 1972). Finally, the coyotes mustered a faster numerical response to rising prey densities than did the wolves.

The Kaibab Plateau Revisited The two previous examples dealt with relatively undisturbed conditions where prey–predator relationships could function naturally. This next example describes conditions disturbed deliberately by humans. It is included here not because it illustrates principles of predation, but instead because it has widely been cited as "proof" that predators regulate deer populations. Recent evaluations have raised some serious and disturbing questions regarding the validity of traditional interpretations. The incident is described here to serve as an illustration of the dangers of oversimplification and overgeneralization when applied to so complex and varied a phenomenon as predation.

Predator control is the earliest form of wildlife management practiced in North America. Since their earliest settlement on that continent, people of European ancestry have killed predators relentlessly, completely eradicating species of larger carnivores from vast areas. Early in the twentieth century, predator control was widely practiced not only to protect livestock, but also in an attempt to increase populations of game animals. The most publicized examples of this practice took place along the north rim of the Grand Canyon in Arizona on the Kaibab Plateau.

From 1906 through 1923, government hunters and trappers took from the Kaibab 781 cougars, 30 gray wolves (the last remnant population), 4,889 coyotes, and 354 bobcats (Rasmussen, 1941). The population of mule deer had been estimated at 4,000 before initiation of predator control. Less than 20 years after reduction began, the deer population underwent a violent irruption, peaking at an estimated 100,000 before crashing during the winter of 1924–1925. The incident became known as a classic example of the shortsightedness of predator control to protect game population. It has also been cited as evidence of the importance of predation in the prevention of prey irruptions.

Caughley (1970) challenged the classical interpretation of the Kaibab incident, developing his conclusions in part from his own studies of the Himalayan thar in predator-free New Zealand. The thar followed the same sort of irruptive population growth that occurred at the Kaibab Plateau, and Caughley concluded that the phenomenon was caused by a dynamic interaction between the herbivores and the vegetation upon which they fed. He criticized earlier interpretations of events at the Kaibab, contending that the census figures were unreliable. The irruption, maintained Caughley, was due more to changes in the vegetation following removal of at least 200,000 head of sheep and 20,000 head of cattle than to removal of predators.

That the deer irruption took place is not at issue; nor is the contention that large numbers of deer overbrowsed the vegetation before starving *en masse.* The question is this: how much of the irruption resulted from predator removal and how much was due to changes in vegetation? In the case of the Kaibab, the answer may never be forthcoming. Certainly Caughley has pointed to what, at the very least, was a confounding variable. While it seems likely that the sudden removal of so many predators from the Kaibab contributed to the population buildup of deer, wildlife managers should use caution in citing the Kaibab as a classic study that "proves" predators regulate prey populations.

Conclusions on the Effects of Predation

The effects of predation are quite variable. Some of the more important components that contribute to these effects include age-specific selection of prey, population characteristics of both predator and prey, and the availability of buffer species. Age-specific selection of prey is more likely to occur whenever the prey is physically larger than the predator, a condition which makes prime-aged animals difficult to take. Predation on territorial prey in a limited environment may more likely be compensatory than predation under other circumstances. Many predator species are themselves territorial and some have dominance hierarchies that inhibit reproduction. These forms of social behavior tend to restrict growth of predator populations, reducing the predator's ability to regulate prey densities.

Of practical necessity most field studies of predators and prey have been done in simple ecosystems with few alternative prey to act as buffers. Predator and prey populations in such simple settings are closely interrelated, but the preponderance of evidence suggests that prey may regulate predators more often than the other way around. Most predators and their prey exist in more complex ecosystems where the effects of competition and buffering most likely dampen population fluctuations of both trophic levels.

Like competition, predation favors species diversity and helps stabilize populations. Predators do not normally overexploit populations of prey with which they have coevolved.

SUMMARY

This chapter reviews two major types of population interactions, competition and predation, as they typically occur under natural conditions. Competition was once viewed as a ruthless process, and it was widely assumed that natural selection could ultimately eliminate species from communities through survival of only the fiercest competitors.

But field studies of competition between similar species within the same communities have revealed a different pattern and allowed a different interpretation of competition. Rather than pursuing the more destructive routes of ever increasing competition, natural selection has instead favored reduction in competition, leading to niche partitioning and, ultimately, to coexistence at lower levels of competition.

Four examples of interspecies competition are presented. Three species of ptarmigans share common ranges in the interior of Alaska by subsisting on slightly different vegetation. Pocket gophers are territorial both within and between species. Four species in Colorado were found to coexist because of differences in both soil tolerance and aggressive abilities. Up to 14 species of large herbivores in East Africa manage to share the same range by using the available vegetation in an array of different ways. Some of them probably benefit one another more than they compete. Five species of large carnivores also inhabit portions of East Africa. These predators manage to coexist through using different activity periods, habitats, and hunting techniques.

The effects of predation are more varied than those of competition. Any given case can be evaluated only when various components of predation, including selection of different age classes of prey, population characteristics of both predator and prey, and the availability of buffer species, are known.

Three examples of prey–predator interactions are presented and evaluated. The first, moose and wolf at Isle Royale, describes the relationship between a prey that is larger than the predator. This size difference limits wolves primarily to very young or very old moose, those vulnerable to predation. Such age-specific selection leaves a larger proportion of moose in the prime-aged classes, the same classes that typically have the highest rate of reproduction. This type of predation is not likely to cause much appreciable decline in prey production. Moreover, the relationship between the number of wolves and the number of moose at Isle Royale is not a stable one; it is varying widely. The two populations clearly influence one another, though there seems little chance that the wolves will ever exterminate the moose.

Coyotes and black-tailed jackrabbits make up a second case study, in which the predator is physically larger than the prey. As a result, jackrabbits are taken by coyotes without regard to age, since an individual coyote is quite capable of killing any jackrabbit. Again, the prey and predator populations are linked together, fluctuating in this case over an even greater range of population sizes than do the wolves and moose at Isle Royale.

A final case study appears in an attempt to clarify and correct a persistent overgeneralization. A large irruption of mule deer on the north rim of the Grand Canyon followed two decades of removal of several species of large predators. The deer population devastated much of the available vegetation and then crashed. Conventional interpretations credit the removal of the predators as the sole cause of the irruption. However, it also appears that during the same period, the habitat was recovering from heavy pressure from both cattle and domestic sheep. No one can therefore determine the extent to which the predator removal caused the irruption and to what extent it was due to range recovery.

Both predation and competition are natural phenomena that, over long periods, enhance the species diversity within a community. Practical problems with either type of interaction typically follow human activity and are described in following chapters.

REFERENCES

Brown, W. L., & E. O. Wilson. 1956. Character displacement. *Syst. Zool.* 5: 49–64.

Caughley, G. 1970. Eruption of ungulate populations, with emphasis on Himalayan Thar in New Zealand. *Ecology* 51: 53–72.

Clark, F. W. 1972. Influence of jackrabbit density on coyote population change. *J. Wildl. Manage.* 36: 343–356.

Cole, L. C. 1960. Competitive exclusion. *Science* 132: 348–349.

Craighead, J. J., & F. C. Craighead, Jr. 1956. *Hawks, owls, and wildlife.* Harrisburg, Pa.: The Stackpole Co., Washington, D.C.: The Wildlife Management Institute.

Errington, P. L. 1963. *Muskrat populations.* Ames, Iowa: Iowa State University Press.

Errington, P. L. 1967. *Of predation and life.* Ames, Iowa: Iowa State University Press.

Gwynne, M. O., & R. H. V. Bell. 1968. Selection of vegetative components by grazing ungulates in the Serengeti National Park. *Nature* 220: 390–393.

Hardin, G. 1960. The competitive exclusion principle. *Science* 131: 1292–1297.

Holling, C. S. 1959. The components of predation as revealed by a study of small mammal predation of the European pine sawfly. *Canad. Entomol.* 91: 293–320.

Hornocker, M. G. 1969. Winter territoriality in mountain lions. *J. Wildl. Manage.* 33: 457–464.

Hornocker, M. G. 1970. An analysis of mountain lion predation upon mule deer and elk in the Idaho Primitive Area. *Wildl. Monogr. No. 21.*

Jordan, P. A., & M. L. Wolfe. 1981. Aerial and pellet-count inventory of moose at Isle Royale. In J. Gougue (Ed.), *Proc. Second Conf. on Scientific Research in National Parks.* American Institute of Biological Sciences and the National Park Service, Washington, D.C.

Kruuk, H. 1972. *The spotted hyena: a study of predation and social behavior.* Chicago: University of Chicago Press.

Kruuk, H., & M. Turner. 1967. Comparative notes on predation by lion, leopard, cheetah, and wild dog on the Serengeti Area, East Africa. *Mammalia* 31: 1–27.

Lamprey, H. F. 1963. Ecological separation of the large mammal species in the Tarangire Game Reserve, Tanganyika. *E. African Wildl. J.* 1: 63–92.

Mech, L. D. 1966. The wolves of Isle Royale. *U.S. Natl. Park. Serv. Fauna Ser. 7, Washington, D.C.*

Mech, L. D. 1970. *The wolf.* New York: Natural History Press.

Miller, R. S. 1964. Ecology and distribution of pocket gophers (*Geomyiidae*) in Colorado. *Ecology* 45: 256–272.

Moss, R. 1974. Winter diets, gut lengths, and interspecific competition in Alaska ptarmigan. *Auk* 91: 737–746.

Murie, A. 1934. The moose of Isle Royale. *Univ. Mich. Museum Zool. Misc. Publ. 25,* Ann Arbor.

Peters, R. H. 1976. Tautology in evolution and ecology. *Am. Nat.* 110: 1–12.

Peterson, R. O. 1977. Wolf ecology and prey relationships on Isle Royale. *Natl. Park Serv. Monogr. Ser. 11,* Washington, D.C.

Pianka, E. R. 1978. *Evolutionary ecology.* New York: Harper & Row.

Rabb, G. B., J. H. Woolpy, & B. E. Ginsburg. 1967. Social relationships in a group of captive wolves. *Am. Zool.* 7: 305–311.

Rasmussen, D. I. 1941. Biotic communities of Kaibab Plateau. *Ecol. Monogr.* 11: 230–275.

Roughgarden, J., & M. Feldman. 1975. Species packing and predation pressure. *Ecology* 56: 489–492.

Rusch, D. H., E. C. Meslow, P. D. Doerr, & L. B. Keith. 1972. Response of great horned owl populations to changing prey densities. *J. Wildl. Manage.* 36: 282–296.

Schaller, G. B. 1972. The Serengeti lion: a study of prey-predator relations. Chicago: University of Chicago Press.

Schoener, T. W. 1968. Sizes of feeding territories among birds. *Ecology* 49: 123–141.

Schoener, T. W. 1982. The controversy over interspecific competition. *Am. Scient.* 70: 586–595.

Spight, T. M. 1967. Species diversity: a comment of the role of the predator. *Am. Nat.* 101: 467–474.

Wagner, F. H., & L. C. Stoddard. 1972. Influence of coyote predation on black-tailed jackrabbit populations in Utah. *J. Wildl. Manage.* 36: 329–343.

MANAGEMENT FOR HARVEST

The killing and subsequent removal of a portion of a game or furbearer population is the "harvest." In part the term is a euphemism, though it is justifiable to apply it to the removal of a portion of any renewable natural resource, not just those with soft brown eyes. That portion removed is the yield.

One of the two major goals in wildlife management is the production of acceptably high yields of game and furbearers while ensuring that enough animals are left to replenish those taken. As shown in Chapter 1, this is the older of the two main approaches, having been practiced in Europe for centuries and in the United States and Canada at least since the 1930s. This chapter introduces the biological basis for harvest, first in a more general sense, then by addressing each of the major groups of terrestrial animals commonly harvested.

This chapter emphasizes the effects of harvest upon game populations. The central question is whether or not hunting mortality is compensatory to natural mortality. In some cases reference is also made to habitat management whenever habitat management has been a primary tool.

PRINCIPLES OF HARVESTING

The term "harvestable surplus" is a popular one among hunters and wildlife managers alike. To most people, the harvestable surplus is that proportion of the game population that would invariably die of other causes if not taken by hunters. Implicit in this definition is that mortality inflicted by hunters is compensatory, not additive to natural mortality. Sometimes, hunting mortality does appear to be compensatory, especially among resident small game. Hunting mortality is much less likely to be compensatory in big game or carnivorous furbearers. The extent to which hunting adds to the mortality patterns of waterfowl is a matter of serious and vigorous debate.

Hunting mortality which is additive is not necessarily bad. Properly regulated, this additive mortality can reduce a game population to a more productive level, thus assuring a higher yield while ensuring the long-term survival of the population.

Whether or not hunting mortality is additive depends in each individual case upon the intensity of the harvest in relation to the ability of the population to compensate. Up to a certain point or threshold, hunting mortality can be expected to be compensatory. Once that threshold is exceeded, however, any more hunting mortality becomes additive. The trick, then, is determining where that threshold lies before setting harvest goals. It is theoretically possible to exceed that threshold for any species including such ubiquitous and *r*-selected species as the starling and the Norway rat (*Rattus norvegicus*). Practically speaking, though, such species are virtually immune from excessive harvests, as the costs for such heavy exploitation would be prohibitive. At the other end of the scale are some of the less fecund large mammals such as most bears. The threshold beyond which hunting adds to the mortality of such large mammals is much lower and, in practice, far easier to exceed.

A relatively new tool in the game manager's kit is the population model. Acceptance of population models grows, but only among those who understand the practical limits to modeling. Some managers reject all models as useless and impractical. Others assume that models will solve all their problems, even to the point of defining management objectives. Both groups are in error.

Before the usefulness of models can be understood, it is essential to know what a model really is. One definition is "a simplified, stylized representation of the real world that abstracts the cause-and-effect relationships essential to the question studied" (Quade, 1966, as cited in Walters and Gross, 1972). Models are both simpler than the "real world" and infinitely easier to manipulate. These are the characteristics that make them so useful.

There is no one all-inclusive model for harvests or for anything else. It is impossible to maximize generality, precision, and realism at the same time; most applied ecologists sacrifice precision and realism for generality (Levins, 1968). The logistic growth equation introduced in Chapter 3 is an example of a model that gets high marks for generality because it can be applied to a wide range of populations. It is less realistic and rates poorly for precision (Green, 1979), weakened in both cases by its simplicity. On the other hand, a more detailed model based upon real data from a population in one place and time may sacrifice generality for precision and realism. Such a model would prove very useful when applied to that particular population. The unique ecological conditions under which that population was living, though, could make it inappropriate for use with other species or with the same species in other regions.

Models are classified in a number of ways (cf. Tipton, 1980). One major division is deterministic versus stochastic. A deterministic model is one in which the population parameters are fixed except for those being tested. In a stochastic model, some or all of the parameters are allowed to vary at random. Stochastic models can be particularly useful in the study of harvesting in fluctuating environments.

Another dichotomy between models is the theoretical versus the empirical. Theoretical models are those derived from one or more theoretical premises, whereas empirical models are developed from real data collected from the field. Walters and Bandy (1972) developed a largely theoretical model of big game populations, concluding that periodic harvests (harvesting at intervals greater than one year) would result in higher yields on average than would annual harvests. Their theoretical premise was that following periodic harvests, fecundity rates would rise substantially in the older age classes. Periodic harvest would then allow more individuals to enter those presumably more fecund age classes, thereby boosting overall yield. McCullough's (1979) empirical model, derived from the George Reserve deer herd in Michigan, permitted him to test Walters and Bandy's conclusions against real data. McCullough demonstrated that periodic harvests would actually reduce yields below those obtained through annual harvests at least for that particular population.

The example above shows how theoretical models and empirical ones can be used in conjunction with one another. Theoretical models establish broader, more general relationships; empirical ones provide the means of testing and refining the more general models. Thus, the combined approach can lead to more detailed, accurate, and precise models.

Harvest models have important uses in wildlife management. They can be used to play "what if" games to test the effects of various harvest schemes or environmental factors on population size and recruitment rate (Walters and Bandy, 1972). A manager can then test the effects of a doubling of the harvest rate by using a model, thereby avoiding potentially disastrous effects of actually doubling harvest in the field.

Models can also show the relative importance of various population parameters upon population growth. For example, Walters and Gross

(1972) used data from the Llano Basin deer herd (Teer et al., 1965) to test their population model. They asked which was the more important for predicting the behavior of the population, accurate data on fecundity or accurate data on mortality? In the case of the Llano Basin population, the model indicated that fecundity data were the more critical.

But perhaps the most important function of harvest models is that they force managers to explain objectives clearly, to quantify relationships within the population, and thereafter to measure performance or progress in meeting management goals (Tipton, 1980; Walters and Gross, 1972). The use of models in wildlife management can therefore ensure more management by objective as opposed to management by tradition. Management by objective can result in greater efficiency and superior long-term planning.

Maximum Sustained Yield

The logistic equation, introduced in Chapter 3, shows that the maximum rate of population growth occurs at or near the inflection point. The inflection point, moreover, is located at half of carrying capacity, or $K/2$. Theoretically, if a harvested population can be reduced to $K/2$ each year, its rate of recovery can be maximized, leading to a productive and efficient management yield that it can continue annually, the maximum sustained yield (MSY).

MSY has been applied to fisheries management, especially for commercial harvests of marine fisheries (Larkin, 1977; Sissenwine, 1978) since the end of World War II. Such applications have not always been completely successful, and some serious overharvests have resulted (Holt and Talbot, 1978; Talbot, 1975). Consequently, the concept of MSY as well as its application, have become controversial in marine fisheries management.

The concept of MSY has found its way into wildlife management only in more recent years. Part of the reason for this delay resulted from the fact that terrestrial wildlife is only rarely harvested commercially. Since wild birds and mammals are typically shot for recreational purposes, the economic pressures for heavy harvests simply have not been present. In addition, to protect their sport, hunters have for the most part been very conservative in their harvest goals, preferring excessive protection over excessive exploitation. But as pressures increase for more recreational hunting in the United States, wildlife managers face the prospect of dealing with heavier harvests. Some will advocate management for MSY.

The application of MSY to big game management was explained in detail by Gross (1969). Citing evidence from several closely studied cervid populations, Gross demonstrated that:

1 Fecundity rates change with density
2 Changes in fecundity rates produce dome-shaped yield curves
3 Yield curves can be used to estimate MSY.

One of the more important points raised in Gross's paper was that maximum productivity occurred not at *K* as had often been assumed, but at about *K*/2. Maximum population size and maximum productivity were found to be biologically incompatible management goals.

Figure 7-1, derived from the logistic equation, illustrates how yield should be greatest at *K*/2. In principle, then, MSY is easy to apply to any game population. If *K* is known, simply estimate the population size and harvest the population until *K*/2 is reached. Repeat the procedure each year and thereby maximize the harvest as well as the subsequent productivity. If *K* equals 1,000 and the prehunt population is 925, harvest 425 so that the population size will be reduced to 500, or *K*/2.

Unfortunately, MSY is more difficult to apply in practice. *K* is only roughly known, except for a few intensively studied populations. Moreover, *K* is likely to change, as it would, for example, as succession advances or is set back. As shown in Chapter 4, methods of population estimation

FIGURE 7-1 A yield curve as derived from the logistic growth equation. Note that maximum productivity occurs when the population is at one-half of carrying capacity.

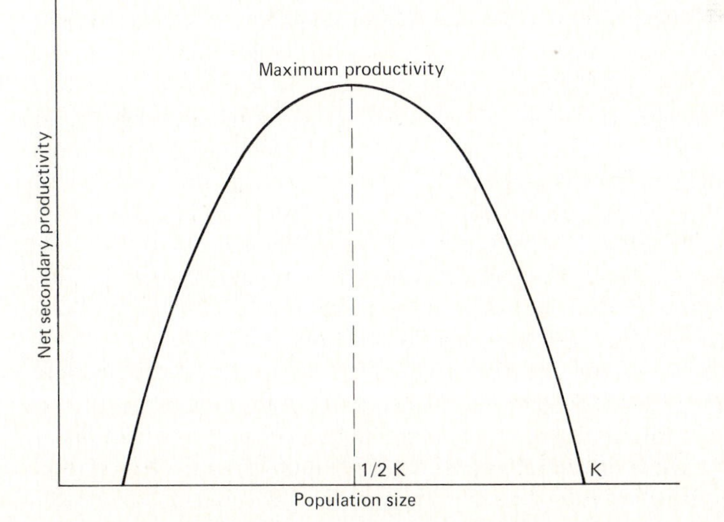

are rarely as precise or as accurate as managers would like, and at times they may be seriously in error. Should errors be made in estimation of either K or N for several consecutive years, populations can be seriously overharvested or even driven to local extinctions.

Another problem with the application of MSY is that the number removed from the population is usually easier to determine than either K or N. Managers quite naturally tend to rely more on the yield than on their estimates of N or K, a practice that leads to harvest by quota. If estimates of K and N are accurate and if environmental conditions remain fairly constant, this approach to achieving MSY is safe. But errors in estimation or fluctuations in the environment can render this technique useless or even dangerous to the resource base.

Finally, to achieve MSY even under the most confident conditions, managers have to be able to regulate the kill very closely. This requires, in most cases, control of hunting pressures and a tally of hunting success rates. Managers must also have authority to stop harvest as soon as MSY is achieved, when the population is reduced to $K/2$. Such conditions could be found only on intensively managed areas with limited access.

Quite aside from these practical problems in the application of MSY, the concept of MSY has received substantial criticism. To begin with, MSY is usually derived from the logistic equation, which is, as previously shown, a general mathematical approximation of growth rates in real populations. The logistic equation is weak in realism. McCullough (1979) compared the observed growth of the George Reserve deer herd with that predicted from the logistic equation (see Figure 7-2). He noted that despite the superficial similarity, the curves were substantially different. Although the logistic equation predicted that the population would take 14 years to reach K, in fact it took only 8 years. McCullough found that the actual population leveled off far more abruptly as K was reached than was predicted by the logistic. This abrupt leveling off implies a tendency to exceed K, a trait for which cervid populations are well known. Most importantly, the principal difference between the logistic model and McCullough's empirical one was that the logistic assumed that r decreased linearly with density. For the George Reserve population it did not, shifting instead to a curvilinear relationship at higher densities. This tendency for r to decrease in a curvilinear fashion at higher densities has been reported for other cervid populations (Gross, 1969).

The basic MSY model allows only for the direct effects of numerical reduction through harvest. Indirect effects, such as alterations in age and sex ratios or disruptions in social organization and behavior, are not taken into account (Larkin, 1977; Talbot, 1975). Even though these effects could be substantial, they have not been estimated for most harvested species and so have remained ignored.

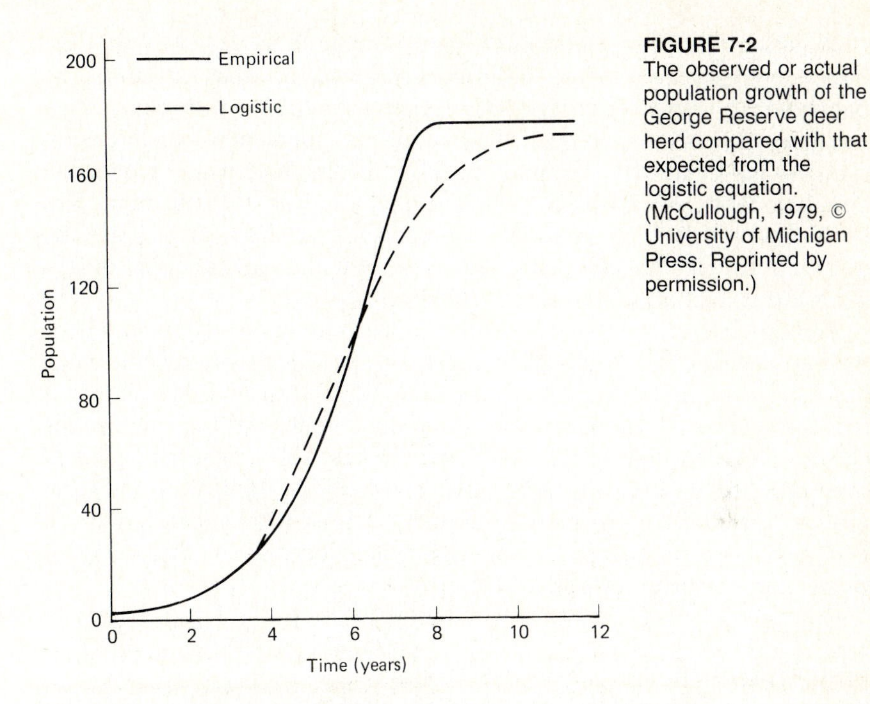

FIGURE 7-2
The observed or actual population growth of the George Reserve deer herd compared with that expected from the logistic equation. (McCullough, 1979, © University of Michigan Press. Reprinted by permission.)

MSY is typically applied to one species at a time, essentially ignoring the effects of other components of the ecosystem for which the harvest is taken. Competitive relationships could be disrupted if only one of the competing species is harvested. Harvest could affect patterns of natural predation that could in turn affect the prey population being harvested.

As it is usually applied, MSY assumes consistent environmental conditions. Such environmental conditions as weather patterns or changes in successional stages typically occur in the real world. What happens to a harvested population when an MSY is taken in a fluctuating environment? Beddington and May (1977) simulated fluctuating environmental conditions in a harvest model and found that populations harvested at MSY take longer to recover. The variability in the yields increased as harvest was intensified. When their simulated population was pushed beyond the MSY level, high harvest produced low yields with high levels of variance between years. This pattern was identical to those observed in several fish and whale populations now thought to be overexploited. These effects proved to be more serious when harvesting was based upon quotas rather than upon given harvesting efforts. Beddington and May concluded that, given the environmental variables of the real world, approaches such as MSY were usually undesirable.

A similar relationship was found in the George Reserve herd. Using the basic deterministic model (with fixed parameters, thus assuming no fluctuations), McCullough (1979) calculated an MSY of 49 deer. Then using a stochastic approach that varied environmental conditions at random, he found that any annual yields in excess of 46 would reduce the population and, if continued, would eventually lead to extinction. This shows not only that fluctuating environments reduce yield, but also that the margin between a harvest that stabilizes a population and one that pushes it toward extinction can be a very thin one.

Perhaps MSY can be effectively applied to harvested populations without endangering them. The concept may be made more reliable when more sensitive, precise, and realistic models replace the logistic equation as its basis. Improved techniques for estimating both K and N will reduce the chances or errors that in turn allow overharvests. As game management intensifies, managers will eventually have both the biological and legal tools to regulate harvest more efficiently, at least within specified areas. But in the meantime, game managers would be wise to explore more conservative approaches to harvest.

Alternatives to Maximum Sustained Yield

There are several alternatives to MSY, all of which result in more conservative harvest levels, thus sacrificing higher yields for greater margins of safety. All are more qualitative than MSY and none require use of any general mathematical formula.

Optimum Sustained Yield

With the right set of data, the wildlife manager can define MSY in strictly biological terms. Optimum sustained yield (OSY) is more complex because it combines both biological and sociological factors (Savidge and Ziesenis, 1980). OSY allows greater flexibility in management decisions, being, as Larkin (1977) pointed out, "anything you want to call it." As it is most commonly applied, OSY is a modification of MSY with built-in safety factors, including a yield deliberately set lower than MSY, as Gross (1969) used the term.

OSY in general, though, has several important advantages over MSY (Larkin, 1977). Proponents of this newer concept admit that it is difficult in practice to achieve MSY. With OSY, there is no implied obligation to harvest a species just because it is there. Finally, unlike some of the more conventional harvest schemes, OSY furnishes a basis for objective documentation of management decisions.

The flexibility of OSY can prove to be a mixed blessing. Most authorities (Larkin, 1977; Sissenwine, 1978) recognize OSY as in improvement over MSY, but because it is less precise than MSY, its application will have more variable results. Larkin (1977) admits that OSY can be "a recipe for achieving heaven or hell."

Conservative Harvests Although it comes as a surprise to many newcomers to wildlife management, hunters in the United States are often more protectionist and more conservative on matters of harvest than are wildlife management agencies. The hunters' sentiment stems from the historical successes in game management that they and their ancestors witnessed during and shortly after the era of protection (see Chapter 1). If strong protective laws led to more game back then, why should similar measures not work now?

State wildlife agencies see game harvests a bit differently. Professional wildlife managers employed by those agencies understand that most game populations are nowadays sufficiently restored to withstand even greater levels of harvest than they typically sustain. Under such conditions, it is futile to try to stockpile game by curbing harvests, when many of today's game populations already exist at or near carrying capacity. Managers also realize that the actual effects of harvest are often compensatory, thus adding less to the total annual mortality of the population than might popularly be supposed.

The result of these different viewpoints is that state wildlife agencies find themselves walking very fine lines. Ideally, they would like to see every licensed hunter satisfied with the agency's results. When game is sufficiently abundant to allow widespread success, it is, as a resource, doing well. By receiving large portions of their funding from hunting license sales, state wildlife agencies have little practical incentives to reduce seasons or bag limits if such action is likely to reduce license sales. But at the same time, that very funding comes from hunters who tend to view any decline in game abundance or harvests as proof that the wildlife agency has granted too little protection. Moreover, disgruntled hunters are likely to complain to their elected representatives or to appointed members of wildlife commissions, either of whom can bring considerable pressure to bear upon a state wildlife agency.

Knowing that many hunters favor strong protection, game managers like to avoid the embarrassing consequences of overharvest. They have for many years worked out empirical schemes that usually result in conservative harvests. One method, usually requiring several years of experience with game populations on a particular area, is to harvest the same conservative quota each year. This quota is set deliberately low and, as it is empirically derived, is a rather safe strategy. Its principal drawback is that it lacks efficiency, tending to underexploit during good years and run some risk of overexploiting in poor years (Savidge and Ziesenis, 1980).

Another method is to harvest by taking the same proportion of the population each year, varying seasons and bag limits in accordance with changes in the population trends. This approach is better than the quota method in that it is more efficient and less likely to overexploit during poor years. It does, however, require reliable estimations of abundance annually. Only with good data on population changes can the manager adjust harvest regulations between years.

In practice, effective regulation of game harvests over areas the size of states or provinces is a tricky and imprecise business. Most state wildlife agencies in the United States have tried setting harvest regulations that simply avoid overharvest. While this strategy may not be as efficient as one computed through OSY, it survives largely through practical necessity. Game managers simply lack the authority to regulate the distribution of hunting pressure throughout large areas. As a result, they adjust hunting seasons and bag limits annually based upon surveys of population abundance and changes in land use practices. The usual assumption about hunting pressures is that they change very little between years.

Despite these problems involved with managing game over large and varied regions, wildlife agencies can still collect valuable information on both game population trends and hunter harvests. Most of the methods are not suitable for rigorous statistical analysis, but they can furnish useful data at nominal cost (Downing, 1980).

Survey of Habitat Conditions The condition of indicator species, introduced in Chapter 2, offers a convenient means of gauging herbivore populations in relation to carrying capacity. The appearance of browse lines should alert managers of too many herbivores, and recovery of heavily used vegetation should signal declining populations. Check stations offer the opportunity to examine the condition of individual animals. Weights, proportions of "trophy-class" animals, and levels of parasitism afford reliable clues to local habitat conditions.

Surveys of Game Populations The frequencies of sign or calls changes along with game populations. Road kills and crop damage complaints involving game animals increase and decrease in accordance with populations. Nest counts, brood counts, and other indexes of reproductive success and recruitment rates can supply useful information on the status of a game population.

Surveys of Hunters Surveys of hunters can help game managers monitor hunting pressure within different parts of their regions. Increases in local hunting pressures often suggest increases in game populations, but they can also imply that the larger number of hunters are having greater impact on game populations. Monitoring of hunter success rates can settle that question in most cases. Some agencies monitor hunter success rates through the use of questionnaires, either sent out after the season or else supplied with the licenses.

Wildlife agencies must also be responsive to public opinions as they relate to the agencies' responsibility (Downing, 1980). Hunters are seldom reticent regarding their opinions on game abundance or the wisdom of proposed changes in hunting regulations. If wildlife agency personnel hear more frequent or louder complaining than usual, they should do something about it. Justifiable complaints should be met by appropriate changes within the agencies. If, as is often the case, complaints are unjustified, the agencies' I and E divisions should attempt public education programs to correct widespread misconceptions.

Adaptive Management The use of models may lead to bolder initiatives in game harvests than the traditional conservative approaches described above. One such method is called "adaptive management," a technique of potential value to applied ecology in general, including game management. Adaptive management attempts to use any form of ecological disturbance as a deliberate field experiment. Its advocates maintain that, if properly done, such experiments could greatly enhance our knowledge and understanding of the ways in which human activity influences ecological systems (Walters and Hilborn, 1978).

Adaptive management has been practiced to a subtle degree in game management for some time. Indeed, most data from which game managers draw their conclusions come from exploited populations, not from protected ones. Nevertheless, game populations could be harvested at varying levels, including quite high ones, in patterns more closely resembling controlled experiments. Ricker (1975) has advocated increasing harvest levels by stages in an effort to define MSY or OSY more clearly. Under such conditions, MSY would be found in the population after harvest, just before the decline in subsequent yield.

Obviously such experimental methods would have to be applied carefully so as to avoid irreversible consequences. They could only be applied under those circumstances in which, like the George Reserve, managers could closely regulate the harvests.

New Principles Recognizing that most exploitation of wild living resources has not been managed in the strict scientific sense of the word, some three dozen scientists and conservationists from the United States and Canada conducted a planning session and two workshops to search for better alternatives. They concluded that MSY and other single-species types of management are inadequate to meet the needs of natural resource management in the future. The primary goal of management, they contended, should be the maintenance of resource systems (as opposed to single species, as though they lived in isolation).

The authors of the final report were especially critical of MSY. While they admitted that some failures of MSY were due to the ways in which it was applied, they added that the concept itself contains a number of serious deficiencies. Among these deficiencies are that MSY:

1 Focuses attention on single species
2 Emphasizes quantity only
3 Depends upon resilience and stability that might not exist
4 Focuses attention on output (yield) without regard for amounts of effort required to meet those yields
5 Allows, or even encourages, overexploitation (Holt and Talbot, 1978).

Holt and Talbot advocated new principles to which they believed exploiters of natural resources should adhere. Their four general principles included:

1 Ecosystems should be maintained so the consumptive and non-consumptive values are maximized, future options are ensured, and risk of adverse or irreversible damage is minimized.
2 Management decisions should include a safety factor to allow for overharvest.
3 Measures to conserve a wild living resource should be formulated so as to avoid wasteful use of other resources.
4 Adequate surveys should precede planned use of resources and should be continued during the period of exploitation in the form of monitoring.

These new principles take a broader view of harvest, a view more compatible with current levels of understanding of ecosystems. Like the methods described under the "conservative harvests" section in this chapter, this, too, is an approach designed first and foremost to avoid overharvest. Other objectives such as efficiency of harvest and maximization of yield receive much lower priority.

LARGE UNGULATES

The management of large hooved mammals or ungulates has been quite successful in North America, as shown by population trends. White-tailed deer, the most widespread of wild ungulates in the United States, increased from fewer than half a million around the turn of the century, to about 12 million by the 1970s (Trefethen, 1975). Of these, about 2 million are taken by hunters annually (Halls, 1978).

Large ungulates are, compared with other game animals, conspicuous, *K*-selected, and for the most part, resident. They are polygynous, such that shifts in sex ratio can have pronounced effects upon population growth. Finally, ungulates have a trait considered troublesome by wildlife managers: they often exceed the carrying capacity of their habitats and damage their own food base.

Given these characteristics, management attention has focused more upon population manipulation than upon habitat manipulation. This is not to imply that habitat management is not important for ungulates. Like all wild species, they have certain habitat requirements and their populations track habitat changes. But most of the controversy in the management of deer and other ungulates centers on the manipulation of populations. The white-tailed deer makes a good case study; that species is emphasized below.

When white-tail numbers reached their all-time low, the principal cause of their decline was widely thought to be unrestricted hunting. Hunters and other conservationists demanded greater protection and state wildlife agencies did their best to provide it. In some cases, seasons were closed either in alternate years or for several years to allow populations to recover. When seasons were opened, only buck or male deer were legal game, a practice that became known as the "buck law." The rationale behind the buck law was easy enough to understand. Since the species was polygynous, the rate of population growth depended only on the number of breeding females, assuming, of course, that there remained enough males to fertilize all of them. Most bucks were "surplus," unimportant to population growth and hence expendable. Besides, they came equipped with antlers, making them nice trophies.

The buck law soon became a tradition in many states, and legislatures commonly passed laws granting protection to does. The crusade to protect female deer probably did more to bring legislatures into wildlife management than did any other single issue. Lucky is the state wildlife agency that escaped the effects of legislative intervention in the management of deer and other ungulates.

Viewed through the magnificence of hindsight, the buck law was a bad idea. Protecting does probably did help many deer populations grow rapidly and become established; the problem was that they kept on growing, aided by the fact that predators had been eradicated from most regions prior to the establishment of the buck law. Range conditions deteriorated, at least temporarily, starvation increased, birth rates declined, and damage to agricultural and forest crops increased. Meanwhile, bucks, the principal trophy and only legal game, became relatively scarce.

Table 7-1 shows why bucks became scarce after several years of "bucks only" hunting. In this case 20 percent of the population, all males, is taken

TABLE 7-1
HOW "BUCKS ONLY" HARVEST CAN SKEW THE SEX RATIO OF A DEER POPULATION, LEADING TO A SHORTAGE OF BUCKS
This simplified model assumes a fixed and rigid carrying capacity so that the number of fawns surviving equals the number of older animals removed. Survival rates are computed from an average fawn production of 1.25 per doe.

	Males	Females	% Fawn mortality
Year 1	250	250	
20% removal (males only)	−100	− 0	
Survivors	150	250	
Fawns surviving	+ 50	+ 50	68
Year 2	200	300	
20% removal (males only)	−100	− 0	
Survivors	100	300	
Fawns surviving	+ 50	+ 50	73
Year 3	150	350	
20% removal (males only)	−100	− 0	
Survivors	50	350	
Fawns surviving	+ 50	+ 50	77
Year 4	100	400	

by hunters each year. The sex ratio becomes more skewed each year until year 4, when all 100 remaining bucks would have to be taken to maintain the quota. Meanwhile, the greater proportion of does in the population means that more fawns are produced; since the population is at or near carrying capacity, however, fawn mortality is quite high, becoming more so as the sex ratio becomes more skewed.

This example is admittedly simplistic. Although a 1 to 1 sex ratio at birth is the general rule, natural mortality seldom affects both sexes equally. In addition, K is not really fixed for most deer populations, but instead tends to fluctuate. This means that the actual population would almost certainly grow beyond the arbitrary 1,000, increasing until the declining food supply forced it downward. Nevertheless, the general comparison is realistic. Populations in which does have been protected for many years tend to have a substantially smaller proportion of bucks and lower fawn survival. They also tend to exceed K, causing at least temporary habitat damage, as well as damage to forest and agricultural crops.

At about the same time that the buck law was gaining widespread acceptance, states began restocking deer into areas from which they had been eradicated. Aided by predator removal and the buck law, these transplants succeeded where habitat conditions allowed, contributing substantially to the recovery of the species. For example, in 12 states in the southeastern United States, some 22,686 deer were transplanted between 1938 and 1967. Before the restocking, populations were estimated to total about 303,500, and the annual kill was about 60,000. Deer populations soared to 2,405,000 by the late 1960s, with an annual yield of 274,000 (Newsome, 1969).

Changes in land use resulted in improved habitat in many places. Small family farms, abandoned during the depression, grew deer browse instead of row crops. Old growth forests were cut, and in their place grew more and better browse characteristic of the early- to midsuccessional stages. The human populations in rural areas declined as people sought economic improvement through city jobs. With the people went dogs and rifles, which had helped keep deer populations low.

By midcentury, most wildlife managers realized that deer populations had become excessive in many regions. Many of them attributed the excessive numbers to the absence of predators (Leopold et al., 1947). Gradually, though, managers began to conclude that the real problem was excessive protection and that the best solution was increased harvest. Unfortunately, much of the public and their representatives in state legislatures were by then convinced that the buck law was the key to deer management. The "unselling" of the buck law has proven to be a difficult task.

Largely because of political opposition to doe hunting, state wildlife agencies have adopted a cautious policy of issuing doe permits in regions where deer populations are especially high. Though far from perfect, this arrangement grants state agencies considerable flexibility. Regions containing deer populations at or near *K* are issued generous numbers of doe permits. Agencies may issue a few permits for regions with modest deer populations and none at all in areas where deer numbers need to be increased.

Pennsylvania has one of the most advanced and sophisticated systems of regulating its deer through issuance of doe permits. The Pennsylvania Game Commission uses a series of equations based on harvest reports, sex ratios, attrition rates, and recruitment rates. Field personnel of the commission collect data during two critical periods. From January 1 through June 30 they determine pregnancy rates from does killed on the highways or killed for crop damage. They also tally frequency of twinning, sex ratio of fetuses, and the age of does.

The bulk of the remaining data comes from the fall harvest, through deer check stations, processing establishments, and mandatory reports from hunters. Once data are collected, the commission calculates the condition of deer populations for groups of similar counties. Next, it calculates through a formula the number of doe permits to be issued for each county class (Lang and Wood, 1976). Clearly, then, the Pennsylvania management plan makes no attempt to obtain maximum sustained yield or even optimum sustained yield from the state deer herd. Its principal management objective, like that of most other agencies responsible for management of large ungulates, is to keep the population within reasonable bounds through harvesting.

SMALL GAME

Resident birds and small mammals regularly hunted for sport are customarily lumped into the rather imprecise category of "small game." Most of these species are fairly *r*-selected, reproducing rapidly under favorable conditions and experiencing high rates of population turnover. In keeping with the usual *r*-selected population traits, most species of small game reach their highest numbers under early-to-midsuccessional stages.

In dealing with small game, managers have usually assumed that because small game populations exhibit such high rates of turnover with or without hunting, that mortality inflicted by hunters must be compensatory. Habitat conditions, rather than harvests, have been assumed to be the most critical determinants of population size and productivity. Both these long-held assumptions seem to be essentially correct.

Using empirical models derived from field data, computer simulation has enabled critical evaluation of the effects of hunting on some small-game populations. Again there is a difference between maximum sustained yield and optimum sustained yield. Roseberry (1979) simulated the effects of different harvest strategies upon populations of America's most popular game bird, the bobwhite quail. According to his harvest model, maximum yields can be sustained from annual harvests of about 55 percent. Harvesting at such a high rate, though, results in depression of the following spring populations by 53 percent below unexploited levels. That leaves little room for error. Roseberry recommended instead an optimum harvest rate of 40 percent to 45 percent, noting that this level resulted in nearly as high a yield without such a severe population reduction.

Thus, hunting mortality even on such an *r*-selected species as the bobwhite is not completely compensatory. Moreover, the extent to which hunting mortality becomes additive increases with higher yields. Eventually, virtually any population could be substantially reduced, even eradicated, with sufficiently great hunting pressure. The great majority of small-game hunting is not of sufficient intensity to reduce the population seriously, though, owing to the animals' rapid reproductive rates and the density-dependent responses of both fecundity and mortality. Hunters favor areas of abundant game over areas with sparse game populations. This "law of diminishing returns" is another reason why most small-game harvests fall short of the additive threshold.

Species that are *r*-selected usually have populations that fluctuate with successional stages, weather variations, and changes in land use. Accordingly, wildlife managers have gone to great lengths to monitor population trends and annual variations in productivity. In the case of the bobwhite quail, auditory indexes (whistle counts) have often been used to estimate population trends. There are two drawbacks with this technique. First, it must be done in June or July when the whistling of unmated males can be heard. A lot of things can happen to a quail population between July and the fall hunting season. Second, whether or not the frequency of whistling among unmated males has any valid relationship to the population size either in summer or the following fall is not conclusively established.

After reviewing several types of survey data collected on bobwhites in Iowa, Schwartz (1974) found a strong linear correlation between roadside counts done in August in marginal habitats and subsequent annual quail harvest and average season bag. He recommended these counts instead of the more widely used whistle counts, at least in Iowa. Counts in marginal habitats do tend to fluctuate more than do those done in prime habitats, so they may be more sensitive indicators of population changes.

Age ratios (see Chapter 4) are commonly used by managers to estimate productivity in quail and certain other small game. Changes in age ratios can be misleading, however, as the implicit assumption is that of a stable between-year breeding population. Roseberry (1974) has shown that percent juveniles correlates with the percent gain in the population over the summer. Age ratios do not correlate with fall densities, nor do they give any reliable estimate of the total productivity of the population as a whole. All that age ratios indicate is the survival of young birds relative to the standing density of adults. Since managers rarely know the latter figure, the utility of age ratios becomes questionable. Anyone contemplating use of the method should carefully consult the simulation of Roseberry (1974) as well as criticisms of age ratios by Caughley (1974).

The practice of releasing pen-raised bobwhite quail by state wildlife agencies has almost completely ceased. For years, this practice was supposed to increase hunting opportunities for sporthunters and was paid for by their license fees. Biologists have known for decades that the stocking was rarely, if ever, cost-effective. Pen-reared quail are generally poorly suited to life in the wild. More importantly, if habitat conditions could sustain higher population densities, then the wild birds would have already responded. It is not possible to raise population densities permanently by adding pen-raised game. A far more reliable and cost-effective alternative is habitat improvement.

Habitat improvement can be accomplished in several ways, but the two most common approaches for small game are improvements in food and cover. On Fort Riley, Kansas, food plots consisting mainly of corn, sorghum, and multiflora rose (*Rosa multiflora*) have for years been maintained for the benefit of bobwhites. Crop contents of birds shot within varying distances of the food plots revealed that quail taken within about 600 meters of the plots obtained food from them. The birds shot farther from the plots also tended to be lighter in weight, having less fat reserves than did those living nearer the plots. Most seed-eating birds take seeds in relation to availability rather than calories. Therefore bobwhites living in an area having an abundance of high-quality seeds will weigh more than those in an area with an abundance of seeds of low quality. Artificial food plots are expensive to install and to maintain. Also some people object to them on the grounds that they detract from the naturalness of the habitat. Alternatives to agricultural crops are ragweed (*Ambrosia* spp.), sunflower (*Helianthus* spp.), and dogwood (*Cornus* spp.); all produce seeds with high metabolizable energy (Robel, et al., 1974).

Increasing certain types of cover can improve habitat for several reasons. Being physically small animals, quail and other resident game often need thermal cover to guard against extreme weather, particularly cold. They also need escape cover from predators. But for most resident game birds, the most critical type of cover is that for nesting. In the case of ring-necked pheasant, the ideal nesting habitats include alfalfa fields, fields of sweet clover, small grains, and fence rows, plus bar ditches with early maturing plant cover (Johnsgard, 1975). Pheasants also need a diversity of habitat types within fairly close proximity. More intensified agricultural practices that require larger fields and mowing of fence rows and bar ditches have substantially reduced ring-necked pheasant populations throughout much of the United States.

Wildlife managers have long believed that game production can be stimulated by increasing the relative amounts of edge between cover types, a phenomenon known as the "edge effect" (see Chapter 2). Although the edge effect has not been conclusively proven in this regard, it seems likely to apply to many small-game species for two reasons. First, most small game have limited mobility, seldom moving over very large areas. Second, they often have varied habitat requirements, such that one habitat or cover type can rarely furnish all their year-around requirements. Thus, increasing the edge or interspersion among different types probably does raise carrying capacity and hence the yield for many small-game species.

At least some species of small game actually decline with increasing amounts of edge. The ruffed grouse, for instance, reaches highest numbers within large tracts of second growth forests. Gullion and Marshall (1968) found that territorial male grouse did not survive as well along forest edges as they did deeper in the forests. Managers need detailed knowledge of the life history of species with which they work, plus a clear set of management objectives, before deciding how they should manipulate the habitat.

WATERFOWL

Ducks, geese, and swans are gregarious, migratory, aquatic game birds of immense popularity. Because their migratory instincts carried them across state boundaries in the United States, waterfowl were among the first groups of wildlife to fall under federal jurisdiction. Management of these birds is necessarily complex, involving programs directed at nesting areas, wintering grounds, and along the migration routes in between. Habitat management, as well as harvest regulation, requires close cooperation between federal and state wildlife agencies, along with involvement of private organizations.

The establishment of annual waterfowl harvest regulations is a complex task shared by federal, state, and private organizations. Biologists conduct annual spring and summer censuses on the breeding grounds. The field data are forwarded to the Flyway Councils for compilation and then sent to the National Waterfowl Council and to the Waterfowl Advisory Council, where they are evaluated along with recommendations from federal biologists and representatives of private conservation organizations. From there the information goes to the U.S. Fish and Wildlife Service and the Secretary of the Interior, whose representatives work out a general framework for seasons and bag limits. State wildlife agencies can then select specific regulations for waterfowl hunting within their boundaries, so long as they fit within the broader federal guidelines (see Figure 7-3).

For decades, waterfowl biologists have studied mortality rates among waterfowl by analyzing data from band returns (see Chapter 4). Not surprisingly, rates of band returns rose or declined with changes in bag limits. This correlation, along with the analytical procedures in use for many years, led many biologists and managers to infer that hunting mortality was in fact additive to the natural mortality on waterfowl. More sophisticated methods have raised some important and intriguing questions regarding the effects of hunting on the population of the most widespread duck in North America, the mallard.

The mallard has received more management and research attention than any other duck. As a result, the U.S. Fish and Wildlife Service and other management agencies have amassed data on hundreds of thousands of mallards banded each year before the hunting seasons. When applied to

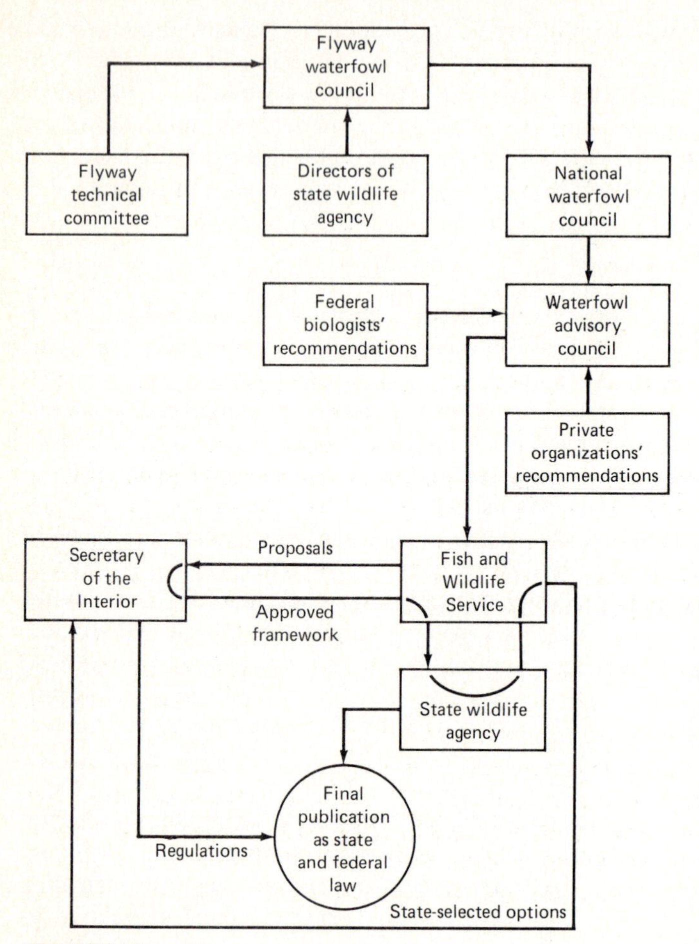

FIGURE 7-3
The steps involved in setting waterfowl regulations annually in the United States. (Giles, 1978, © W. H. Freeman & Co. Reprinted by permission.)

these data, the new analytical techniques have yielded surprising results. To get at the question of the effects of hunting mortality on mallards, researchers first compared recovery rates of bands with annual harvest regulations. They found that band return rates were significantly higher in years of more liberal bag limits than during years of more conservative bag limits. This was certainly not surprising. Next, they used the band return data to determine that survival rates did indeed vary significantly between years. Finally, they compared survival rates with harvest rates during the restrictive and liberal years. There was no correlation (Anderson and Burnham, 1978; Rogers et al., 1979).

These results imply, of course, that hunting mortality on the mallard is largely compensatory, and that mallards cannot be effectively stockpiled through several consecutive years of restrictive hunting regulations. The researchers have emphasized that their findings are not totally conclusive, pointing out that important breeding areas in northern Canada have not been represented by banding samples.

Other researchers have disagreed with the results of the mallard mortality studies. Trauger and Stroudt (1978) studied long-term trends in waterfowl production from three widely separated regions in south central Canada. They compared fluctuations in water conditions (and hence habitat conditions) over a 25-year period with fluctuations in breeding success for a number of waterfowl species. The authors concluded that there was no apparent correlation between habitat conditions and production of broods, adding that historically, major declines in waterfowl populations occurred before the loss of many wetlands and the alterations of others. All this evidence suggests that factors other than habitat conditions on breeding grounds influence waterfowl populations. Trauger and Stroudt suggested that hunting might play a major role in holding down populations of at least some species of ducks. They suggested that conservative harvests be imposed for several years following major droughts to allow populations to recover.

Most management-oriented research on waterfowl has been done on the more productive breeding grounds or "duck factories," certainly the prudent place to begin. Yet the controversy over the effects of hunting mortality upon mallards and other ducks points out the need to learn more about mortality during migrations and, especially, on the wintering grounds.

Another warning about concluding that hunting mortality is compensatory came from Patterson (1979). While disagreeing with neither the analysis nor the guarded conclusions of Anderson and Burnham, he questioned whether or not the population dynamics of the mallard are comparable to those of other ducks. The mallard is widespread, opportunistic, and reproduces rapidly, making it one of the more *r*-selected of North American ducks. Since *r*-selected species sustain high natural rates of mortality, hunting mortality is likely to be at least partly compensatory. *K*-selected species, on the other hand, are generally more vulnerable to the additional mortality from hunting. Patterson established a scale on which he ranked 10 species of ducks based upon their places on an *r*–*K* continuum (see Figure 7-4). The mallard was among the most *r*-selected, while the canvasback and other divers were more *K*-selected. Patterson warned against using the mallard as the yardstick for other waterfowl species,

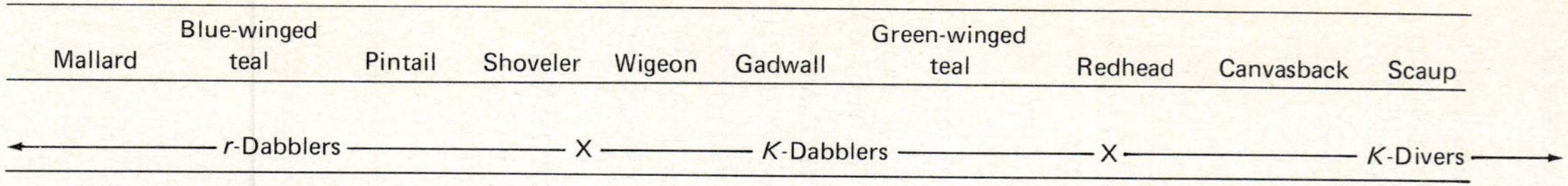

Mallard	Blue-winged teal	Pintail	Shoveler	Wigeon	Gadwall	Green-winged teal	Redhead	Canvasback	Scaup

← *r*-Dabblers —— X —— *K*-Dabblers —— X —— *K*-Divers →

FIGURE 7-4
North American waterfowl compared on an *r–K* continuum. (Patterson, 1979, © Wildlife Management Institute. Reprinted by permission.)

contending that the more *K*-selected have, in all probability, much lower harvest thresholds.

The wetland habitats that produce and sustain waterfowl are among the most fragile and rapidly disappearing of habitat types (see Chapter 9). Sanderson (1976) reported that the original wetlands in the United States totaled an estimated 49.6 million hectares (127 million acres). By 1968, at least 20.3 million hectares (52 million acres) had been lost to various forms of drainage or filling, leaving only about 20.3 million hectares (75 million acres). Of these remaining wetlands, only 3.5 million hectares (9 million acres) could be classed as high-quality, and the rest of either low or negligible value. Clearly, the amount of high-quality waterfowl habitat has declined seriously. To help offset these losses, waterfowl specialists and responsible agencies have directed their attentions toward habitat acquisition and habitat manipulation.

Habitat acquisition has been carried out to a large extent by the U.S. Fish and Wildlife Service through their National Wildlife Refuge System. Beginning in 1903 with President Theodore Roosevelt's executive order protecting Pelican Island, the system received an important boost in 1929 with passage of the Migratory Bird Conservation Act. This act authorized purchases of sanctuaries for migratory birds. In 1934, the Duck Stamp Act furnished funding for such acquisition. A 1958 amendment to the Migratory Bird Conservation Act extended authority for acquisition to waterfowl production areas. The acquisitions, concentrated mainly in the prairie pond or pothole region of the north central United States, were purchased through funds from the Wetlands Loan Act of 1961, to be repaid through future duck stamp sales (U.S. Fish and Wildlife Service, 1976).

State wildlife agencies often have their own acquisition programs for waterfowl habitat, with state funds obtained through sales of state waterfowl hunting stamps or licenses. By 1975, states had acquired a grand total of over 2 million hectares (5 million acres) of waterfowl habitat (Bellrose, 1976).

The National Wildlife Refuge System devotes 276 of its 367 refuges to waterfowl, with a total area of more than 1.5 million hectares (4 million acres). In addition, the U.S. Fish and Wildlife Service holds title to nearly 156,000 hectares (400,000 acres) of waterfowl production areas plus 388,670 hectares (995,000 acres) maintained under easements (U.S. Fish and Wildlife Service, 1976).

Ducks Unlimited, a private organization comprised mainly of waterfowl hunters, has for years worked to acquire and to improve wetlands habitat. The organization owns no land but develops waterfowl habitat through easements from landowners. Being private, Ducks Unlimited can secure waterfowl habitat in Canada and Mexico, making its efforts truly continental in scope. The organization has, since 1932, spent over $32 million to lease 849,000 hectares (2.2 million acres) of land under lease and has developed through water projects 540,500 hectares (1.4 million acres) and 15,856 kilometers (9,910 miles) of shoreline (Stahr and Callison, 1978). Ducks Unlimited has been extremely successful as a fund raiser, obtaining its operating funds from members and other contributors.

Another important private organization concerned with wetland habitat is the Nature Conservancy. The conservancy, which employs specialists in real estate, business, and law, as well as wildlife specialists, is well-known for its effective, businesslike approaches to acquiring habitat. Sometimes when state or federal agencies are unable to purchase a tract of habitat to save it from development, the conservancy buys the land. It may then maintain it as a sanctuary or nature reserve, or else deed it over to state or federal agencies when the agencies are able to purchase it.

Once suitable habitat is acquired, the next step may be habitat manipulation, an effort to improve habitat quality. One particularly dramatic way of increasing breeding habitat in the northern prairies of North America is through blasting new small, shallow ponds (potholes). Managers can create new potholes by using ammonium nitrate, a commercial fertilizer. The ammonium nitrate is soaked in fuel oil and detonated. The size and depth of the resulting pothole depends upon the soil characteristics and the amount of ammonium nitrate used (Burger and Webster, 1964).

Water-level control is very important for intensively managed waterfowl habitat to provide the best conditions for feeding, nesting, or resting. This is why so many of the national wildlife refuges have artificial dikes and devices for altering water levels. They may detract from the scenic or esthetic values of some refuges, but they are essential for intensive waterfowl management.

Control of vegetation is another important component of habitat management. Grazing and prescribed burning are used in some cases to reduce vegetative cover and open up feeding areas. Depending upon the circumstance, either seeding or prescribed burning can be used to improve nesting habitat (U.S. Fish and Wildlife Service, 1976).

For many years, agricultural crops were planted on national wildlife refuges as an important source of supplemental food for waterfowl. The practice has declined somewhat, in part because of rising costs of farming and in part because supplemental feeding leads to unnatural, and often unhealthy, concentrations of ducks and geese (Greenwalt, 1978).

FURBEARERS

Several important differences exist between management of furbearers and management of game animals hunted for sport. Harvest of furbearers involves commercial motives because the hides of these animals, unlike the meat from game species, can legally be sold on the free market. This difference gives rise to a unique and frustrating problem for managers. Changes in the market values of the pelts can be substantial and occur quite independently of the availability of the animals from which they come. Trapping pressure for many species shifts roughly in accordance with changes in the market prices. This shifting makes it very difficult to protect the populations adequately when numbers are low and prices high. Conversely, it makes it virtually impossible for managers to use harvests to reduce populations of abundant (and sometimes destructive) species when prices are low.

Furbearer harvests are more controversial than are the harvests of game animals (Payne, 1980; Reiger, 1978). Some methods of capture, especially the steel leg-hold trap, are viewed by many persons as cruel and inhumane, so much so that some nations and some states have outlawed their use. Many people also have strong convictions against the sale and display of products made from animal furs, particularly those from rarer species.

Reliable estimates of many furbearer populations are often very difficult to obtain. This is especially true of the more cryptic species of carnivores. Moreover, carnivore populations tend to fluctuate from annual variations in the survival rates of juveniles. Although there have been valiant and imaginative attempts to develop reliable and practical methods of monitoring population trends and survival rates, they still lag behind those used for game animals.

One extensive means of estimating population trends is the scent station survey (see Chapter 4). Originally developed for monitoring trends in coyote populations, this method does attract many other carnivorous furbearers. One difficulty with the scent station survey is that the chemical attractant used was developed especially for coyotes. Almost certainly the attractant does not lure all species of carnivores equally, and to date there are no reliable estimates of differential attractability. In addition, serious questions have been raised regarding the validity of making between-area comparisons with this technique, particularly if the areas vary in terms of habitat types and cover (Roughton, 1979).

Another way of estimating population trends has been through use of questionnaires distributed to resident landowners and people who can report sightings of furbearers (Hatcher and Shaw, 1981). One interesting variation on this theme was the use of rural mail carriers who reported sightings of red foxes in North Dakota (Allen and Sargent, 1975).

More often, wildlife agencies use various types of harvest data to monitor furbearer populations. Most states require annual reports from fur buyers and also require trappers to file catch reports. These furnish some estimate of the yearly harvests. Occasionally, trappers are interviewed or else questionaires are mailed to samples of them in an effort to obtain more detailed information.

Another important means of evaluating furbearer harvests is through analysis of skinned carcasses. Biologists collect skinned carcasses from trappers and fur buyers. They are then able to determine age structures, sex ratios, fecundity rates, and the general physical condition of the populations from which the carcasses were obtained.

Erickson (1981) compared the variables affecting harvests of muskrat, beaver, raccoon, and coyotes in Missouri. In more *r*-selected species such as muskrat, he found that annual variations in recruitment (most likely reflecting differences in survival rates) were the major correlates of numbers harvested. Variation in season lengths was also important, particularly for beaver harvests, a finding that underscores the misfortune of states whose legislatures have set season length by statute. Changes in demand resulting from shifting prices were the main influence for raccoon and coyote harvests.

The effects of harvest on aquatic rodents is then quite different from those on terrestrial carnivores. This difference is illustrated through comparison of two important furbearers in North America, the muskrat and the bobcat. Muskrat populations are very sensitive to drought, which, of course, shrinks the marshes and shallow lakes that constitute their habitat. After several years of drought, six marshes in northwest Iowa were closed for a year to muskrat trapping. Five shallow lakes within the same general area remained open to trapping. The number of muskrat harvested yearly was recorded for both areas following the closed season in an effort to measure the effects of the closed season on subsequent harvests. Table 7-2 shows that harvests per hectare remained nearly the same on both areas. From these data, Neal (1977) concluded that closing the season had no real effect on muskrat populations. Droughts, along with their effects upon production of emergent aquatic vegetation, seemed to be

TABLE 7-2
RESULTS OF A STUDY SHOWING THE EFFECT OF A CLOSED SEASON UPON MUSKRAT POPULATIONS IN NORTHWEST IOWA
(Modified from Neal, 1977, © The Wildlife Society. Reprinted by permission).

Year	Harvest/ha (experimental)	Harvest/ha (control)
1961	Closed	12.2
1962	26.8	23.5
1963	34.9	55.8
1964	36.6	38.6
1965	38.8	30.4

the critical factor influencing harvests. Interestingly, Neal added that there was no significant correlation between the average price paid for muskrat furs and the annual harvests.

Bobcat harvest, on the other hand, like those of other upland carnivorous furbearers, is more closely tied to demand. Following enactment of legislation and international treaties restricting imports of spotted cat hides, fur buyers sought substitutes in bobcat hides. Bobcat prices rose sharply during the 1970s, so much so that the species was placed on Appendix II of the Convention on International Trade in Endangered Species (CITES). This action required special export permits from the country of origin. The federal government, acting through the newly created Office of Scientific Authority (OSA), irritated many state wildlife agencies by imposing quotas upon a ubiquitous, resident species whose jurisdiction had previously rested solely with states.

Concern for the effects of harvest on bobcat populations led Crowe (1975) to develop a model based on data from exploited bobcat populations from Wyoming. Crowe used 161 carcasses to determine age structure (by dental characteristics), sex ratios, and fecundity rates. Analysis of the resulting data indicated that survival rates were about 67 percent for all age classes, the sex ratio was 1 to 1, and all females bred and produced litters that averaged 2.78 kittens. These basic data permitted calculation of the juvenile survival rate necessary to keep the population stable, through use of the following formula: lambda = $S_a + F(S_y)$, where S_a = survival rate of adults (in this case presumed to average .67), F = rate of production of kittens per adult (one-half the litter size with a 1 to 1 sex ratio), and S_y = the survival rate of juveniles.

A stable population has a lambda value of 1.00 (see Chapter 3). Solving, then, for S_y gives the survival rate of juveniles necessary for keeping the population stable. Hence:

$$1.00 = .67 + 1.39(S_y) \qquad S_y = 0.24$$

The population could remain stable with an adult survival rate of 0.67, so long as juveniles survived at a rate of 0.24, or 24 percent.

Being considerably more *r*-selected, muskrat populations can sustain themselves with much lower survival rates. Each female muskrat gives birth to an average of 2.8 litters totaling over 18 young (Smith et al., 1981). If the population has a 1 to 1 sex ratio, the resulting F value from the equation above becomes 9.18, more than six times as great as that of the bobcat. If the survival rate of adults is a mere 20 percent, then:

$$1.00 = 0.20 + 9.18\ (S_y) \qquad S_y = 0.087$$

A muskrat population would thus remain stable with only a 20 percent survival rate for adults and an 8.7 percent survival rate for the young.

As a general rule of thumb, then, muskrat management should emphasize preservation of marshland habitat with the right combination of water level and emergent aquatic vegetation. Beavers require adequate stream conditions and, if pelt prices remain low, little or no harvest regulation. Should harvest pressures rise, or else populations increase to troublesome levels, season lengths can be reduced or increased, respectively. Harvest restrictions for most of the carnivorous furbearers should rise and fall along with pelt prices.

GAME RANCHING

A relatively new form of wildlife management, "game ranching" is the controlled exploitation of game animals for profit on a sustained yield basis. The profit motive distinguishes game ranching from most other wildlife management practices. Profit can be derived from the direct sale of game meat or hides, from the sale of trophy animals to sport hunters willing to pay for the privilege, and even from admission tickets for game viewing.

Most commonly practiced in southern Africa and Texas, game ranching is almost always carried out on private lands, many of which are also used for raising livestock as well as game. Proponents of game ranching (cf. Dasmann, 1964; Mossman, 1975; Mossman and Mossman, 1976) argue that the profit motives provide incentives to landowners to maintain wildlife habitat and game populations. Without such direct financial returns, the landowners are likely to destroy wildlife habitat by converting their lands to other profitable use.

Africa is blessed with a wide variety of wild herbivores. Natural selection has reduced competition among species (see Chapter 6), resulting in partitioning of feeding niches. Such partitioning means that native vegetation will be used more efficiently by several species than it ever could by any single species. Where habitat is still sufficiently diverse, several species of wild herbivores can yield substantially more biomass per unit area than can cattle. Wild herbivores are also far better adapted to water conservation in their hot, seasonally arid environment. The eland (*Taurotragus oryx*), for example, passes very little water in its feces; this ability to conserve water, along with its unique metabolism, allow it to exist virtually without surface water (Jewell, 1969). Adding to this difference is the fact the livestock cannot inhabit much of Africa because of sleeping sickness carried by the tsetse fly. Wild species have natural resistance to the disease.

Texas game ranching differs from that in Africa in two key respects.

First, while African game ranching draws on numerous species of native wildlife, game ranching in Texas and other parts of the United States includes several exotic introductions. Most of the successful exotic ungulates are Eurasian, perhaps because they are more tolerant of temperate zone climates than are African ungulates.

Second, the main objective of African game ranching has been meat production obtained through periodic culling. Texas game ranching, on the other hand, derives most of its income from sport hunters, who pay for hunting privileges. In more recent years, this distinction has been blurred somewhat as some South African game ranchers have discovered higher profits from safari sales than from culling (Mossman and Mossman, 1976).

Compared with conventional livestock ranching, African game ranching offers several advantages. More efficient use of native vegetation and greater disease resistance have already been mentioned. Estimates of meat production potentials are often quite high. Cole and Ronning (1974), for example, calculated that more than 70 percent of the present total world meat production could be met through careful cropping of African wildlife. Another advantage of game ranching is that it represents a "gentler ecological intrusion" than do other forms of land use (Mossman, 1975).

Wild game is generally higher in protein and much lower in fat than is meat from domestic stock (Table 7-3). This difference offers important nutritional advantages to modern American diets and may result in higher yields of protein for Africans. Yet some authorities have questioned whether the Africans need more protein or more calories (due to higher fat content) from livestock (Coe, 1980).

For the wildlife manager, African game ranching has a special pragmatic appeal. As Africa's human population continues to grow, placing greater demands on the environment (Table 7-4), game ranching may be the only way to justify survival of many species. If game ranching can really provide

TABLE 7-3
NUTRIENT VALUE OF MEAT

Item	Protein (%)	Fat (%)	Calories/100 g
Fresh caribou	27.2	4.7	120
Fresh moose	25.0	0.9	123
Standard grade beef	29.4	15.8	225
Prime grade beef	13.6	41.0	428
Eland	23.0	1.9	125

Source: Modified from Novakowski and Solman (1975, © American Society of Animal Science. Reprinted by permission) from original sources in *Alaskan Sportsman,* May 1972, and from Cole and Ronning, 1974.

TABLE 7-4
A LIST OF SPECIES RAISED COMMERCIALLY IN SOUTHERN AFRICA (ZIMBABWE AND REPUBLIC OF SOUTH AFRICA)
(Mossman and Mossman, 1976, © International Union for Conservation of Nature and Natural Resources. Used by permission)

Species
Warthog *(Phacochoerus aethiopicus)*
Springbok *(Antidorcas marsupialis)*
Bushbuck *(Tragelaphus scriptus)*
Greater kudu *(Tragelaphus strepsiceros)*
Eland *(Taurotragus ornx)*
Nyala *(Tragelaphus angasi)*
African buffalo *(Syncerus caffer)*
Waterbuck *(Kobus ellipsprymnus)*
Reedbuck *(Redunca arundinum)*
Sable *(Hippotragus niger)*
Brindle gnu (Wildebeest) *(Connochaetus taurinus)*
Impala *(Aepyceros melampus)*
Zebra *(Equus aethiopicus)*
White rhino *(Ceratotherium simum)*
Ostrich *(Struthio camelus)*

abundant meat and also allow wild species to persist in natural environments, then it surely merits a high priority in Africa.

Texas game ranching is justified on somewhat different grounds. Advocates point out the loss of species diversity resulting from Pleistocene extinctions, especially large mammals. Wild species of exotics are therefore advocated on the grounds that they will fill some of the "empty" niches left by extinct species. (The deliberate introduction and release of exotic wildlife, especially on public land, is one of the most controversial areas in wildlife management. It is discussed in detail in Chapter 9.)

Another justification of game ranching in Texas is that, just as in Africa, landowners receive an incentive to retain wildlife habitat. Indeed, the fact that almost all of Texas is privately owned, combined with the state's effective trespassing statutes, has been largely responsible for development of hunting leases in general and game ranching in particular (Teer, 1975). Table 7-5 shows the economic value and profitability from commercial game ranches in the United States.

Game ranching has not been without substantial opposition, as well as some legal and logistic problems. Much of the opposition comes from the livestock industry, notably veterinarians, who fear that diseases and parasites will be transmitted from wild to domestic species. Many livestock growers tend to be conservative in general and resist the changes that would come through game ranching. A good deal of opposition comes

TABLE 7-5
AVERAGE TROPHY FEE AND INVESTED COST PER ANIMAL FOR U.S. COMMERCIAL HUNTING RANCHES
(Attebury et al., 1977, © The Wildlife Society. Reprinted by permission)

Species	$\bar{x}$ trophy value (fee)	$\bar{x}$ invested cost	Gross profits (%)
Aoudad sheep	$ 632	$293	53.6
Axis deer	$ 621	$308	50.4
Barbado sheep	$ 180	$ 37	79.4
Blackbuck antelope	$ 600	$295	50.8
Corsican sheep	$ 236	$ 55	76.7
Elk	$1,167	$639	45.2
Fallow deer	$ 398	$212	46.7
Mouflon sheep	$ 338	$149	55.9
Red deer	$1,054	$585	44.5
Sika deer	$ 408	$246	39.7

from wildlife enthusiasts who lean toward the protectionist side of the conservation spectrum. Such people are skeptical of deliberate, systematic slaughter of wild animals in the name of profit and feel that the temptation for greater short-term gains will lead to excessive exploitation.

There are some legal difficulties with game ranching. In the United States and in some African nations, native wildlife is regarded as public property, even on private land. This arrangement limits the abilities of private landowners and their clients to harvest wild game. The most successful game ranching programs in Africa have been in South Africa, where game on private lands can be treated as private property (Parker and Graham, 1971). Although native wildlife in Texas is still managed as public property, exotic species of game are regarded as property of the landowner. Exotic game can then be harvested at the landowner's discretion.

Government assistance and tax laws often favor the livestock rancher over the game rancher, especially in southern Africa. Mossman and Mossman (1976) have pointed out the need for reform of tax laws and for assistance programs to place game ranching on a more competitive footing with livestock ranching.

SUMMARY

Up to a certain level or threshold, hunting mortality tends to be compensatory to natural mortality. Once that threshold is exceeded, however, hunting mortality becomes additive. Harvest models have aided wildlife managers in understanding such thresholds and in furnishing rough guidelines for setting harvests. These models can also be used to play "what if" games, reveal the relative importance of various population parameters upon a population's growth, and help formulate clear management objectives.

The number of animals that can safely be taken from a population can be estimated in several ways. The most basic is maximum sustainable yield (MSY), thus far used more widely in commercial fisheries than in terrestrial game harvests. MSY is easy to understand in principle but difficult and, at times, even hazardous to apply. Practical problems include the need for accurate estimates of both numbers and carrying capacity, tendencies to rely more upon quotas than upon actual population sizes, and the need to regulate the harvest precisely. Some conceptual problems associated with MSY are that the logistic equation lacks precision and realism; that it fails to consider indirect effects of harvests, competitive relationships, or both; and that it inherently assumes stable environmental conditions.

One alternative to MSY is optimum sustainable yield (OSY). OSY uses both biological and sociological criteria that are more complex to integrate. But for managers, OSY is much more flexible and can be implemented with a margin of safety.

Traditional game managers often rely on conservative harvests such as taking small annual quotas or proportions set deliberately low. Over areas the size of states or provinces, seasons and bag limits, adjusted in accordance with estimated trends in game populations, may afford the only practical means of regulating harvests. These conservative measures rarely allow harvests to approach the levels attainable through MSY. Usually, the harvests are far lower and less efficient, but these management practices do protect against overharvests.

A radical alternative to conservative harvests is adaptive management, in which populations are deliberately and seriously disrupted. If done systematically, adaptive management could help pinpoint levels of MSY or OSY. It would, however, be risky and hence unlikely to be implemented on any large scale.

Other biologists have recommended new and broader principles for harvesting wildlife populations. These principles reject MSY and place greater importance on maintaining natural ecosystems than on achieving certain quotas or population levels.

The threshold beyond which hunting mortality becomes additive is quite low in ungulates and other big game. The reduction of ungulate populations in the past has led to strong protective measures such as the buck law. But under such protection, many ungulate populations exceeded carrying capacity, inflicting serious habitat and crop damage. As a consequence of these radical changes in ungulate numbers, management has concentrated on population manipulation more than on habitat management.

Small game populations are closely tied to habitat conditions. Being more fecund than ungulates, small-game species quickly respond to habitat improvement and seldom exceed carrying capacity. Most small-game management has thus been directed at habitat improvement rather than at population regulation. The threshold beyond which hunting mortality becomes additive is quite high and, practically speaking, difficult to exceed.

Waterfowl harvests present special difficulties. For years hunting mortality was assumed to be additive. Research on the mallard, the species for which the most data are available, suggests that hunting mortality is largely compensatory. Some authorities question the interpretations of the mallard research and others caution against its application to other waterfowl. But authorities have long agreed on the importance of wetlands to waterfowl, so much management attention has been directed at habitat. Acquisition of habitat has been carried out through the U.S. Fish and Wildlife Service and through private organizations such as Ducks Unlimited and the Nature Conservancy. Manipulation consists mainly of regulation of water levels and control of aquatic and adjacent terrestrial vegetation. Waterfowl hunting regulations are set each year based on field surveys and followed by decisions by the Flyway Councils, the U.S. Department of the Interior, and state wildlife agencies.

Furbearer harvests differ from the harvests of game in that they have a profit motive. This difference can mean that the demand for furbearers changes radically with shifting fur markets. Managers face a considerable challenge because the demand for furbearers changes independently of furbearer population trends. While management of aquatic furbearers often considers habitat conditions, most efforts to manage upland furbearers have been directed at the still elusive goal of developing reliable census methods.

Game ranching also features a profit motive, for the sale of game meat, trophies, or both. Most African game ranching is done primarily for meat production from native wildlife whereas most Texas game ranching emphasizes sale of trophy animals, most of which are exotics. Proponents of game ranching argue that it offers private landowners a financial incentive to maintain wildlife. Opponents cite potential disease risks, concerns about the ecological impact of exotic introductions, and the ethics of slaughter for profit.

REFERENCES

Allen, S. H., & A. Sargeant. 1975. A rural mail-carrier index of North Dakota red foxes. *Wildl. Soc. Bull.* 3:74–77.

Anderson, D. R., & K. P. Burnham. 1978. Effect of restrictive and liberal hunting regulations on annual survival rates of the mallard in North America. *Trans. N. Am. Wildl. and Natur. Resour. Conf.* 43: 181–186.

Attebury, J. T., J. C. Kroll, & M. H. Legg. 1977. Operational characteristics of commercial exotic big game hunting ranches. *Wildl. Soc. Bull.* 5: 179–184.

Beddington, J. R., & R. M. May. 1977. Harvesting natural populations in a randomly fluctuating environment. *Science* 197: 463–465.

Bellrose, F. C. 1976. *Ducks, geese, and swans of North America.* Harrisburg, Pa.: Stackpole Books.

Burger, G. V., & C. G. Webster. 1964. Instant nesting habitat. In J. P. Linduska (Ed.), *Waterfowl tomorrow.* Washington, D.C.: U.S. Government Printing Office.

Caughley, G. 1974. Interpretation of age ratios. *J. Wildl. Manage.* 38: 557–562.

Coe, M. 1980. African wildlife resources. In M. Soule and B. Wilcox (Eds.), *Conservation biology.* Sunderland, Mass.: Sinauer Associates, Inc.

Cole, H. H., & M. Ronning (Eds.). 1974. *Animal Agriculture; the biology of domestic distribution and their use by man.* Chap. 16. San Francisco, Calif.: W. H. Freeman & Co.

Crowe, D. M. 1975. A model for exploited bobcat populations in Wyoming. *J. Wildl. Manage.* 39: 408–415.

Dasmann, R. F. 1964. *African game ranching.* London: Pergamon Press.

Downing, R. L. 1980. Vital statistics of animal populations. In S. Schemnitz (Ed.), *Wildlife management techniques manual* (4th ed.). Washington, D.C.: The Wildlife Society.

Erickson, D. W. 1981. Furbearer harvest mechanics: An examination of variables influencing fur harvests in Missouri. In J. Chapman and D. Pursley (Eds.), *Worldwide furbearer conference proceedings. Vol. II.*

Green, R. H. 1979. *Sampling design and statistical methods for environmental biologists.* New York: John Wiley & Sons.

Greenwalt, L. A. 1978. The National Wildlife Refuge System. In H. P. Brokaw (Ed.), *Wildlife and America.* Washington, D.C.: Council on Environmental Quality.

Gross, J. E. 1969. Optimum yield in deer and elk populations. *Trans. N. Am. Wildl. and Natur. Resour. Conf.* 34: 372–387.

Giles, R. H. 1978. *Wildlife management.* San Francisco, Calif.: W. H. Freeman & Co.

Gullion, G. W., & W. H. Marshall. 1968. Survival of ruffed grouse in a boreal forest. *The Living Bird* 7: 117–167.

Halls, L. K. 1978. White-tailed deer. In J. P. Schmidt and D. L. Gilbert (Eds.), *Big Game of North America: Ecology and management.* Harrisburg, Pa.: Stackpole Books.

Hatcher, R., & J. Shaw. 1981. A comparison of three indices to furbearer populations. *Wildl. Soc. Bull.* 9: 153–156.

Holt, S. J., & L. M. Talbot. 1978. New principles for the conservation of wild living resources. *Wildl. Monogr. No. 59.*

Jewell, P. A. 1969. Wild mammals and their potential for new domestication. In P. Ucko and G. Dimbleby (Eds.), *The domestication and exploitation of plants and animals.* London: Gerald Duckworth and Co.

Johnsgard, P. A. 1975. North American game birds of upland and shoreline. Lincoln, Neb.: University of Nebraska Press.

Larkin, P. A. 1977. An epitaph for the concept of maximum sustained yield. *Trans. Amer. Fish. Soc.* 106: 1–11.

Lang, L., & G. W. Wood. Manipulation of the Pennsylvania deer herd. *Wildl. Soc. Bull.* 4: 159–165.

Leopold, A., L. Sowls, & D. Spencer. 1947. A survey of over-populated deer ranges in the United States. *J. Wildl. Manage.* 11: 162–177.

Levins, R. 1968. Evolution in changing environments: some theoretical explorations. Princeton, N.J.: Princeton University Press.

McCullough, D. R. 1979. The George Reserve deer herd. Ann Arbor, Mich.: University of Michigan Press.

Mossman, A. S. 1975. International game ranching programs. *J. Anim. Sci.* 40: 993–999.

Mossman, Sue L., & A. S. Mossman. 1976. Wildlife utilization and game ranching. *IUCN Occas. Paper No. 17.* International Union for Conservation of Nature and Natural Resources. Morges, Switzerland.

Neal, T. J. 1977. A closed trapping season and subsequent muskrat harvests. *Wildl. Soc. Bull.* 5: 194–196.

Newsome, J. D. 1969. History of deer and their habitat in the south. *Proc. of the symposium white-tailed deer in the southern forest habitat.* USDA Southern Forest Experimental Station, Nacogdoches, Texas.

Novakowski, N. S., & V. E. F. Solman. 1975. Wildlife as a protein source. *J. Anim. Science* 40: 1016–1019.

Parker, I. S. C., & A. D. Graham. 1971. The ecological and economic basis for game ranching in Africa. In E. Duffey and A. S. Watt (Eds.), *The scientific management of animal and plant communities for conservation.* Oxford: Blackwell Scientific Publications.

Patterson, J. H. 1979. Experiences in Canada. *Trans. N. Am. Wildl. and Natur. Resour. Conf.* 44: 130–139.

Payne, Neil F. 1980. Furbearer management and trapping. *Wildl. Soc. Bull.* 8: 345–348.

Reiger, George. 1978. Hunting and trapping in the New World. In H. P. Brokaw (Ed.), *Wildlife and America.* Council on Environmental Quality. Washington, D.C.: U.S. Government Printing Office.

Ricker, W. E. 1975. Computation and interpretation of biological statistics of fish populations. *Fish. Res. Board Canada Bull.* 191: Ottawa, 382 pp.

Robel, R. J., R. M. Case, A. R. Bisset, & T. M. Clement, Jr. 1974. Energetics of food plots in bobwhite management. *J. Wildl. Manage.* 38: 653–664.

Rogers, J. P., J. D. Nichols, F. W. Martin, C. F. Kimball, & R. S. Pospahala. 1979. An examination of harvest and survival rates of ducks in relation to hunting. *Trans. N. Am. Wildl. and Natur. Resour. Conf.* 44: 114–126.

Roseberry, J. L. 1974. Relationships between selected population phenomena and annual bobwhite age ratios. *J. Wildl. Manage.* 38: 665–673.

Roseberry, J. L. 1979. Bobwhite population responses to exploitation: real and simulated. *J. Wildl. Manage.* 43: 285–305.

Roughton, R. D. 1979. Developments in scent station technology. In *Proc.*

Midwest Furbearer Conf. Coop. Ext. Serv. Manhattan, Kan.: Kansas State University.

Sanderson, G. C. 1976. Conservation of waterfowl. In F. C. Bellrose (Ed.), *Ducks, geese, and swans of North America.* Harrisburg, Pa.: Stackpole Books.

Savidge, I. R., & J. S. Ziesenis. 1980. Sustained yield management. In S. D. Schemnitz (Ed.), *Wildlife management techniques manual.* Washington, D.C.: The Wildlife Society.

Schwartz, C. C. 1974. Analysis of survey data collected on bobwhite in Iowa. *J. Wildl. Manage.* 38: 674–678.

Sissenwine, M. P. 1978. Is MSY an adequate foundation for optimum yield? *Fisheries* 3(6): 22–24; 37–38; 40–42.

Smith, H., R. Sloan, and G. Walton. 1981. Some management implications between harvest rate and population resiliency of the muskrat (*Ondatra zibethicus*). In J. Chapman and D. Pursley (Eds.), *Worldwide furbearer conference proceedings. Vol. I.*

Stahr, E. J., & C. H. Callison. 1978. The role of private organizations. In H. P. Brokaw (Ed.), *Wildlife and America.* Washington, D.C.: U.S. Government Printing Office.

Talbot, L. M. 1975. Maximum sustainable yield: an obsolete management concept. *Trans. N. Am. Wildl. and Natur. Resour. Conf.* 40: 91–96.

Teer, J. G., J. W. Thomas, & E. A. Walker. 1965. Ecology and management of white-tailed deer in the Llano Basin of Texas. *Wildl. Monogr. No. 15.*

Teer, J. G. 1975. Commercial uses of game animals on rangelands of Texas. *J. Anim. Sci.* 40: 1000–1008.

Tipton, A. R. 1980. Mathematical modeling in wildlife management. In S. D. Schemnitz (Ed.), *Wildlife management techniques manual.* Washington, D.C.: The Wildlife Society.

Trauger, D. L., & J. H. Stroudt. 1978. Trends in waterfowl populations and habitats on study areas in Canadian parklands. *Trans. N. Am. Wildl. and Natur. Resour. Conf.* 43: 187–205.

Trefethen, J. B. 1975. *An American crusade for wildlife.* New York: Winchester Press.

U.S. Fish and Wildlife Service. 1976. *Final environmental impact statement: operation of the National Wildlife Refuge System.* Washington, D.C.: U.S. Department of the Interior.

Walters, C. J., & P. J. Bandy. 1972. Periodic harvest as a method of increasing big game yields. *J. Wildl. Manage.* 36: 128–134.

Walters, C. J., & J. E. Gross. 1972. Development of big game management plans through simulation modeling. *J. Wildl. Manage.* 36: 119–128.

Walters, C. J., & R. Hilborn. 1978. Ecological optimization and adaptive management. *Ann. Rev. Ecol. & Syst.* 9: 157–188.

CHAPTER 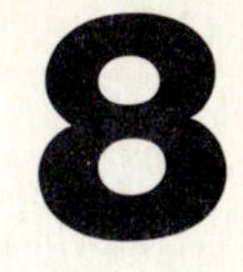

MANAGEMENT FOR PRESERVING NATURAL DIVERSITY

Extinction is nothing new. More than 99 percent of all species that ever lived have vanished, unable to cope with changing environments. The dinosaurs of the Mesozoic are gone as are the giant mammals of the Pleistocene. So why are conservationists and resource managers so concerned over the 1,000 or more vertebrate species officially recognized as threatened with extinction?

Besides those values of wildlife reviewed in Chapter 1, the concern stems from the greatly accelerated *rates* of species extinctions. By all accounts, human activity greatly increases the rate at which species disappear. Of course estimating the "natural" extinction rate is necessarily imprecise, but it can be done by emergences and disappearances in the fossil record. Myers (1976), for example, estimated that the rate of the "great dying" of the last dinosaurs at the end of the Cretaceous was about one species every 1,000 years. Currently the official estimate of extinction rates is about one "wildlife" species per year, and when plants and invertebrates are included it could be as high as one species per day (Eckholm, 1978).

Biologists still do not know how many species live on earth. The tropics, the most species-rich realms of all, remain poorly known biologically. The number of known species, together with the rate at which new species are being discovered, allow estimates of between 3 and 10 million species. Of these, somewhere between 563,000 and 1,875,000 are projected to become extinct by the end of the twentieth century (Lovejoy, 1980). Very roughly speaking then, the earth may lose between about 20 and 33 percent of its species very soon.

Recognizing the lack of precision in estimating extinction rates, Ehrlich et al. (1977) restricted their estimates to an order of magnitude. They computed that between the years 1600 and 1975, the rate of extinction was between 5 and 50 times the natural rate. Furthermore, they estimated that between the years 1980 and 2000, the rate would climb to between 40 and 400 times the usual rate. Even allowing for the rough nature of these estimates, it is quite evident that extinction is a lot more common than it used to be.

Since all the estimates of projected extinction rates are crude, authorities must exercise extreme care in qualifying their use. They are only guidelines for establishing the magnitude of the potential loss. Moreover, as Harwood (1982a) rightly pointed out, such statistics take on a life of their own when crude estimates become cited as "facts." Because these estimates are little better than educated guesses, they are subject to severe criticism by those who oppose conservation measures for reasons of economic self-interest, an aversion to pessimistic predictions, or both.

As species disappear the ecosystems in which they lived become simpler. Since species within a community are interdependent, the extinction of one is certain to affect others in unpredictable ways. An intriguing example of how the disappearance of one species may affect another comes from the island of Mauritius. While conducting field studies on endangered parakeets, Temple (1983) learned of a large species of forest tree, the calvaria (*Calvaria major*). The only calvaria trees left were very old and there had been no evidence of regeneration despite ample fruit production.

Noting the thick coating surrounding the seed and knowing that passage through digestive tracts of birds increases germination rates for many other fruit-producing species, Temple wondered if the disappearance of the dodo (*Raphus cucullatus*) led to the decline of this tree. The dodo, a large flightless bird endemic to Mauritius, had become extinct in 1681 after early colonists and their dogs found them to be easy game. No other bird on the island was large enough to ingest the entire pit of the calvaria.

Temple tested his hypothesis first by force-feeding some fresh calvaria pits to turkeys. He found that 3 of the 10 pits that passed through the turkeys' digestive tract eventually sprouted. But dodos were larger than turkeys and their gizzards undoubtedly were more powerful. Temple estimated that dodo gizzards could exert about 800 pounds of pressure and wondered if the pit would be completely pulverized by such force. So he tested the pits under a laboratory pressure gauge and discovered that the sturdy pits could withstand nearly 1,400 pounds of pressure, far more than the dodo's gizzard could have produced.

While Temple's evidence is admittedly circumstantial, it seems quite likely that the demise of the dodo, presumably a fruit-eating species, was linked to the subsequent decline of calvaria. The thick coat was probably an adaptation to withstand the pressures of the large bird's gizzard. But in this coevolutionary arrangement, that same protective coat became, if not eroded by the journey through the dodo's gizzard, an impenetrable barrier to germination.

Thus if Temple's interpretation is correct, the loss of the dodo would, if not corrected by other means, lead to the extinction of a large species of tree. As the tree disappeared, so might yet another species. The sequence could continue indefinitely, given long enough time and sufficient interdependency.

HOW SPECIES BECOME ENDANGERED

The four major causes of endangerment appear below. Many of today's endangered species, however, reached their status through a combination of causes (see Fig. 8-1).

Excessive Exploitation

The last dodo died late in the seventeenth century. The great auk (*Pinguinis impennes*), another flightless bird, was gone by the mid-nineteenth. The bison barely escaped a similar fate and will never again roam the plains in herds numbering in the millions. These and many more species and subspecies were pushed to the brink by direct, deliberate killing in numbers beyond the thresholds that their populations could withstand.

Motives for this excessive exploitation have varied. Commercial value led to the decline of the sea otter (*Enhydra lutris*), the whooping crane, the bison, the blue whale (*Balenoptera musculus*), and five species of rhinocerous. Other species were greatly reduced because they competed with human beings, particularly for livestock, as did the gray and red wolves, the grizzly bear, and the cougar.

Not all forms of excessive exploitation result from direct killing. The international market in the exotic pet trade has, in recent years, been increasing pressures on declining wildlife populations, especially in the tropics. According to figures compiled by the U.S. Fish and Wildlife Service, more than 95 percent of wild birds and between 30 and 45 percent of all wild mammals imported into the United States are destined for the pet trade (King, 1978). The worldwide trade in live birds is estimated at at least 7.5 million annually and rising (Nilsson, 1981). The real impact of the trade upon wild populations is even more severe when mortality rates are

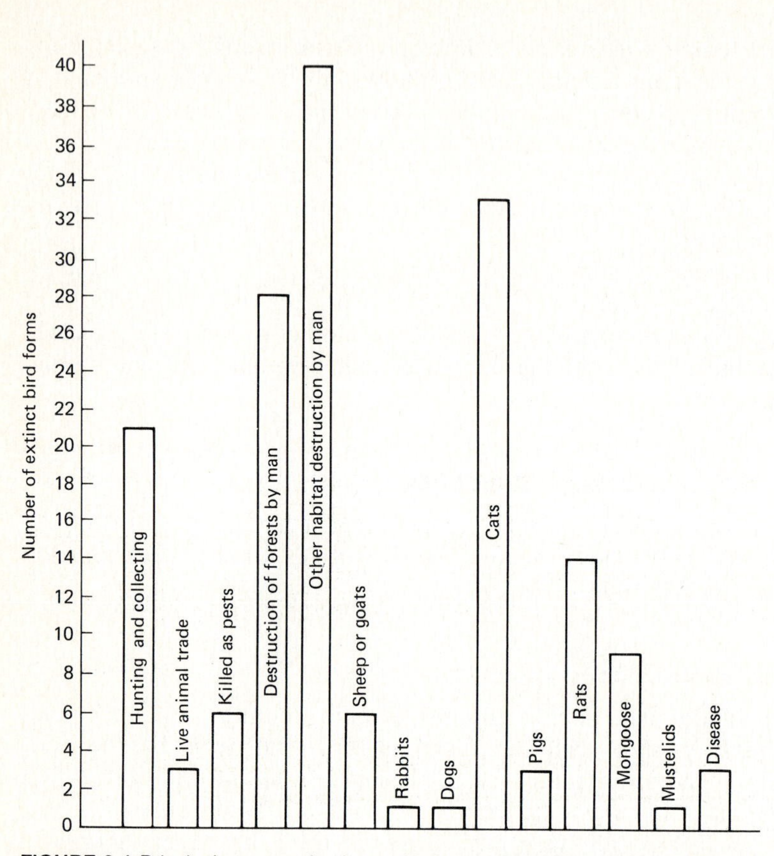

FIGURE 8-1 Principal causes of avian extinctions worldwide (after Jackson, 1978, based on data from Ziswiler, 1967 © Univ. of Wisc. Press and Springer-Verlag. Used by permission).

added. Roughly speaking, at least 40 to 50 percent of the birds captured die before they are exported (Domalain, 1977; Nilsson, 1981). In addition, many of the capture techniques for young nestlings require the cutting of valuable and increasingly rare nest trees, thus adding habitat damage to that imposed by direct removals. Finally, mortality rates remain high after export, with an estimated 12 percent overall mortality for the 509,874 birds arriving in the United States during 1980 (Nilsson, 1981).

Introduced Species

Birds dwelling on islands far from continents enjoyed a special luxury. They lived and reproduced in environments free of mammalian predators. There was no need for these birds to nest in trees or along cliffs so they simply laid eggs in exposed, accessible areas. So secure were these distant islands that some species, such as the dodo, completely lost the ability to fly.

Then the sailing ships appeared and their crews landed on the islands, searching for fresh water and food and claiming these islands for distant nations. No doubt they killed some of these birds to supplement salt pork and sea biscuits, but what really devastated the populations was what the mariners left behind. Some of the rats (*Rattus* sp.) that inevitably stowed away in the holds jumped ship. In some cases a few of the cats carried on board also remained behind. The cats and rats found the unprotected eggs easy pickings and so began the decline of many island-dwelling birds. Sailors also left domestic goats on many of the islands, hoping to exploit their descendants for fresh meat on future voyages. The goats became numerous and, as goats typically do, inflicted severe damage on the native vegetation. Such damage undoubtedly affected many native species indirectly through habitat changes.

Other introduced species contributed to the decline of insular birds. Mongooses (*Herpestes auropunctatus*), adaptable and agile predators, were sometimes introduced. Dogs became feral. Besides goats, other livestock damaged the natural vegetation, causing irrevocable habitat changes. There is even evidence that the accidental introduction of mosquitoes into Hawaii in 1826 contributed to the demise of several bird species by providing an agent for disease transmission. A testimony to the vulnerability of island species comes from the fact that of the 33 taxa of birds driven to extinction in the United States within historical time, 24 were Hawaiian (Fischer et al., 1969).

Environmental Contaminants

As recently as the 1950s, brown pelicans (*Pelicanus occidentalis*) were common sights along the Gulf Coast. A few years later they were almost extinct. Other predatory birds including the osprey and the peregrine falcon also declined sharply at about the same time. The timing provided one clue—the populations first began diminishing after the end of World War II, concurrent with widespread use of DDT and related pesticides to combat insects. Another discovery was large-scale nesting failure, found to be due to a sharp increase in the proportion of broken eggs. Experiments followed, demonstrating that DDT (or DDE, as its persistent form is called) had reached such high levels in these birds that it caused a hormonal imbalance, interfering with calcium metabolism and resulting in thinner eggshells. Shells became so thin that they broke under the weight of incubating parents and in a few cases eggs were produced with no shells at all.

The osprey, the peregrine falcon, and the brown pelican all had one habit in common—they lived at the end of long food chains. As DDE passed through the food chain it reached higher concentrations at each step. These top predators received the heaviest doses, almost causing their extinctions. Most species affected by DDT have increased in regions where the pesticide use has been curtailed or banned. But DDT is only one of thousands of synthetic chemicals entering natural environments in recent years. The effects of most of them remain unknown.

Habitat Destruction

The Kirtland's warbler (*Dendroica kirtlandii*) always nests in jack pines (*Pinus banksiana*) of about 1 to 3 meters in height. This small songbird found sufficient nesting habitat in Michigan each year before migrating to winter in the Bahamas. But jack pines cannot regenerate without fires. Since most modern forestry practices in the region exclude fire, young jack pines of the right size for nesting have become very scarce. Only by setting aside special reserves of jack pine have conservationists been able to save this bird from extinction.

Habitat changes were once a relatively minor factor in the decline of species, being overshadowed for centuries by excessive exploitation and introduced predators. During the twentieth century the relative importance of habitat changes began to shift until in the last decade they have become clearly the most severe threat to wildlife survival. The greatest problems lie in tropical regions, where conditions are so serious that they require a complete change in management strategy.

AGENCIES INVOLVED

In the United States, state wildlife agencies have played relatively minor roles in the preservation of endangered species, partly because of their problems in obtaining funds for nongame projects. Some states, however, have passed their own endangered species acts, making their wildlife departments eligible for federal matching funds to aid preservation.

The U.S. Fish and Wildlife Service, operating under authority from the various Endangered Species Acts of 1966, 1969, and 1973, has assumed major responsibilities for endangered species. A laboratory for research and captive propagation of endangered species has been established at Pautuxent, Maryland. Under the authority of the Secretary of the Interior, the U.S. Fish and Wildlife Service has maintained and published periodically in the *Federal Register,* an official list of threatened and endangered species. The service appoints recovery teams of scientists and administrators to advise in matters related to management or recovery programs. Importation of species officially recognized as endangered is severely restricted and enforcement of federal endangered species laws is a responsibility of the U.S. Fish and Wildlife Service.

Alarmed by threats posed to wildlife by the international animal trade, a number of nations signed the Convention on International Trade in Endangered Species (CITES) beginning in 1973. Each signatory is required to establish its own agencies to regulate import and export of threatened species or parts thereof. In the United States, the regulations are set by the Endangered Species Scientific Authority (later known as the Office of Scientific Authority, or OSA), a semiautonomous federal agency comprised of seven members from other federal agencies. OSA has authority to regulate exports of all threatened species and imports of species endangered abroad. The U.S. Fish and Wildlife Service is charged with enforcement of OSA regulations.

There are several important international organizations concerned with endangered species preservation. One of these is the International Union for Conservation of Nature and Natural Resources (IUCN), a private organization based in Gland, Switzerland. The IUCN has for years published the *Red Data Book,* a collection of volumes each representing rare and endangered members of a particular taxonomic group. The *Red Data Book* continuously updates, lists the status, reasons for decline, and management measures proposed or undertaken for each species. The IUCN also promotes the cause of endangered species preservation, provides guidelines for establishment and operation of national parks and reserves, and disseminates information through its newsletters.

Closely affiliated with the IUCN is another private organization, the World Wildlife Fund. Supported by private contributions, the World Wildlife Fund furnishes direct financial support for research and restoration involving endangered species.

The United States, through its Educational, Scientific, and Cultural Organization (UNESCO) has become involved in preservation of endangered species. UNESCO's Biosphere Reserve Program, for example, began establishing reserves representative of the earth's various natural communities.

STATUS CLASSIFICATIONS

There are two basic levels at which disappearing species are commonly classified. A species is endangered if it is in imminent danger of extinction. Less critical, though still serious, are species that are threatened or rare. Rare or threatened species are in no immediate danger of extinction but have declined enough to merit close attention. These last two terms are often used interchangeably, but since the 1973 Endangered Species Act, "threatened" has been used more commonly than "rare."

These classifications can be applied to either species or subspecies. Among conservation agencies, endangered species justify a higher level of concern than do endangered subspecies, provided that other subspecies exist safely elsewhere.

For classifying species, OSA uses the criteria developed by CITES. Appendix I lists species threatened with extinction and actually or potentially threatened by international trade, a status corresponding approximately with "endangered." Both import and export of Appendix I species are highly restricted. Species listed in Appendix II are not immediately threatened by extinction but may become so unless trade is closely regulated, a status roughly equivalent to "threatened" or "rare." Exports of Appendix II species are regulated. In the United States, permits to export Appendix II species or to import Appendix I species must be granted by the Federal Wildlife Permit Office under advisement by OSA.

There are no fixed criteria with which to decide whether a particular species is safe, threatened, or endangered. Because of the many biological and economic variables involved, the best that can be hoped for is a set of useful guidelines. One such set (see Table 8-1), designed to help set management priorities, listed four classes of guidelines: population status, vulnerability, recovery potential, and special attributes (Sparrowe and Wight, 1975).

SINGLE-SPECIES APPROACHES

Wildlife management emerged largely as a science applied to solving problems of one species at a time. This single-species approach was quite naturally carried over in North America, Europe, and elsewhere to try to save endangered species. Some species clearly benefitted. The whooping crane and Kirtland's warbler are examples of species saved from extinction by single-species management.

Survey, Protection, Reserves

Once conservation agencies or organizations suspect that a species is having trouble, a series of steps can be taken. The first step is usually a status survey to try to establish more clearly the actual remaining numbers, geographic distribution, and causes of the decline. Legal protection generally follows if the status survey shows that the species is endangered or threatened. This should be succeeded by the establishment of suitable

TABLE 8-1
SOME CRITERIA FOR EVALUATING STATUS OF ENDANGERED SPECIES
(After Sparrowe and Wight, 1975 © Wildlife Management Institute. Used by permission.)

Criteria
Population status
Index of population size
Index of population trend
Vulnerability
Amount of habitat reduction
Rate of reduction of remaining habitat
Population concentration and spatial distribution
Reproductive rate
Unusual mortality factors
Recovery potential
Status of protected habitats
Successional stages of available habitat
Potential for population growth
Species characteristics
Taxon level threatened (species, subspecies)
Hybridization threat
Isolation from other populations
Ecological specialization
Taxonomic uniqueness
Security of related species and subspecies

reserves of critical habitat. If these steps are carried out soon enough and with sufficient vigor, most species can be saved in the wild.

Captive Propagation

A comparatively new method of preserving endangered species that is gaining wide attention is captive propagation. In fact, many of the larger zoological parks are devoting more and more space and facilities to breeding species that are threatened or endangered in the wild. This propagation is being done with the hope of eventually reintroducing descendants of the captive animals into the wild. Some rather impressive results have been obtained in species ranging from the bison to the peregrine falcon. Larger zoos are now seeing themselves as "gene banks" for perpetuation of wild animals otherwise doomed to extinction.

Captive propagation is quite appealing to many people. Zoos welcome the opportunity to serve in the fight against extinction, and they appreciate the favorable publicity received from such efforts. Conservation agencies, both national and international, are pleased to have qualified zoos to help propagate seriously endangered fauna as a last ditch option. The viewing public gets a chance to see some very rare creatures, at least when public displays do not interfere with breeding success.

Captive animals are shielded from predation, competition, severe weather, many diseases and parasites, and hunting. Thus, mortality can be drastically reduced below natural levels. Fecundity rates can, under certain conditions, be raised through the use of manipulative techniques. "Double-clutching," for example, has been used successfully to increase fecundity in many species of birds. Zoologists simply remove the first clutch of eggs and use artificial incubation to hatch them. Meanwhile, removal of the first clutch stimulates the birds to lay another. This process can double or even triple fecundity (Conway, 1980a). Obviously, captive propagation can help build up endangered species populations quite rapidly, a highly desirable accomplishment in the perpetuation of vanishing wildlife.

Yet captive propagation alone has some severe drawbacks. In view of available space and monetary support, breeding endangered animals in captivity will be limited mainly to "showcase species." There are about 4,100 species of mammals on earth, yet Campbell (1980) estimated that space would allow only 100 to 150 species to be maintained for captive breeding in all the zoos in the United States.

Moreover, the costs to propagate some species are truly astronomical. Against all odds peregrine falcons have been bred in captivity, thanks to the heroic efforts of Tom Cade of Cornell University and some of his associates. Falcons have been reared in captivity at Cornell and more recently at Fort Collins, Colorado, and Santa Cruz, California. But the cost by 1982 was about $1,500 to $3,000 for each young falcon reintroduced into the wild, and the annual budget had grown to some $600,000 (Harwood, 1982b).

Large mammals can be even more expensive to propagate in captivity. William Conway and George Rabb calculated that the annual cost of feeding the 750 Siberian tigers living in captivity is $1.2 million. Adding veterinary, curatorial, and maintenance costs, the upkeep on these tigers totals at least twice the food cost, bringing the bill to $2,432,000 per year (Conway, 1980b).

One of the most serious and persistent problems with captive propagation is the loss of genetic variability. For practical reasons, the number of "founders" taken from the wild to begin a captive population must be small, especially for endangered species. This small number can at best contain only a portion of the species' remaining gene pool, a restriction known as the "founder effect." Thus, the ability of the species to adapt to future conditions is impaired as its genetic options are lost.

Once the founders begin to breed, another potentially more serious problem emerges. Unless the facilities are sufficiently great to sustain large numbers of animals per generation, the propagated species, already suffering from the founder effect, begins to lose more genetic diversity through inbreeding. Zoos typically have precious little extra space and are often forced to keep the numbers per generation low. After a few generations, reproductive success may decline markedly, even critically, as inbreeding depression (see Chapter 3, Ralls et al., 1979; Senner, 1980) sets in.

After simulating the effects of inbreeding depression using an elaborate computer model, Senner (1980) concluded that maintenance population size was much more critical in retaining genetic variability than was the size of the founding population. This might explain why many wildlife biologists have been reluctant to accept the existence of inbreeding problems in wild populations. They have seen hundreds of successful transplants of reintroduced game species, beginning with perhaps a half-dozen individuals, and watched while their populations grew and became established without evident difficulty. But these founder groups were unconfined, and habitat conditions and legal protection allowed them to "breed up" to high numbers rapidly. The greater the population size, the lower the rate of inbreeding per generation.

Another genetic problem stems from selection, either deliberate or subconscious, on the part of those charged with captive propagation. Any selection increases the risk of permanent genetic loss. Behavioral traits which are determined at least partially through genetics are particularly susceptible to selection (Kleiman, 1980). For example, individual animals that are particularly fearful of or aggressive toward humans are not likely to be retained for breeding purposes.

Other behavioral problems can interfere with captive propagation. Two animals may be ideally suited for mating from a genetic standpoint yet be behaviorally incompatible. Hand-rearing, often practiced in conjunction with captive propagation, may ensure survival of genes that would have been selected against in the wild. Human interference, probably accidental, in normal mate selection and dispersal patterns (Kleiman, 1980) could disrupt these patterns, with deleterious results.

Since space and funding will not permit captive propagation of all endangered species, how should species be selected? Several sets of criteria have been proposed. Myers (1979) and Lovejoy (1976) have accepted the view that conservation cannot possibly save all endangered species. Myers proposed using the "triage." During World War I, battlefield surgeons could not possibly attend promptly to the needs of all wounded soldiers. So the physicians devised the triage scheme for the most efficient salvage job possible. Wounded were divided into three groups: those likely to die even after medical attention, those likely to survive even without medical aid, and those for whom medical treatment was likely to spell the difference between life and death. The limited medical personnel gave priority to the last group.

If a triage approach were devoted to the world's endangered species, those most critically endangered, such as the California condor (*Gymnogyps californianus*), would be left to their all but inevitable fate. Conservation agencies and private organizations would then be free to devote all their resources toward saving those species, such as the jaguar (*Felis onca*) or the Kirtland's warbler, for which the outlook is a bit brighter.

Perhaps a better approach, in part advocated by Sparrowe and Wight (1975) and by Conway (1980b), is to grant first priority to those endangered species which are monotypic or otherwise quite unique. This would clearly put a premium on preservation of natural diversity.

Most of the criteria described above apply to all methods of saving endangered species, not just to captive propagation. So attached to the other conditions should be those for which captive breeding is likely to make a real difference. For example, species endangered through severe, yet reversible, habitat loss would make good candidates for captive breeding. Another group might be species, such as the black rhino (*Diceros bicornis*), seriously threatened by poaching because of an unusually high market demand. In any case, captive propagation might well be the only realistic hope for survival in the short run. This characteristic, combined with reasonable prospects for availability of habitat in the future, should make captive propagation realistic as a management tool.

How long will a particular endangered species have to be kept in captivity? This question, posed more in terms of generations than in years, is critical. It is far easier to introduce first-generation captive-bred animals to the wild than it is to introduce animals maintained for generations in captivity. The more generations passed in captivity, the greater the loss of genetic variability and the greater the chances that inbreeding depression will arise.

Minimum Effective Population Size

In establishing a captive breeding population, especially if it is to be the last chance for an endangered species to survive, zoologists need to estimate the number of animals necessary. The size of the founder group is critical for two reasons. First, the population must be large enough so that the chances of complete extinction through a few ill-timed deaths are minimal. Second, the population must be large enough to maintain the maximum possible genetic diversity. For this latter category, mere numbers of animals within a captive breeding program are alone insufficient. This is because different species and different populations vary considerably in terms of sex ratios, breeding patterns, and population fluctuations. All these characteristics affect the gene pool of a population and, within a small, finite population, the rate at which genetic diversity is lost.

Some simple formulas allow estimation of effective population size (Kimura and Crow, 1963; Franklin, 1980). An important parameter is variation in progeny number per family unit. For monogamous species, this family unit would consist of a breeding pair. If the objective is to maintain the maximum possible genetic variability within a small, closed population, then all family units should contribute equally to the next generation. This arrangement minimizes the chances of losing certain genes through differential reproductive success. If carried out properly, allowing each unit to contribute equally will slow evolutionary change dramatically, virtually "freezing" evolution. By freezing evolution, zoologists would be ensuring perpetuation of the maximum number of genes, thereby granting the species the best chance of being able to adapt to the wild once again.

The formula for computing the effect of variance in progeny number upon effective population size is:

$$N_e = \frac{4N_a}{2 + \sigma^2}$$

where N_a is the actual population size and σ is the variance in the number of progeny produced. For example, if the variance is 3, the N_e equals four-fifths of the actual population. The effective population size for a population of 50 individuals would then be 40. On the other hand, if all family units contributed absolutely equally so that there were no variance, then N_e would equal $2N_a$. That same population of 50 animals could then be genetically the equivalent of 100.

Suppose that the sex ratio is unequal in a captive population. Alternatively, assume that in a polygamous species, only a small proportion of the males actually breed. In either case, the effective population size is reduced. This reduction is approximated by the formula:

$$N_e = \frac{1}{\frac{1}{(4N_m)} + \frac{1}{(N_f)}}$$

where N_m is the number of males breeding and N_f is the number of females breeding. If the sex ratio is exactly even and there is no differential reproductive success within either sex, then N_e equals N_a. Conversely, in a population of 50 animals with 10 males and 40 females, N_e would equal only 32.

Wild populations almost always fluctuate to some extent. Even those kept in captivity will likely undergo some changes. Any fluctuation in numbers between generations reduces the effective population size. The formula of estimating these effects is:

$$\frac{1}{N_e} = \frac{1}{t}\left(\frac{1}{N_1} + \frac{1}{N_2} + \ldots \frac{1}{N_t}\right)$$

which effectively reduces to:

$$N_e = \frac{t}{(1/N_1 + \ldots .1/N_t)}$$

A completely stable population would be one in which $N_e = N_a$. However, if a population maintained at 50 individuals for two generations suddenly plummeted to 20 for the third generation, the effective populations size at the time of the third generation would then be 33.3.

So, how many animals are enough to constitute a minimum effective population size? At present, zoologists have only very rough guidelines. If the objective is to avoid inbreeding depression for a few generations so that a population might some day be returned to the wild, then 50 is the best rule of thumb (Franklin, 1980). This estimate of effective (not actual) population size is inferred from livestock breeding records. As domestic stock have been subject to selective breeding for quite a long time, they may offer a misleading approximation, but it is the best available. Moreover, there is now evidence that wild species vary considerably in their tolerance to inbreeding (Ralls et al., 1979). Still, the population parameters discussed above should have about the same relative effects upon populations of different species, and thus should be followed as closely as possible.

Some Special Cases

The outlook for some endangered species has improved in recent years, thanks in part to newer and more innovative techniques. Here are some avian examples.

The Peregrine Falcon Breeding populations of the peregrine falcon declined throughout the United States and Canada and disappeared completely from the eastern United States by the late 1960s. Canada banned the use of DDT in 1970 and the United Stated followed with similar action 2 years later, giving wildlife managers reason to hope that the species could be restored. The problem was that the species was thought to be extremely difficult to rear in captivity and no one knew if captive-reared birds could really adapt to the wild and themselves reproduce there.

Researchers at the Cornell University Laboratory of Ornithology solved the first problem and have successfully bred hundreds of young peregrines since 1973. The next problem was getting the young birds introduced into the wild. Two methods were used successfully: hacking and augmentation of wild nests. Hacking consists of caring for young at selected sites to help them adjust gradually to the wild. Typically the young falcons are placed in "hacking boxes" a few days before fledging, where they are furnished with food in such a way that they will not associate it with humans. Augmentation is done by placing young 2- to 3-week-old captive hatched falcons in the nests of wild peregrines. Experiments in two falcon nests in Colorado resulted in the restoration of five young peregrines where no more than one would have been produced naturally (Burnham et al., 1978).

Thanks to these captive breeding and reintroduction techniques, wild peregrine falcons in the eastern United States produced the first young in that region in 1980. Two years later there were 10 breeding pairs in the east, and the outlook for further restoration is quite good (Harwood, 1982b).

Sandhill Cranes as Foster Parents The whooping crane has long been the symbol of America's endangered wildlife. This large, conspicuous bird has a very low reproductive rate, and only through painstaking protective efforts have its numbers grown from 15 in 1941 to over 70 by 1977. Furthermore, whoopers stay together, breeding in Wood Buffalo National Park, Alberta, and wintering at the Aransas National Wildlife Refuge, Texas. This cohesiveness leaves the whole population vulnerable to disasters such as hurricanes or environmental contaminants.

To establish a second wild population, the U.S. Fish and Wildlife Service has begun an elaborate program of placing whooping crane eggs in the nests of sandhill cranes farther west. This technique, called "cross-fostering," depends upon the foster parents' acceptance of a different species' young. The sandhills accepted the young whoopers. Although mortality of eggs and young birds has been high, at least a half-dozen or so whoopers now migrate with their foster parents from the breeding grounds at Gray's Lake National Refuge (Zimmerman, 1978). The real test will be whether the whoopers mate among themselves or attempt to mate with sandhills. As they approach sexual maturity, the whoopers have shown an increasing tendency to segregate, a good sign that a second breeding population of wild whoopers will eventually be established.

The Trumpeter Swan Only rarely does an endangered species increase in numbers to the point at which it is no longer endangered. Such a pleasing change took place for the trumpeter swan (*Cygnus buccinator*). In 1932, the known population of trumpeters was only 69 (Bellrose, 1976; Harwood, 1982c) and the species was thought to be critically endangered. By 1980, the total population was known to be 10,000 (Harwood, 1982c).

Management can claim only a small part of the credit for this change in status. Some national wildlife refuges have undoubtedly helped, particularly the Red Rock Lakes National Wildlife Refuge in southwestern Montana. Two warm springs keep ponds open year-round, thus permitting the trumpeters to stay for the winter. Numbers at Red Rock have in recent years ranged from 200 to 300 (Bellrose, 1976; Harwood, 1982c). More important, the refuge has served as a source for trumpeters to be transplanted to other protected areas, where they have since become established.

Meanwhile, other populations of trumpeter swans were discovered in more remote parts of south central Alaska. (Although swans had been known to nest in such regions for a long time, they had been presumed to be the more common whistling swan (*Cygnus columbianus*). The tally of trumpeters in Alaska was known to be 7,696 by 1980 (Harwood, 1982c). More populations have been discovered in Alberta, Saskachewan, the Yukon, and British Columbia, and more trumpeters undoubtedly await discovery.

ECOSYSTEM MANAGEMENT

While single-species approaches will no doubt continue to be important in assuring survival for many species, they are inadequate as a management strategy in the tropical, so-called underdeveloped countries. There the tide of human activity is fast colliding with diverse, fragile ecosystems. By some estimates (Eckholm, 1978; Myers, 1976) only a small fraction of the species in tropical forests have ever been classified, much less studied in even the most rudimentary ways. The complexities of tropical ecosystems make predictions of the effects of disturbance a guessing game. Most of all, time is running out. Some projections (e.g., Myers, 1979; Barney, 1980) indicate clearly that the fate of many, perhaps most, tropical species will be decided by the end of the twentieth century.

The only real hope for preserving species under such conditions is through preservation of large reserves representing the different biogeographical provinces. An effort toward saving representatives of the world's biogeographic provinces has begun through the United Nations Educational, Scientific, and Cultural Organization's (UNESCO) Biosphere Reserve Program. Under this program, protected areas are managed by the countries in which they occur. By late 1979 there were 162 protectorates in 40 nations, with many more planned (IUCN, 1980).

In establishing biological reserves, conservationists are faced with difficult practical problems. How large must reserves be to ensure that most species will survive? What guidelines are available regarding shape and proximity of parks and other reserves? One promising set of guidelines comes from island biogeography or, as its applied version is known, insular ecology.

Insular Ecology

Traditionally, wildlife conservationists have established refuges, parks, and other nature reserves when and where they could. There were no guidelines for minimum size or shape or the proximity to other protected places. More important, once protected areas were established, most conservationists assumed that the species contained within them were forever safe from extinction, barring special intrusion such as poaching. The truth is that species often disappear from protected reserves, even though the actual causes are not readily apparent.

The number of species that a reserve can hold and the rate at which species are lost are now at least partially predictable. Predictions come from the theory of island biogeography (MacArthur and Wilson, 1963, 1967). Island biogeography originally attempted to answer the question: "Why do some islands have so many more species than others?" MacArthur and Wilson concluded that the size of any given island is a major factor, as is its proximity to the mainland or to other islands. Smaller islands have higher extinction rates, while islands closer to mainlands have greater colonization rates. Eventually, an equilibrium is reached between extinctions and colonizations, setting the maximum number of species that may be carried on any given island.

As humans continue to modify drastically most of the earth's surface, the few natural areas remaining will become more and more like islands in a sea of agriculture, intensive forestry, and other land uses. Theories of island biogeography can, along with empirical evidence, furnish a useful basis for estimating the necessary size, shapes, and proximities of protected natural areas.

Size Based upon studies of species diversity, principally among birds, on islands of different sizes, biologists have found that the ability of an island (natural or controlled) to hold species can be calculated through the following formula:

$$S = CA^z$$

Where S = number of species
C = a constant for a given species group in a given community
A = area
z = power function relating species diversity to area

C is not especially important for comparing the effects of size within a given community type. The really critical value is z. Although z values have been shown to range from 0.18 to 0.48 (Wilcox, 1980), the overall average is between .27 and .30. If, for simplicity, C is assumed to be 1.0, and z is 0.30, the relative holding power of different sizes of reserves can be estimated by the formula. Hence, for a reserve of 250 square kilometers

$$S = 1 \times 250^{.30} = 5.24$$

If the reserve were 2,500 square kilometers, then

$$S = 1 \times 2{,}500^{.30} = 10.46$$

According to the formula, then, a tenfold increase in area is necessary to double the holding power of a biological reserve.

The formula also provides guidelines for the total proportion of a biogeographical province that must be protected to ensure survival of a given proportion of species. One begins with the rather conservative view that if 100 percent of the natural habitat were conserved, 100 percent of the species would survive (other than natural extinctions) indefinitely. Then, applying the formula shown above, 50 percent of the area preserved would permit survival of 81 percent of the species, 10 percent would allow 50 percent, and 1 percent would grant survival to 25 percent. As the curve in Figure 8.2 shows a sharp rise of between 10 to 20 percent of the land area preserved, this range may be close to optimum.

The guidelines from insular ecology are quite crude, but they provide the best available estimates. By the mid-1970s, just over 1 percent of the earth's land area was being effectively protected by parks and reserves formally recognized by the United Nations–IUCN list (Myers, 1979). Apparently, at least ten times that area, distributed more evenly than at present, may be needed to save even half the world's wild species.

The land bridge islands such as Sumatra, Borneo, Bali, Java, New Guinea, Trinidad, and Tobago lie close to continental mainlands and are separated by fairly shallow waters. During the Pleistocene, when much of

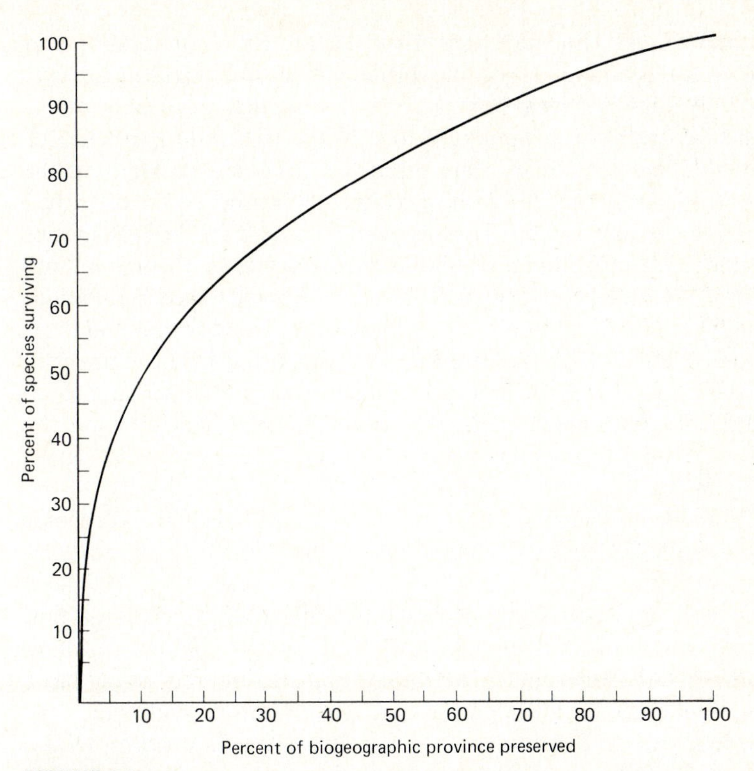

FIGURE 8-2
A theoretical relationship between the percent of a particular biogeographical province preserved and the percent species saved (see text for explanation).

the earth's water was tied up in glaciers, these islands were connected to the mainland. Ten thousand years or so ago the glaciers subsided and sea levels rose, cutting the islands from the mainlands. These land bridges thus act as a 10,000-year experiment in the effects of insularization. When land bridges existed, the faunal diversity on the archipelagos was approximately the same as that on mainlands. Because the geological record clearly provides a reasonably accurate estimate of the time since separation, scientists then can measure the rate of species loss after insularization.

But do such estimates furnish valid models for establishing biological reserves? If they do, then most of the earth's parks and reserves are too small and many, perhaps most, of their species will eventually disappear. The empirical evidence so far suggests that small, isolated reserves lose species rather rapidly. Perhaps the best example is Barro Colorado Island. Located in Gatun Lake, Panama, Barro Colorado Island was, until 1914, merely rain forest atop a mountain. Gatun Lake was constructed to provide water for the Panama Canal. By 1923, the 15.7–square kilometer (6 square mile) island had become protected as a biological reserve. Scientists, including the ornithologist Frank Chapman, carefully determined the bird diversity, beginning in the 1920s. By 1970, 45 of the 208 species of breeding birds had disappeared from the island. Many of the disappearances clearly resulted from successional advance as the disturbed areas grew into mature forests. But between 13 and 18 of the lost species were known to be inhabitants of undisturbed rain forest. Moreover, no "new" forest species have colonized Barro Colorado since separation, even though some occur as close as .5 kilometers away on the mainland (Wilson and Willis, 1975). A careful reassessment of avian extinctions indicated that the real extinction rate was higher than previously supposed. Only about half the number of species were caught in mist nets on the island as were caught in similar habitat on neighboring mainland (Karr, 1982).

Other examples come from East Africa. The 3,276–square kilometer (1,265 square mile) Mkomazi Game Reserve in northern Tanzania contained 43 species of large mammals in 1952. Twenty-five years later, 4 of them had disappeared. From this rate of decline, Miller (1978) calculated that the reserve would eventually lose up to 21 species before reaching equilibrium, about 300 years after separation. Using the same basic approach, Soule et al. (1979) estimated that Nairobi National Park (Kenya) would lose over 80 percent of its species within a few centuries and that Serengeti National Park (Tanzania) would lose about 70 percent of its species within 1,500 years.

Once an island or biological reserve becomes isolated, the species are not lost at random. On Barro Colorado Island, for example, most of the forest-dwelling species that disappeared were those that nested or fed upon forest floors or those that were very large species for their particular feeding association or guild (see Table 8-2). Actual causes of disappearance of ground-living species were unclear. Larger members of guilds, however, are always vulnerable, as they typically exist at lower population densities and, within any given area, are more likely to suffer extinctions.

Different vertebrate classes fare differently with insularization. Other than bats, mammals generally do poorly, particularly among the larger species. Their greater metabolic requirements, the resulting limits to population density, combined with limited dispersal work against many mammals. Birds also suffer declines due to their metabolic requirements but can offset this by their greater dispersing abilities, enabling recolonization. Amphibians and reptiles, with their much lower metabolic demands, should probably do much better than mammals. Wilcox (1980) compared these vertebrate groups on a series of West Indian islands and derived z values of .48 for nonflying mammals, .38 for reptiles and amphibians, and .24 for birds and bats.

TABLE 8-2
CHARACTERISTICS OF ENDANGERED OR "EXTINCTION-PRONE" SPECIES AND RECOMMENDATIONS FOR THEIR CONSERVATION
(After Terborgh, 1974 © American Institute of Biological Sciences. Used by permission.)

Characteristics	Recommendations
Top trophic level and largest members of guilds	Protect from excessive hunting; establish large parks
Species with poor dispersal and colonization ability	Establish medium-to-large parks; provide maximum diversity of habitat types
Continental endemics (unusually restricted distributions)	Identify pockets of endemism; define habitat needs; establish small, well-located reserves
Insular endemics	Establish reserves; prohibit or limit exotic introductions; control or eradicate feral predators and competitors
Colonial nesters	Protect nesting grounds; remove predators from offshore islands
Migratory species	Preserve adequate breeding and wintering ranges; establish international cooperative programs where appropriate

Land bridge islands smaller than about 2,500 square kilometers (965 square miles) contain substantially fewer species than do larger islands. Below about 250 square kilometers (9,650 square miles), land bridge islands have no more species than oceanic islands that never had a land bridge (Terborgh, 1974).

Shape and Proximity As biological reserves and parks become increasingly like islands in a sea of intensive land use, questions of shape and the proximity of these islands to one another become crucial (see Figure 8-3). Both of these criteria are important because they determine chances of recolonization. Should a given species become extinct in one portion of a large reserve, it could easily recolonize that portion from populations in other regions. The smaller the reserve, on the other hand, the greater the likelihood of extinction for any given species.

If it is feasible to establish several smaller reserves, the closer they can be to one another, the greater the chances of recolonizing from one to another. Unfortunately, the species most likely to become extinct throughout a series of unconnected reserves are often those with poorest dispers-

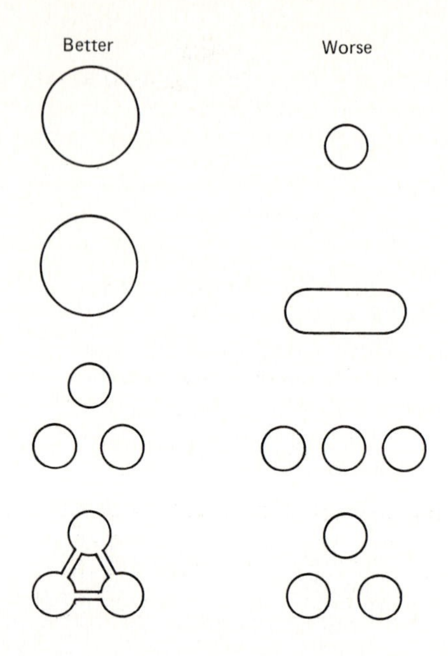

FIGURE 8-3
Ideal shapes and proximities for biological reserves. (Diamond, 1975 © Elsevier Applied Science Publishers, Ltd. Used by permission.)

ing abilities. For this reason, corridors of natural habitat connecting the different reserves are recommended to facilitate dispersal. Many birds inhabiting tropical rain forests, for example, will not disperse across expanses of open habitat. In some cases, a mere dirt road may serve as a barrier against dispersal.

The ideal shape for a biological reserve is circular. Such a shape minimizes dispersal distances within it. As shown in Chapter 2, the circle is the geometric figure with the least edge per unit area. Since threats and disturbances to the biota of a reserve typically come from outside it, a circular shape minimizes the zone of contact. This tenet, of course, is quite the opposite of the much-touted edge effect in wildlife management. But a moment's reflection should clarify the apparent contradiction. Managers who maximize edge generally have as their objective improving habitat conditions for resident game species. Most of these "edge" species are fairly *r*-selected and do well in disturbed early-successional conditions.

The management of biological reserves and parks has quite a different set of objectives. Reserves protect maximum diversity rather than produce maximum harvests. Those species typically most restricted and endangered are *K*-selected, and the greater edge created by increased disturbance is about the last thing they need.

Some Conclusions on Insular Ecology

The guidelines from island biogeography are not universally accepted (cf. Simberloff and Abele, 1975, and rebuttals by Diamond, 1976, Terborgh, 1976, and Whitcomb et al., 1976). Critics are reluctant to accept a theoretical basis for conservation without greater empirical support. The size estimates generated by insular ecology are, some feel, too large to be either biologically realistic or politically and economically acceptable. In particular, Simberloff and Abele (1975) argued that several small reserves could contain more species than a single large one of equivalent area. Diamond (1976) countered this point by stating that the primary objective is not to maintain the greatest number of species but rather to maintain the greatest number of species otherwise doomed to extinction. If this indeed is the case, a single large reserve would likely do a better job of preserving these "extinction-prone" species.

Nevertheless, two criticisms of insular ecology raised by Simberloff and Abele are valid. One was the danger that a disaster could wipe out an entire single reserve, analogous to that of putting all the eggs in one basket. Another criticism, implied in this case, was that superior competitors within a single reserve might eventually eliminate inferior ones (Diamond, 1976). Several smaller reserves might allow more competitors to persist because competitive outcomes might differ between them.

Despite these problems, insular ecology now offers the best set of systematic guidelines for establishing biological reserves. No doubt the basic model will become more refined as more empirical evidence becomes available. Insular ecology will likely become an increasingly important basis for wildlife management.

Although community or ecosystem approaches to conservation must of necessity assume a greater role than single-species approaches, they are unlikely to displace single-species management completely. Under the best of conditions, biological reserves will not become large enough nor well distributed enough to function passively. Larger mammals, in particular, as well as other "showcase" species among extinction-prone types, will have to be aided. For such species, active reintroduction programs (preferably from wild-caught stock) will remain necessary both to guard against extinctions in individual reserves and to ensure against inbreeding depression. Captive propagation will, in the short run, be the only hope for some highly endangered species found temporarily without suitable habitat.

Besides furnishing guidelines for biological reserves in the rapidly developing tropics, applied insular ecology has found other uses. Two such uses include wilderness wildlife management and its antithesis, urban–suburban wildlife management.

Wilderness Wildlife Management

One of the areas in which wildlife management emphasizes natural diversity as an objective is in the maintenance of wilderness areas. Wilderness areas are protected from nearly all forms of human destruction, particularly in terms of habitat disturbance. As such they serve as biological reserves for all sorts of organisms, including those commonly called "wildlife."

Definitions of wilderness vary widely, but they generally stipulate that wilderness must contain no permanent human inhabitants, be of a certain minimum size or larger, and have few, if any, signs of human activity. The Wilderness Act of 1964 in the United States, for example, specified that a wilderness area must be of at least 1,930 hectares (5,000 acres), roadless, and devoid of human inhabitants. Most potential wilderness (areas meeting the basic criteria yet not yet formally recognized as wilderness areas by Congress) lies in the vast domain of the Bureau of Land Management (BLM). Of the more than 450 million acres of land administered by the bureau, nearly 90 million acres qualify for wilderness status. None of the BLM lands have yet received wilderness protection.

The U.S. Forest Service administers the greatest amount of wilderness, with about 4.9 million hectares (12.6 million acres) under wilderness act protection as of 1978 (Hendee and Schoenfeld, 1978). Another 16.4 million hectares (42 million acres) are potential wilderness areas but have not yet been officially designated. The National Park Service and the U.S. Fish and Wildlife Service manage a combined total of just under 780,000 hectares (2 million acres) of official wilderness. The U.S. Fish and Wildlife Service also maintains nearly 5.5 million hectares (14 million acres) of potential wilderness.

Leopold (1949) was one of the earliest advocates of wilderness. He championed the establishment of wilderness areas for three basic reasons. Wilderness areas were necessary for certain types of outdoor recreation. In Leopold's time such recreation included mainly pack trips and canoe trips. More recently these activities have been joined by armies of hikers and backpackers who take advantage of new lightweight foods and equipment to travel deep into the wilderness on foot.

Leopold also urged wilderness areas for science. Because they were regions least disturbed by human beings, wilderness areas served as a "control," against which the effects of human activities elsewhere could be compared. The justification has, if anything, increased along with levels of human disturbance outside wilderness areas in recent years.

Last, but not least, Leopold championed wilderness for species of wildlife unable to survive elsewhere. He recognized clearly that species requiring late-successional conditions or else inhabiting fragile environments were doomed without wilderness. Likewise, large predators persecuted for destroying livestock needed the protection afforded by large tracts of undisturbed land.

Time and events have shown that Leopold's advocacy was justified. Today the justification for wilderness areas is even greater than it was then.

Yet wilderness wildlife management presents special problems. One thorny question is whether or not sport hunting should be permitted in keeping with the pristine and natural conditions of wilderness. In practice, the National Park Service prohibits hunting on its wilderness areas just as it generally does on all its other lands. The U.S. Forest Service, on the other hand, generally allows hunting on its wilderness in accordance with state laws. However, use of mechanized vehicles is prohibited, so that anyone hunting in those areas must be willing and able to pack the game out on foot or by horseback.

Considering hunting and other potentially controversial issues, Schoenfield and Hendee (1978) have proposed a list of general guidelines for wilderness wildlife management:

1 Seek natural distributions, numbers, and interactions of native wildlife.

Allow natural processes to control wilderness ecosystems and wildlife.

2 Keep wildlife wild, with behavior altered as little as possible by human influence.

3 Permit viewing, hunting, and fishing where such activities are biologically sound and legal.

4 Whenever appropriate, favor preservation of rare or endangered species.

5 Seek the least possible degradation of qualities that make wilderness-solitude, and absence of human activities.

A list of wilderness-dependent species appears in Table 8-3. Notice that species often associated with wilderness, such as mule deer and elk, are not wilderness-dependent. They, like many other wild species, can thrive in disturbed areas as well as in wilderness.

Insular ecology also applies to wilderness areas. One investigation by Picton (1979) compared occurrences of 10 species of large mammals on 24 mountain ranges of varying sizes in central Montana. Picton found a definite species–area relationship which had increased with human disturbance. The historical distribution revealed a z value of only about .15, showing quite weak insularity for large mammals. Between about 1880 and 1920, when large mammals were most severely exploited, the z values reached a high of .39 for larger ranges. Restocking along with improved law enforcement and better cooperation among hunters allowed the large mammals to become partially restored, with z values of .25 to .29.

TABLE 8-3
WILDERNESS-DEPENDENT AND WILDERNESS-ASSOCIATED WILDLIFE FROM PANHANDLE NATIONAL FOREST, IDAHO
(Hendee and Schoenfeld, 1979 © The Wildlife Society. Used by permission.)

Wilderness-dependent	Wilderness-associated
Grizzly bear (*Ursus arctos*)	Black bear (*Ursus americanus*)
Canada lynx (*Lynx lynx*)	Marmot (*Marmota* spp.)
Wolverine (*Gulo gulo*)	Pika (*Ochotona princeps*)
Fisher (*Martes pennanti*)	Mule Deer (*Odocoileus hemionus*)
Pine marten (*Martes americana*)	Elk (*Cervus elaphus* or *canadensis*)
Mountain caribou (*Rangifer caribou*)	Moose (*Alces alces*)
Mountain goat (*Oreamnos americanus*)	Mountain sheep (*Ovis canadensis*)
Richardson's Blue Grouse (*Dendragapus obscurus*)	Clark's Nutcracker (*Nucifraga columbiana*)
Bald eagle (*Haliaeetus leucocephalus*)	Gray jay (*Perisoreus canadensis*)
Northern white-tailed ptarmigan (*Lagopus leucurus*)	

Another interesting finding from the central Montana study was that the z values tended to increase abruptly for areas larger than about 710 square kilometers (274 square miles). This shift implies a threshold effect, with mountain ranges of 710 square kilometers or greater being large enough to withstand human impact more effectively.

Recreational use of wilderness areas has been increasing sharply in the United States since the early 1960s. Some wilderness enthusiasts as well as professional researchers in the field have recommended use of strict quotas to regulate the impact of recreationists upon wilderness areas. One of the most critical issues in wilderness wildlife management will be the extent of human activities in those areas and their effects upon animals that require wilderness for survival.

Urban Wildlife Management

Natural diversity is also a desirable wildlife management objective within the expanding limits of municipalities. Until the 1970s, little thought was given to the deliberate management of wildlife within towns and cities. But by the mid-1970s, 7 out of 10 Americans lived in cities of 50,000 or more; by the year 2000, at least 90 percent of Americans will be residing in cities which will cover only about 2 percent of the nation's land area (Allen, 1974). Energy shortages may curtail peoples' abilities to travel away from cities. All of which means that efforts to perpetuate wildlife within towns and cities will become increasingly justifiable in the future.

Most wildlife within cities occurs within parks, greenbelts, and cemeteries. Quite often, a great many more species live within the limits of our larger cities than people realize. This fact was clearly illustrated by studies of "cemetery ecology" by Jack Ward Thomas and his colleagues. Based in urban Massachusetts, Thomas and friends sought nearby natural areas within which to have lunch. The nearest thing to a convenient natural area turned out to be a local cemetery. While the biologists enjoyed their lunches, they noticed that quite a large array of wild species inhabited the cemetery; they set about to learn more about these species. Their studies led them to the greater Boston area, where cemeteries constituted some 35 percent of the available open land. After walking some 200 miles of transect lines in Boston-area graveyards, Thomas and Dixon (1974) tallied 95 species of birds, nearly 1,200 nests of at least 34 species, 20 species of wild mammals, and 12 species of reptiles and amphibians. The species included such "rural" animals as great blue heron (*Ardea herodias*), sharp-shinned hawk (*Accipiter striatus*), bobwhite quail, ring-necked pheasant, raccoon, red fox, cottontail rabbit, muskrat, opossum, garter snake (*Thamnophis*), plus several species of turtle, frog, and salamander.

Despite these impressive occurrences, the fact remains that urban and suburban environments do generally support fewer species than do nearby rural habitats. (Important exceptions occur whenever cities and suburbs are surrounded by large tracts of agricultural monocultures. See Chapter 10.) Moreover, the changes in fauna that take place with development usually involve shifts from more desirable species toward less desirable or pest species. This change is best illustrated by avian studies. Table 8-4 shows the changes in index values for bird species in Maryland before and after suburban development. Such species as the mourning dove (*Zenaida macroura*) declined sharply; others, such as the starling, increased dramatically. Although interpretation of the significance of these changes necessarily involves some value judgment, most people would probably rather live around doves and thrushes than around starlings and house sparrows (*Passer domesticus*). Similar studies, done in other parts of the United States confirm this general trend, with a reduction in overall species numbers and a domination by omnivorous species and those adapted to nesting within human-built structures (Davis and Glick, 1978). Records of birds nesting within Cambridge, Massachusetts, showed a decline of 65 percent between 1860 and 1964. This decline followed a reduction of some 68 percent of the native vegetation within the same area (Walcott, 1974).

The biological characteristics of species that disappear with development are quite consistent. Broad-niched, highly competitive species tend

TABLE 8-4
CHANGES IN BREEDING BIRD POPULATION INDEX VALUES BEFORE AND AFTER SUBURBAN DEVELOPMENT NEAR COLUMBIA, MARYLAND
(After Geis, 1974 Used by permission.)

Species	Before development	After development
Decreasing		
Bobwhite quail (*Colinus virginianus*)	37.0	9.0
Mourning dove (*Zenaida macroura*)	17.0	4.3
Wood thrush (*Hylocichla mustelina*)	16.0	2.3
Eastern meadowlark (*Sturnella magna*)	16.0	2.3
Red-winged blackbird (*Agelaius phoeniceus*)	9.0	1.3
Indigo bunting (*Passerina cyanea*)	14.0	2.3
Grasshopper Sparrow (*Ammodramus savannarum*)	15.0	2.0
No change		
Cardinal (*Richmondena cardinalis*)	4.0	5.0
Increasing		
House wren (*Troglodytes aedon*)	0.0	2.7
Mockingbird (*Mimus polyglottos*)	1.0	10.0
Starling (*Sturnus vulgaris*)	1.0	31.7
House sparrow (*Passer domesticus*)	0.0	20.0
Chipping sparrow (*Spizella passerina*)	0.0	12.3
Song sparrow (*Melospiza melodia*)	4.0	12.3
Total birds per coverage	207	191

to dominate at the expense of more specialized ones. Edge species win out over species requiring larger tracts of contiguous habitat.

In a general way, then, changes in fauna with urban–suburban development tend to follow the patterns predicted by insular ecology. Comparing cities with islands, Davis and Glick (1978) concluded that (1) small cities function like islands that are large or near mainlands or both, and that (2) larger cities operate more like smaller, more distant islands. Smaller cities tend to have higher turnover rates, though most cities support lower levels of diversity than do surrounding countrysides, with extinction rates higher than colonization rates. Interestingly, species well-adapted to urban conditions tend to colonize cities not from the adjacent rural area, but rather from other urban regions.

Successful urban species often reach very high levels of population density and biomass. Robins, for example, have been found to nest at densities of up to 6.2 pairs per hectare (2.5 pairs per acre) (Howard, 1974) in some residential areas. In a suburb of Cincinnati, Ohio, Schinner and Gauley (1974) used a combination of mark-and-recapture and radio telemetry to estimate a raccoon population of 1 per .55 hectares (1 per 1.4

acres). Nuorteva (1971) calculated that the total avian biomass within Helsinki was about 10 times that of the surrounding countryside.

Some of the species that decline most sharply with increasing development are those most sensitive to habitat fragmentation. Such "area-sensitive" species form North America are listed in Chapter 10, along with some minimum estimates of the size of contiguous habitat necessary for sustaining nesting.

Two principles have been recommended for wildlife management in areas slated for development as cities and suburbs (Leedy et al., 1978). First, wildlife management should be considered in detail when plans for development are first drafted. This action ensures that the impact of the proposed development on wildlife is considered right along with elementary economic concerns. Early inclusion of wildlife in the planning process also prevents the obstructionist image that emerges whenever conservation concerns are voiced only after development is well under way.

The second principle is that wildlife management in suburbs and cities, as almost anywhere else, is primarily habitat management. Accordingly, the natural diversity and composition of local vegetation should be favored. Whenever possible, developers should spare native trees and shrubs, particularly those with high values to wildlife. Steps should be taken to protect against soil erosion on development sites, both to maintain fertility and to protect local waters from excessive siltation. Natural wetlands should be retained for flood control as well as for wildlife (see the section on wetlands in Chapter 9).

Spatial arrangements of both buildings and natural features of the site are also of prime importance. Clustered housing tracts featuring large communal open spaces are especially well-suited for wildlife habitat management. Natural corridors of woody vegetation can be maintained for wildlife. Minimum-impact nature trails, complete with boardwalks over wet areas and interpretive signs, can change a brushy and ill-kept eyesore into a valued and educational natural area.

Pest species can often be controlled through habitat manipulation. Aelred Geis, observing that some neighborhoods in Columbia, Maryland, were plagued by dense populations of starlings, pigeons, and house sparrows, set out to discover why. The answer lay in neighborhood architecture. Homes and apartment buildings with unscreened eaves and vents made ideal nesting sites for troublesome birds. Neighborhoods without these artificial nest sites had no real problem (Durham, 1981).

Some of the more adaptable game animals may reach damagingly high levels within some suburbs. An exclusive neighborhood on the outskirts of Minneapolis–St. Paul, for example, began suffering damage to gardens and ornamental plants from white-tailed deer. Aerial reconnaissance showed that more than 15 deer per square kilometer (40 per square mile) lived in this development (Jordan, 1980). Freed from pressures of hunting and natural predation, and aided by supplemental food provided by residents, these deer had increased sharply. Local residents, charmed at having these attractive wild animals in their backyards, became less enchanted with them once the damage grew serious.

To sum up, there are two basic approaches to management of wildlife in urban and suburban areas. The first is to maintain open areas (parks, greenbelts, etc.) as large as possible. This will minimize losses of area-sensitive species, particularly if the open spaces are predominantly wooded. The second strategy is to maintain as much diversity within each of the open spaces as possible, avoiding a "manicured look." By combining these methods, the less common urban–suburban species will be favored, and, as a general rule, the more abundant pest species will not find conditions suitable for achieving very high numbers.

SUMMARY

This chapter presents a newer form of wildlife management that differs fundamentally from that described in the preceding chapter. It is applicable to a much larger range of species and conditions. The primary objective of this form of management is the avoidance of extinction or, put another way, the preservation of the world's genetic resources from wild species.

Although current and projected estimates of extinction rates vary, there is no question that human activity has greatly accelerated them. The most obvious reason for species becoming either endangered or extinct is excessive exploitation beyond the threshold that their populations can withstand. Such excessive harvests have taken place for food and furs or for other marketable goods such as plumes or hides and for the wild animal trade. Introduced species have precipitated declines and disappearances of many native species, especially on islands. As more and more chemicals are added to the world's food chains, environmental contamination will threaten more and more species. But by far the leading threat to wild species has in recent years been habitat destruction. The rate of habitat destruction continues to increase worldwide, with the highest rates in the tropics, where the highest species diversity occurs.

Several agencies and organizations devote considerable efforts in the fight to avoid extinctions. The U.S. Fish and Wildlife Service has become a leader in the management and recovery of endangered species programs. The International Union for Conservation of Nature and Natural Resources (IUCN) and the World Wildlife Fund, both based in Switzerland, are private organizations that work on behalf of endangered species. The Convention on International Trade in Endangered Species (CITES), an international treaty, has been an important step in promoting international cooperation aimed at reducing the rates of extinction.

Species that decline to low levels are classified in two broad ways. The more serious category is "endangered," defined as those species likely to become extinct without positive management steps. Somewhat less serious are the categories of "rare" or "threatened." These terms apply to species in no immediate danger of extinction, but whose declines are sufficiently serious to justify close attention.

To date most of the management of endangered species has been aimed at saving one species at a time. Legal protection, together with maintenance of sufficiently large reserves of habitat, are often enough to ensure the survival of many species. Captive propagation is a management tool of last resort, as the method is expensive, labor-intensive, and plagued with the question of whether or not captive-reared animals will ever be able to adapt to the wild. Biologists can now use field data to estimate effective population size of endangered species. This method allows maintenance of populations of sufficient size to ensure enough genetic diversity for long-term preservation.

But single-species management has its limitations, especially in the species-rich tropics, where ecosystems are threatened on a large scale. A more realistic strategy under such conditions is the preservation of representative samples of each major habitat type. Insular ecology, an application of island biogeography, furnishes a set of guidelines for determining minimum size, shape, and proximity for such habitat reserves.

Insular ecology and other management techniques can be applied to a wide range of conditions. Two rather extreme conditions are described in this chapter; (1) urban–suburban wildlife and (2) wildlife in officially designated wilderness areas.

REFERENCES

Allen, D. L. 1974. Philosophical aspects of urban wildlife. In J. Noyes and D. Proguske (Eds.), *Wildlife in an urbanizing environment.* Cooperative Extension Service, Amherst, Mass.: University of Mass., U.S. Dept. Agriculture and County Extension Services Cooperating.

Barney, G. O. 1980. *The global 2000 report to the president of the U.S. Vol. 1: the summary report.* New York: Pergamon Press.

Bellrose, F. C. 1976. *Ducks, geese and swans of North America.* Harrisburg, Pa.: Stackpole Books.

Burnham, W. A., J. Craig, J. H. Enderson, & W. R. Heinrich. 1978. Artificial increase in reproduction of wild peregrine falcons. *J. Wildl. Manage.* 42: 625–628.

Campbell, Sheldon. 1980. Is reintroduction a realistic goal? In M. Soule and B. Wilcox (Eds.), *Conservation biology.* Sunderland, Mass.: Sinauer Associates, Inc.

Conway, W. G. 1980a. An overview of captive propagation. In M. Soule and B. Wilcox (Eds.), *Conservation biology.* Sunderland, Mass.: Sinauer Associates, Inc.

Conway, W. G. 1980b. Where we go from here. *Int. Zoo. Yrbk.* 20: 184–189.

Davis, A. M., & T. F. Glick. 1978. Urban ecosystems and island biogeography. *Env. Cons.* 5: 299–304.

Diamond, J. M. 1975. The island dilemma: lessons of modern biogeographic studies for the design of natural reserves. *Biol. Cons.* 7: 129–145.

Diamond, J. M. 1976. Island biogeography and conservation: strategies and limitations. *Science* 193: 1027–1029.

Domalain, J. 1977. *The animal connection: the confessions of an ex-wild animal traffiker.* New York: William Morrow & Co.

Durham, Megan. 1981. Urban wildlife expert makes research a family affair. In *Fish and Wildlife News* (special ed., research) Washington, D.C.: U.S.D.I. Fish and Wildlife Service (April–May).

Eckholm, Erik. 1978. Disappearing species: the social challenge. *Worldwatch Paper 22.* Washington, D.C.: Worldwatch Institute.

Ehrlich, P., Anne Ehrlich, & J. Holdren. 1977. *Ecoscience: population, resources, environment.* San Francisco, Calif.: W. H. Freeman & Co.

Fischer, J., N. Simon, & J. Vincent. 1969. *Wildlife in danger.* New York: Viking Press.

Franklin, I. R. 1980. Evolutionary change in small populations. In M. Soule and B. Wilcox (Eds.), *Conservation biology.* Sunderland, Mass.: Sinauer Associates, Inc.

Geis, A. D. 1974. Effects of urbanization and types of urban development on bird populations. In J. Noyes and D. Progulske (Eds.), *Wildlife in an urbanizing environment.* Cooperative Extension Service, Amherst, Mass.: University of Massachusetts.

Harwood, M. 1982a. Math of extinction. *Audubon* 84(6): 18–21.

Harwood, M. 1982b. Peregrine redux. *Audubon* 84(5): 9–10.

Harwood, M. 1982c. Unmiraculous comeback of the trumpeter swan. *Audubon* 84: 32–41.

Hendee, J., & C. Schoenfeld. 1979. Wildlife management for wilderness. In R. Teague and E. Decker (Eds.), *Wildlife conservation—principles and practice.* Washington, D.C.: The Wildlife Society.

Howard, Deborah V. 1974. Urban robins: a population study. In J. H. Noyes and D. R. Progulske (Eds.), *Wildlife in an urbanizing environment.* Cooperative Extension Service, Amherst, Mass.: University of Massachusetts.

IUCN. 1980. *World conservation strategy.* International Union for Conservation of Nature and Natural Resources, Gland, Switzerland.

Jackson, J. A. 1978. Alleviating problems of competition, predation, parasitism, and disease in endangered birds. In S. A. Temple (Ed.), *Endangered birds: management techniques for preserving threatened species.* Madison, Wisc.: University of Wisconsin Press.

Jordan, P. A. 1980. Problems with suburban white-tailed deer. *Paper presented at the Midwest Wildlife Conf.* St. Paul, Minn.

Karr, J. 1982. Avian extinction on Barro Colorado Island, Panama. *Am. Nat.* 119: 220–239.

Kimura, M., & J. F. Crow. 1963. The measurement of effective population number. *Evolution* 17: 279–288.

King, F. 1978. The wildlife trade. In H. Brokaw (Ed.), *Wildlife in America.* Washington, D.C.: Council on Environmental Quality.

Kleiman, D. G. 1980. The sociobiology of captive propagation. In M. Soule and B. Wilcox (Eds.), *Conservation biology.* Sunderland, Mass.: Sinauer Associates, Inc.

Leedy, D., R. Maestro, and T. Franklin. 1978. *Planning for wildlife in cities and suburbs.* Washington, D.C.: USDI Fish and Wildlife Service Office of Biological Services. FWS/OBS-77/66.

Leopold, A. 1949. *A Sand County almanac.* London: Oxford University Press.

Lovejoy, T. E. 1976. We must decide which species will go forever. *Smithsonian* 7 (4): 52–59.

Lovejoy, T. E. 1980. A projection of species extinctions. In G. Barney (Dir.) *The global 2000 report to the president of the U.S. Vol. 1. the summary report.* New York: Pergamon Press.

MacArthur, R. H., & E. O. Wilson. 1967. *The theory of island biogeography.* Princeton, N.J.: Princeton University Press.

Miller, R. 1978. Applying island biogeographic theory to an East African reserve. *Env. Cons.* 5(3): 191–196.

Myers, N. 1976. An expanded approach to the problem of disappearing species. *Science* 193: 198–202.

Myers, N. 1979. The sinking ark. New York: Pergamon Press.

Nilsson, Greta. 1981. *The bird business: a study of the commercial cage bird trade.* Washington, D.C.: Animal Welfare Institute.

Nuorteva, P. 1971. The synanthropy of birds as an expression of ecological cycle disorder. *Ann. Zool. Fenn.* 8: 547–553.

Picton, Harold. 1979. Application of insular biogeography theory to conservation of large mammals in the northern Rocky Mountains. *Biol. Cons.* 15(1): 73–79.

Ralls, Katherine, Kristin Brugger, & Jonathan Ballou. 1979. Inbreeding and juvenile mortality in small populations of ungulates. *Science* 206 (4422): 1101–1103.

Robbins, C. 1979. Effect of forest fragmentation on bird populations. In Management of north central and northeastern forests for nongame birds. *Gen. Tech. Rep. NC-51.* St. Paul, Minn.: USDA Forest Service, North Central Forest Experiment Station.

Schinner, J. R., & D. L. Gauley. 1974. The ecology of urban raccoons in Cincinnati, Ohio. In J. H. Noyes and D. R. Progulske (Eds.), *Wildlife in an urbanizing environment.* Cooperative Extension Service, Amherst, Mass.: University of Massachusetts.

Schoenfeld, C. A., & J. C. Hendee. 1978. *Wildlife management in wilderness.* Pacific Grove, Calif.: Boxwood Press.

Senner, J. W. 1980. Inbreeding depression and the survival of zoo populations. In M. Soule and B. Wilcox (Eds.), *Conservation biology.* Sunderland, Mass.: Sinauer Associates, Inc.

Simberloff, D. F., & L. G. Abele. 1975. Island biogeography theory and conservation practice. *Science* 191: 285–286.

Soule, M. E., B. A. Wilcox, & C. Holtby. 1979. Benign neglect: A model of faunal collapse in the game reserves of East Africa. *Biol. Conserv.* 15: 259–272.

Sparrowe, R. D., & H. M. Wight. 1975. Setting priorities for the endangered species program. *Trans. N. Am. Wildl. and Natur. Resour. Conf.* 40: 142–156.

Temple, S. 1983. The dodo haunts a forest. *Animal Kingdom* 86(1): 20–25.

Terborgh, J. 1974. Preservation of natural diversity: the problem of extinction-prone species. *Bio Science* 24: 715–722.

Terborgh, J. 1976. Island biogeography and conservation: strategy and limitations. *Science* 193: 1029–1030.

Thomas, J. W., & R. A. Dixon. 1974. Cemetery ecology. In J. H. Noyes and D. R. Progulske (Eds.), *Wildlife in an urbanizing environment.* Cooperative Extension Service, Amherst, Mass.: University of Massachusetts.

Whitcomb, R. F., J. F. Lynch, P. A. Opler, & C. S. Robbins. 1976. Island biogeography and conservation: strategy and limitations. *Science* 193: 1030–1032.

Wilcox, B. 1980. Insular ecology and conservation. In M. Soule and B. A. Wilcox (Eds.), *Conservation biology.* Sunderland, Mass.: Sinauer Associates, Inc.

Wilson, E. O., & E. O. Willis. 1975. Applied biogeography. In M. L. Cody and J. M. Diamond (Eds.), *Ecology and evolution of communities.* Cambridge, Mass.: Harvard University Press.

Walcott, C. F. 1974. Changes in the birdlife in Cambridge, Mass., from 1860–1964. *Auk* 91: 151–160.

Zimmerman, D. R. 1978. To help save the endangered whooping crane, biologists put endangered whooper eggs in the nests of sandhills, which then hatch the eggs and rear the young as their own. *Smithsonian* 9(6): 52–63.

Ziswiler, V. 1967. *Extinct and vanishing animals* (2d ed.). New York: Springer-Verlag.

CHAPTER 9

SPECIAL PROBLEMS: PREDATOR CONTROL, EXOTIC INTRODUCTIONS, PARASITES AND DISEASES, AND WETLANDS PRESERVATION

When Aldo Leopold and his associates published their North American game policy in 1930, two of the more controversial areas cited were predator control and the introduction of exotic wildlife. Although knowledge about both these problems has expanded considerably since, they still remain controversial. Predator control is an emotional topic, with neither side especially disposed to moderation or objective analysis.

During the last century, attitudes about predators were almost universally negative. These attitudes stemmed from economic concern liberally spiced with prejudice and ignorance. On one side such traditional attitudes largely remain. The other side, however, maintains with equal conviction the view that predators virtually never cause livestock damage. Opponents of predator control argue that predators keep harmful species, such as many of the more destructive rodents, in check. Given these polar views, a quote from Leopold, "Only the mountain has lived long enough to listen objectively to the howl of the wolf (now read coyote)," still seems appropriate (Leopold, 1949).

Debates over exotic introductions are somewhat less emotional and less widespread. Many wildlife professionals cite serious ecological consequences resulting from exotic introductions: starlings in the United States, European rabbits (*Oryctolagus cuniculus*) in Australia, and mongoose in the Caribbean, to name but a few. They then caution against further introductions on the grounds that they might lead to similar disasters. The other side of the debate argues that game species introduced onto continents have often enriched the local fauna. Moreover, as more species are threatened through habitat destruction, especially in the developing nations, their survival outside of zoos may be contingent on their introduction into other parts of the world. Should they accept the risk to native flora and fauna to help save endangered species? Can the risks posed by the exotic species be adequately assessed beforehand?

In addition to these old problems, wildlife managers face a new one that is less well understood: diseases and parasites. Despite Leopold's (1933) contention that the role of diseases in wildlife conservation had probably been radically underestimated, wildlife managers have shown neither the interest nor the technical expertise needed to gain a better empirical understanding. Yet there is abundant evidence that diseases and parasites can play crucial roles in affecting wildlife populations. Worst of all, modern wildlife management will almost certainly increase the risks associated with parasites and disease.

Finally, this chapter reviews an especially important aspect of habitat conservation: the preservation and management of natural wetlands. Wetlands are among the richest and most productive of wildlife habitats as well as the most fragile. They are also rapidly and permanently disappearing; they have been developed into lakes and channelized waterways, and they have been filled in for suburbs, agricultural lands, airports, and roadways.

PREDATOR CONTROL

As shown in Chapter 1, the first form of wildlife "management" practiced by the colonists along the eastern seaboard of North America was predator control. In 1630, the Massachusetts Bay Colony began paying bounties on wolves. Two years later, the Virginia Colony instituted similar practices. Payment of bounties on predatory animals was by then an established practice in England. The colonists merely followed tradition in trying to rid themselves of predators.

Various methods were developed for disposing of large carnivores. The animals were shot whenever they were seen within range. All sorts of traps, including pitfalls and deadfalls, were used, as was pursuit by dogs. During the nineteenth century, wolf hunters on the plains laced animal carcasses with strychnine, returning later to skin the poisoned wolves that had scavenged them. The steel leg-hold trap, developed first to catch beaver, was later modified to catch wolves and other large carnivores. Poisoning and trapping virtually exterminated the gray wolf from the continental United States. The red wolf's numbers were left so low that the species became susceptible to interbreeding with coyotes. Grizzly bears, once common throughout the west, were eradicated from all but two remote, protected areas. The more ubiquitous cougar and the black bear were substantially reduced, particularly in most of the eastern and central states. Against them, the predator war had largely been won by the mid-twentieth century.

In the decades following eradication of the larger carnivores, the coyote greatly expanded its geographic range, moving from its western range eastward into the Atlantic states and northward into Alaska. There is some circumstantial evidence that the increase in both numbers and geographic distribution resulted from removal of the larger predators. Whatever the cause or causes, the success of the coyote makes it undoubtedly the number one target of predator control practices in the United States.

Methods

The technology of predator control has advanced considerably since the end of World War II. A common practice of the 1950s and 1960s was the use of "1080 bait stations." Compound 1080, known chemically as sodium monofloroacetate, is a stable, toxic substance to which canids are quite susceptible. Each bait station consisted of a portion of a livestock carcass, usually a hindquarter of a horse, into which compound 1080 was injected. Coyotes scavenging the carcasses would then ingest the poison and die. Following an extensive review of predator control practices (Cain et al., 1972), President Nixon banned 1080 for predator control early in 1972.

Other popular tools for killing coyotes include the powder-charged "humane coyote getter" and its offspring, the spring-loaded M-44. Both devices are concealed under a scent bait placed along coyote runs. Part of each device consists of a hollow pipe pounded partway into the ground. Into the top of the pipe fits a cyanide charge. When the coyote, attracted to the scent bait, tugs at the charged portion, the cyanide explodes into its mouth with fatal results. One difficulty with coyote getters and the M-44 is that they are not selective. Cyanide works equally well on any animal impelled to tug at the bait. The same executive order which banned 1080 also banned M-44's, which had by then completely replaced the older "getters." Shortly after the executive order, the Environmental Protection Agency canceled registrations for compound 1080, strychnine, and sodium cyanide, further curtailing their use. The M-44's, however, have gradually been finding their way back into service, first under emergency conditions and later for operational use (U.S. Fish and Wildlife Service, 1978). Early in 1983 President Reagan rescinded Nixon's executive order and the Environmental Protection Agency authorized limited use of 1080 once again.

Another widely used method employed by agents of the U.S. Fish and Wildlife Service's Division of Animal Damage Control (ADC) is aerial gunning. Agents fire heavy shot charges from helicopters at coyotes. Although expensive, aerial gunning is more selective than other techniques, as it involves killing only coyotes and does so within the immediate area where damage has been reported.

Aversive conditioning is a fairly new and nonlethal technique that holds some promise in reducing predator damage. Early experiments showed that captive coyotes fed hamburger, rabbit, or lamb treated with lithium chloride became violently ill. Thereafter they tended to avoid eating the type of meat that had made them ill, apparently associating the smell and taste of the meat with the emetic experience. Some of the tested coyotes, after one or two treatments, would even refuse to attack live rabbits or lambs (Gustavson et al., 1974). Further tests by the same research team showed that gray wolves and even an aged cougar could be aversively conditioned (Gustavson et al., 1976).

Field tests of aversive conditioning have been, as might be expected, more difficult than tests conducted on captive predators. Gustavson and his colleagues enticed wild coyotes to consume both sheep-flavored baits and sheep carcasses containing lithium chloride on a ranch in southwestern Washington. Thereafter sheep losses were rather crudely estimated as having declined from between 30 and 60 percent of those in previous years (Gustavson et al., 1976).

Other researchers have obtained ambiguous results with aversive conditioning. Individual coyotes seem to vary, due perhaps to age or experience, in their susceptibility to such conditioning. Extinction rates (the time required for the conditioned response to subside once the stimulus is removed) can also vary greatly. Coyotes apparently learn to avoid the chemical itself, eating portions of baits without the salty emetic (Griffiths et al., 1978). Perhaps they can cue in on the chemical itself and disregard association with the type of bait. Finally, experienced coyotes may learn to avoid carrion and take only live prey instead. Aversive agents such as lithium chloride or poisons such as 1080 may thus miss the very individuals most likely to kill livestock. Only after further evaluation can the real feasibility of aversive conditioning as a tool in predator control be determined.

Another relatively new method of predator control has been the toxic collar. The method is quite simple in principle. Sheep are tethered on the perimeters of flocks where they are most likely to be attacked by coyotes. Around the sheeps' necks are fitted collars containing sodium cyanide or other toxic solutions. Since coyotes usually kill sheep by attacking the throat, they are quite likely to puncture the polyethylene collar and to receive a mouth full of poison. Toxic collars seem to work best whenever most of the flock can be placed in protective pens at night, leaving only collared sheep or lambs exposed. Obviously this nightly penning is far more practical for small farm flocks than for large range operations.

The toxic collar has received a great deal of publicity as the most selective of the lethal forms of predator control. The device would kill only those coyotes which actually attacked sheep. Tests of the collars have been encouraging in experiments on captive coyotes. Savarie and Sterner (1979) observed 12 coyotes attack collared sheep. All coyotes punctured a collar and 9 of the 12 received lethal doses. Field tests against free-ranging coyotes were less conclusive, partly because coyotes did not attack collared sheep frequently enough to permit a thorough evaluation. Moreover, some coyotes avoid the collar by not attacking the throat (Connolly et al., 1978).

Despite these problems, losses of sheep to coyotes have declined in most cases where toxic collars have been in use. The design is undergoing refinement and it seems quite likely that some form of toxic collar will see widespread use in the future. The principal drawbacks include the need to sacrifice collared sheep and the usual hazards associated with use of toxic chemicals (Connolly et al., 1978).

For centuries, Europeans have used large breeds of dogs to protect their sheep from wolves. The komondor, a Hungarian breed, has been tested against coyotes in the United States. In one such test (Linhart et al., 1979) the komondors were first obedience trained, familiarized with sheep, and trained to be aggressive toward captive coyotes. They were then taken to sheep ranches in Montana and North Dakota for 60-day field trials. Sheep losses to coyotes were monitored for the first 20 days before the dogs were put to use. The monitoring continued through the next 20 days while the dogs stayed with the sheep and for 20 days after following the removal of the dogs.

Sheep losses dropped substantially during these preliminary tests, from an average of 15 coyote kills to one of only 6, and an average of only 5 after the dog phase. Not all the results, however, were quite so encouraging. The dogs are expensive (up to $1,200 each) and require careful training and care. A pair of komondors on one of the three test ranches harassed, attacked, and even killed some of the sheep, illustrating the need for careful screening and selection of dogs. Pending more testing and refinement of training and screening methods, komondors and other breeds such as the Great Pyrenees will probably become established tools in predator damage control.

The Effects of Control on Coyote Populations

Coyotes are almost as prolific as they are adaptable, with adult females producing litters of up to seven pups each year. As is the case with many carnivores, rates of fecundity and survival of young can vary markedly from year to year, with profound effects upon population growth. There is also good evidence that coyote reproduction is compensatory, responding positively to reductions in the population size. Knowlton (1972) compared variations in the average litter size for coyotes in areas with varying intensities of control. Where control (and subsequent population reduction) was light, the average litter size was 3.65. In areas where control levels were moderate, litters averaged 4.75. Where control was most intensive, the average litter was 6.56. These results suggest that intensive predator control can be expected to stimulate greater production of coyote pups per female. This response can allow the population to rebuild rapidly once control is reduced or terminated.

Given the reproductive capacities of the coyote, how much control would be needed to ensure that a coyote population would be kept to, say, half its carrying capacity? To answer this question, Connolly and Longhurst (1975) developed a population model. They found that reductions on the level of about 70 percent per year would be required to hold the coyote populations below carrying capacity. Even with reductions of 75 percent, the calculated threshold below which the population could no longer compensate, it would take some 50 years of control to drive a population to extinction. By comparison, the U.S. Fish and Wildlife Service estimates that all its ADC activities kill less than 5 percent of the coyote population in the 17 western states where its programs are operational (U.S. Fish and Wildlife Service, 1978). Obviously, then, this level of control could hardly suppress the coyote populations.

Predator Control and Livestock

Historically, the primary reason for predator control has been protection of livestock. Larger predators can kill cattle as well as sheep and goats; coyotes occasionally kill cattle, but inflict heaviest damage on sheep. In assessing the extent of such losses, analogies between predators and their natural prey are of little value. Natural prey species coevolved with their predators, with predation shaping, through natural selection, the abilities of wild prey to protect themselves against predators. Livestock, in contrast, are the products of selective breeding for traits deemed desirable by their human owners. They are thus relatively helpless when attacked by predators.

Sheep are especially vulnerable to predation by both coyotes and domestic dogs. Experiments have shown that even though not all coyotes will attack and kill sheep even when deprived of food (Connolly et al., 1976), most of them will. Lambs and smaller ewes are the most common victims. Even the sheep's behavior works against it, as its tendency to run from a coyote acts to release the coyote's predatory responses (Lehner, 1976). In addition, the tendency for coyotes to attack sheep is also influenced by the predator's sex and age. Connolly et al. (1976) learned that adult male coyotes almost invariably attack sheep when both animals are placed within an enclosure. Adult females would not initiate an attack, but they would assist males in making a kill. Younger male coyotes would occasionally initiate an attack. No yearling females were tested, though presumably they would not attack sheep alone.

Coyote damage patterns differ sharply through space and time. One livestock owner can suffer serious damage while his neighbor incurs none at all. This difference may in part be due to differences in individual coyotes and in how they teach their young to hunt. It may also be due to differences in terrain, vegetation, or how individual landowners care for their livestock. This variation has inspired different conclusions about the real threat that coyotes pose to livestock, thus fueling the controversy further.

One alternative to conventional predator control that has often been proposed is some form of depredation insurance for livestock owners. This proposition has been evaluated in some detail (U.S. Fish and Wildlife Service, 1978). Practical problems exist, such as dealing with fraudulent or exaggerated claims. These could, however, be minimized by compensating exceptionally high losses at less than 100 percent of the market value and by making certain approved management practices mandatory. As with automobile insurance, individual premiums could be adjusted over time based upon the owners' loss record. Such adjustment should furnish additional incentive for management practices that minimize risk of predator losses. Finally, there already exist some administrative alternatives or supplements to strictly private insurance coverage. The Federal Crop Insurance Corporation (FCIC) has had extensive experience in providing similar coverages and could either reinsure private companies or else furnish direct coverage. Either option would require legislative authority from Congress.

Despite these advantages, some problems might result from the use of insurance payments as a substitute for predator control. Suppose one rancher relaxed pressure on coyotes, let them take whatever they wanted, and then collected insurance. Would that rancher be in essence training coyotes to take a neighbor's livestock? Finally, what sort of documentation should be required to verify losses, and how many federal or state employees might be needed to carry out this task?

Another alternative, one actually in use in both Kansas and Missouri, is that of extension education for livestock owners. It is far less expensive than ADC Cooperative programs. Extension programs teach landowners to minimize losses through sound management of stock. When losses to predators occur, owners are taught to identify the predator and to eliminate offending animals. The primary method taught is the use of steel leg-hold traps. One drawback to extension education is that the actual control would be done by livestock owners, who would not and could not be as skillful as trained professionals. However, against this disadvantage should be weighed the fact that extension programs offer a built-in incentive for livestock owners to minimize their losses (U.S. Fish and Wildlife Service, 1978). Although the programs have been quite successful in both Missouri and Kansas, serious questions exist as to whether or not they could work farther west, where private land holdings are larger and where livestock are commonly ranged on large tracts of federal lands. Under such large-scale conditions, the task of controlling predators might be met only by skilled, full-time professional agents. If control under such conditions were left entirely up to the owners, they might revert to ecologically damaging practices such as widespread use of poisons (U.S. Fish and Wildlife Service, 1978).

Livestock losses attributed to predators can be quite high. According to figures compiled by the U.S. Department of Agriculture, the total losses for livestock in 22 western states exceeded $170 million in 1973 (Terrill, 1977). Cattle and calf losses ($80 million) exceeded those of sheep and lambs ($53 million). The balance was comprised of chickens ($32 million) and pigs ($4 million).

Since these losses are not evenly distributed, some individual stock owners can incur severe losses. On one sheep ranch near Missoula, Montana, for example, coyotes took 20.8 percent of the herd including nearly 30 percent of the lamb crop within a year (Henne, 1977). Bart O'Gara (personal communication) reported that some livestock owners who have tried insurance have had their policies canceled following a year of heavy losses.

Given the dynamic and controversial nature of predator control to protect livestock, the future of the practice is difficult to predict. It seems safe to conclude, though, that publicly funded predator control against coyotes will not seek to eradicate or even to reduce populations substantially. Such an undertaking would be far too expensive and of dubious value, since the relationship between coyote population size and actual depredation loss is still poorly understood. Instead, control will likely become even more selective. It will likely employ nonlethal methods whenever possible and will strive to supplement agent involvement with landowner involvement where practical through extension education. Depredation insurance is almost certain to be tried within the near future and, if successful, may greatly alter the practice of predator control.

Predator Control to Protect Game Populations

During the first half of the twentieth century, predator control was routinely used to protect game populations, particularly big game. Big-game populations, meanwhile, increased where habitat conditions permitted, in part because of restocking, increasingly effective legal protection, and habitat improvement. Larger predators were by then greatly reduced but their role in regulating ungulate populations remains controversial (Connolly, 1978; Peek, 1980). As deer and other big-game populations rose, justification for further predator control waned. Moreover, some deer populations irrupted, such as the celebrated mule deer herd at the Kaibab (see Chapter 6). Although conclusive data were lacking, many biologists concluded that the irruptions were largely due to release of ungulate populations from predation (Leopold et al., 1947). Since then, predator control as a game management tool has virtually ceased, with the prevailing opinion among wildlife biologists that it is unnecessary and wasteful. Society as a whole has begun to recognize the value of predators and in general now realizes that predators pose no threat to populations of their natural prey.

In the 1970s, researchers began taking a closer look at the interplay between predators and game populations. Some of their findings suggested that if the management objectives call for maximizing harvest and if the habitat can support more of the game population, then short-term, local predator control may be justifiable. Of course, such a practice would undoubtedly raise ethical questions. Biologically, though, the outlook for increasing local game populations through predator control seems real enough. As demand for more intensive harvests grows, so will increasing pressure for predator control in an attempt to meet that demand. This section reviews the outlook for predator control in game management. Because of the controversial and emotional nature of the subject, it serves as an excellent example of management by carefully structured objectives.

Beasom (1974a,b,) showed that intensive predator control could stimulate higher game production in south Texas. He selected two study areas of essentially identical habitat, 23 square kilometers (8.8 square miles) each and separated by 8 kilometers. Coyotes, bobcats, raccoons, skunks (*Mephitis mephitis*), and a few other predators were removed from one area while the other served as a control. Game populations in the area where predators were removed, including white-tailed deer, turkey, and bobwhite quail, all averaged higher levels of density, productivity, or both. Weather was also a factor, because rainfall and productivity in both areas were higher in 1972 than in 1971 (Table 9-1).

Is such predator control economically justifiable? Hunting leases, especially for deer and wild turkey, are quite common in south Texas. Using the standard leasing fees as guidelines, Beasom estimated the dollar value for both deer and turkey, taking into account the greater yield obtainable through predator control. He then measured this income against the actual costs of predator control. His conclusion: predator control is cost-effective only at higher levels of harvest. In other words,

TABLE 9-1.
EFFECTS OF EXPERIMENTAL PREDATOR CONTROL ON SOUTH TEXAS GAME POPULATIONS
(From Beasom, 1974a,b © Wildlife Management Institute and The Wildlife Society. Used by permission.)

	Experimental area (predators removed)	Control area (predators not removed)
1971 (dry year)		
No. coyotes removed	129	0
No. bobcats removed	66	0
No. raccoons removed	31	0
No. striped skunks removed	22	0
Deer fawn: doe counts	0.47	0.12
Turkey poult: hen counts	1.3	0.0
% Increase in bobwhite quail	98.7	39.2
1972 (normal year)		
No. coyotes removed	59	0
No. bobcats removed	54	0
No. raccoons removed	34	0
No. striped skunks	24	0
Deer fawn: doe counts	0.82	0.32
Turkey poult: hen counts	4.7	1.5
% Increase in bobwhite quail	213.8	154.6

hunters have to be able to take a large proportion of the "surplus" that would have otherwise disappeared down the throats of the predators. In this particular case, Beasom estimated that if 100 percent of the "surplus" deer and turkey were harvested, the net return to the landowner, after deducting the cost of control, would average \$6.40 per hectare (ca 1970 figures). If half the surplus were taken, the net profit would have been \$1.53 per hectare. On the other hand, if only 10 percent of the game crop were harvested, the landowner would suffer a net loss of \$2.89 per hectare.

If intensive predator control really will allow greater survival of game animals, the next question becomes whether or not the habitat can support the additional game. If it cannot, then efforts toward predator control are both biologically and economically wasteful. It makes no sense to promote survival of young game animals that would have died from predation only to have them starve the following winter. An example of this problem comes from the Welder Wildlife Refuge, also in south Texas. Researchers there constructed a predator-proof fence around 3.9 square kilometers (1.5 square miles), then removed virtually all coyotes from inside it. They then compared white-tailed deer population trends inside the enclosure with those outside. Within 2 years the population density within the enclosure reached 84 per square kilometer, and the average density of deer outside remained at 34.6 per square kilometer (Kie et al., 1979). The work was experimental; aside from a few deer taken for research, the population at the Welder Refuge is not harvested. With neither predators nor rifles to hold the population in check, it gradually began to show signs of stress. Deer inside the enclosure conceived later, retained antler velvet longer, and shed antlers earlier than did those outside. Adult mortality began to increase some 4 to 5 years after predator control began, thus reducing the protected population (Kie et al., 1979).

Increases in game populations following intensive predator control are not restricted to south Texas. South Dakota researchers experimentally reduced populations of red foxes, raccoons, badgers (*Taxidea taxus*), and skunks on 11 study areas, each 16 by 16 kilometers (256 square kilometers or 99 square miles) for 5 years. They also measured changes in populations of ring-necked pheasant, jackrabbits, cottontail rabbits, and small rodents. The pheasant population increased by 132 percent, with other prey showing smaller, but significant, increases (see Table 9-2). Despite their consistent results, the researchers cautioned against the temptation to substitute predator control for habitat improvement, since the latter is definitely needed. They also suggested that population reductions of predators could often be satisfactorily accomplished through increasing the fur harvests, because the predators studied were also furbearers (Trautman et al., 1974).

Wildlife management has come full circle, from widespread predator control to protect game populations, to a complete retreat from control, to the use of short-term, local predator control to boost specific game populations. But there is as much difference between the two types of control as there is between indiscriminate use of a meat axe and the precise (and perhaps reluctant) use of a scalpel. The more that managers learn about predator–prey relationships, the more precisely control can be employed.

So when should wildlife managers decide to use predator control? They would be well-advised to follow the careful guidelines of Connolly (1978) (see Figure 9-1). Connolly's approach stresses the management objective. The first two objectives on the left-hand side are the ones most likely to be encountered by the game manager. Predator control should be undertaken only if:

1 Hunters will take the extra game produced.

TABLE 9-2.
EFFECTS OF EXPERIMENTAL PREDATOR REDUCTION UPON PREY POPULATION IN SOUTH DAKOTA
(Trautman et al., 1974 © Wildlife Management Institute. Used by permission.)

	Experimental area (predators removed)	Control area (predators not removed)	Difference (%)
Predator populations			
(% Change)	−74	−41	33
Red fox	−69	+62	131
Badger	−58	+24	82
Raccoon	−66	+45	111
Prey populations			
(% Increase over control area populations)			
Ringneck pheasant	+132		
Cottontail rabbit	+50		
Jackrabbit	+63		
Small rodents	+18		

2 The habitat will support more game.
3 Predator control will actually increase game.
4 Control is economical and acceptable.

If the answer is "no" for any of these four questions, as it might well be in the majority of cases, managers should not control predators.

EXOTIC INTRODUCTIONS

Two of the more common exotic introductions into North America—the ring-necked pheasant and the starling—illustrate both sides of the controversy. The starling was first successfully introduced (after several failed attempts) into Central Park in New York City in 1890, thanks to the efforts of one Eugene Scheifflin. Mr. Scheifflin had two passions, the works of William Shakespeare and ornithology. It became his mission in life to introduce into the United States all the birds mentioned in Shakespeare's plays (Laycock, 1966). One of them was the starling.

Starlings have since spread all over the United States and into parts of Canada. Aggressive competitors, starlings have displaced many native hole-nesting birds, including the bluebird (*Sialia sialis*) (deVos and Pet-

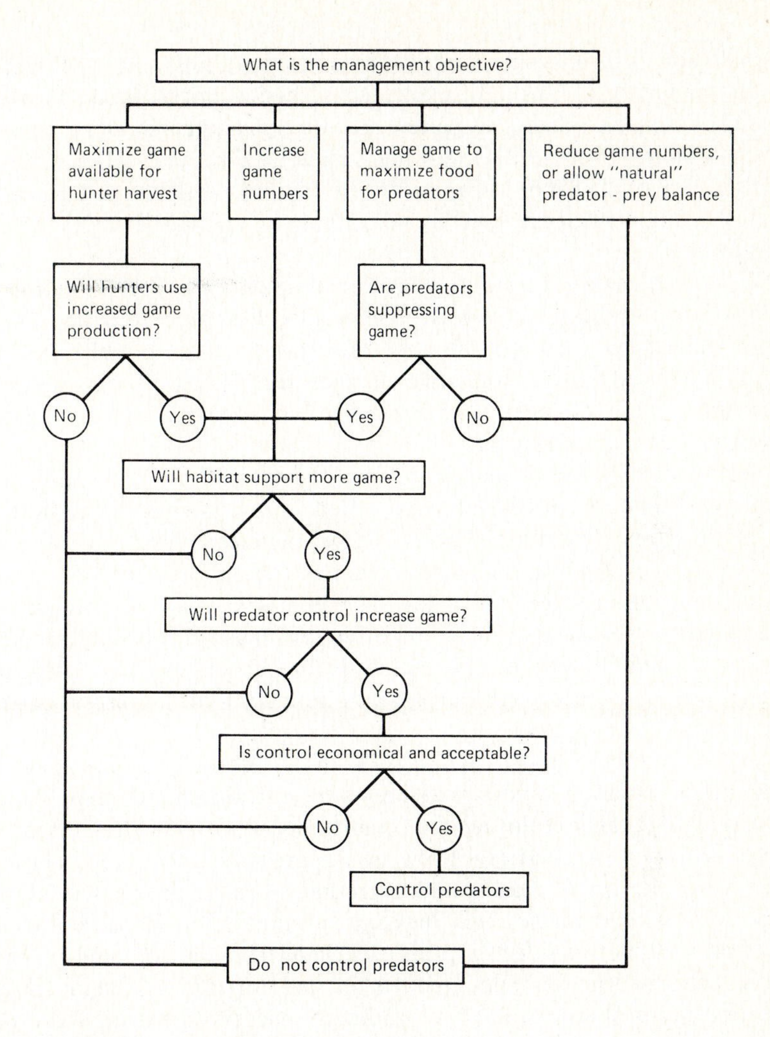

FIGURE 9-1 Decisions to be made when considering the use of predator control to aid game populations. (*Connolly, 1978* © Wildlife Management Institute. Used by permission.)

rides, 1967). They have caused extensive crop damage. Starlings move in large flocks; they commonly choose roosting sites around towns and cities and may disturb residents. By leaving tons of droppings, they may even pose a health hazard. Finally, flocks of starlings around airports have been sucked into the intakes of jet engines, causing fatal crashes. Millions of dollars are spent on starling control each year.

Eight years before Scheifflin liberated starlings, Owen Denny shipped

10 male and 18 female ring-necked pheasant from China to the Willamette Valley in Oregon. This transplant established a prized game bird that would eventually inhabit 18 states. In South Dakota pheasant numbers reached their highest levels, allowing annual legal harvests of more than 3 million (Laycock, 1966). Proponents of exotic introductions have often cited the ring-necked pheasant as an example of a beneficial exotic with no negative effects.

Without question, the ring-necked pheasant is an attractive and desirable species. Perhaps this is why people failed to look closely for adverse ecological effects. At about the same time that pheasant populations built up in the United States, large-scale changes were taking place in the use of the land. These changes may have masked some of the ring-necked pheasant's ecological impact.

Yet the ring-necked pheasant is as aggressive as it is adaptable. Moreover, it can be a nest parasite, often laying its eggs in the nests of other ground-nesting birds (Vance and Westemeier, 1979).

As ring-necked pheasant populations rose, those of the prairie chicken or pinnated grouse declined. Most of the prairie chicken's decline was attributed to the fact that more intensive farming and ranching rendered habitat unsuitable. But it now seems likely that competition with ring-necked pheasants also contributed to the reduction of the prairie chicken populations in many areas.

Jasper County, Illinois, lies outside of the contiguous range of the ring-necked pheasant. These pheasants were introduced into Jasper County in 1969, affording biologists the opportunity to observe their effects on prairie chicken sanctuaries. They saw aggressive encounters between the two species on chicken booming grounds year-round, peaking during the chickens' breeding season. Pheasants dominated in 78 percent of the encounters, sometimes driving prairie chickens from the booming grounds. These actions disrupted both courtship and territorial systems of the chickens and probably reduced reproductive success substantially.

But the pheasants did more than merely harass the prairie chickens on their booming grounds. Pheasant hens sometimes parasitized prairie chicken nests. Only 24 percent of the parasitized nests were successful for the chickens, whereas 51 percent of the unparasitized nests succeeded in producing chicks (Vance and Westemeier, 1979). The combined effects of both harassment and nest parasitism could gradually but steadily reduce prairie chicken populations. Not even so prized an exotic as the ring-necked pheasant is without adverse ecological impact.

Another exotic introduction that has done widespread damage in the United States is the wild pig (*Sus scrofa*). All members of the pig family (Suidae) are native only to the Old World and all those living in North America are descended from exotic introductions. Wild boar, a European forest dweller prized as a game trophy, were introduced into several eastern states, from North Carolina to New Hampshire. The wild pigs in the southeastern and southern Appalachian states are a mixture of wild boar and feral domestic pig. In all of the United States more than 100,000 wild pigs are taken legally by hunters each year (Wood and Barrett, 1979).

Despite some value as a game animal, wild pigs pose severe ecological and health problems. They damage both natural vegetation and agricultural crops, including forest tree seedlings. Some wildlife managers have suspected that they take eggs of ground-nesting birds, although their impact upon bird populations remains unknown. Wild pigs are serious competitors with native wildlife, particularly for hard mast, and they may pose significant problems for deer, turkey, black bear, and other mast-dependent wildlife. Finally, wild pigs can carry serious infectious diseases, including trichinosis, foot and mouth disease, and swine brucellosis (Wood and Barrett, 1979).

Texas can boast of the dubious distinction of containing the largest collection of free-living exotic big game in the United States. By 1979, the Texas Parks and Wildlife Department had tallied 72,147 exotic big-game animals representing 51 species. One reason for the popularity of these exotics has been that the state of Texas legally treats exotic wildlife on private land as though it were livestock. For years there was little evidence of harmful effects, but this lack of evidence may have been due primarily to the fact that few people bothered to look closely enough.

The Texas Parks and Wildlife Department became concerned about the effects of exotics on native white-tailed deer. The department began a series of experiments to measure competition. In one experiment 6 exotic sika deer (*Cervus nippon*) were placed in a 37-hectare (92 acre) enclosure with 6 white-tailed deer. Three years later, the populations were roughly equal, 16 sika and 15 white-tailed deer. But sika deer have broader feeding niches than white-tailed deer, and once the browse and forbs were exhausted the sika deer proved better able to feed on grass. Accordingly, the population of sikas reached 32 by the fifth year and the white-tailed deer declined to only 6. But the eighth year there were 62 sika and only 3 white-tailed deer, all 3 of which died the following winter (Armstrong and Harmel, 1981). Clearly the sika deer proved to be better competitors than the native white-tailed deer.

Reasons for Exotic Introductions

Advocates for exotic introductions have justified their views on many different grounds. Most of the reasons fall into one or more of the following categories.

Esthetics A leading justification, especially popular during the nineteenth century, was esthetics. Immigrant peoples, whether on the coasts of North America or in the valleys of New Zealand, sometimes longed for the familiar birds and beasts they had known in the old country. So they set about importing them, with the deliberate aim of establishing wild populations. Introductions were carried out haphazardly, in almost total ignorance of what are now called ecological principles. Usually these experiments ended somewhere between disappointment and disaster.

Economic Benefit Some exotics arrived through the efforts of entrepreneurs hoping to gain wealth. The nutria (*Myocastor coypus*), a large, aquatic rodent from South America, was sent to hopeful breeders in the southeastern United States. When the fur value of nutria proved to be less than expected, many breeders simply released their animals into local waterways. Other nutria escaped. These destructive rodents are highly fecund and have become established in at least 18 states (Laycock, 1966). Nutria reach very high population densities, inflicting serious damage upon rice and other crops and hampering irrigation and other water control projects by tunneling into levees. They have also altered native aquatic vegetation (Courtenay, 1978).

One very risky practice is the introduction of carnivores that are supposed to control another introduced pest. Sugar cane growers in the West Indies suffered serious crop damage from black rats (*Rattus rattus*), which, as stowaways on ships, had introduced themselves to the islands. E. B. Espeut thought he had the solution when he introduced 4 male and 5 female Indian mongooses on Jamaica in 1872. From Espeut's early and optimistic reports on how efficiently these predators reduced rats, cane growers on Cuba, Puerto Rico, Barbados, and other islands demanded and received mongooses. Although the mongoose did kill rats, the predator's diurnal habits and high rate of reproduction caused it to turn its attention to other prey. These included poultry, puppies, kittens, and a variety of native wild species. Mongooses severely reduced populations of short-tailed capromys (*Capromys brachurus*), ground dove (*Columbigallian passerina*), Jamaican petrel (*Aestrelata caribboea*), plus at least 5 species of snake and 20 species of lizard (Laycock, 1966).

Sport Hunting Besides the ring-necked pheasant, the gray or Hungarian partridge (*Perdix perdix*) and the chukar partridge (*Alectoris graeca*) were successfully introduced into the United States as sport animals. So were the fallow deer (*Dama dama*), the sika deer, the axis deer (*Axis axis*), the barbary sheep (*Ammotragus lervia*), and many other exotic big game. The potential benefits to private landowners are considerable. For minimum operating costs, the landowner can lease hunting rights for exotic animals. Where state laws permit, landowners may actually sell them to hunters on a per-head basis. Sport hunting has replaced esthetics as the number one justification for importing and liberating exotic birds and mammals. Table 9-3 lists a recent estimate of the numbers of exotic big game in the United States.

TABLE 9-3.
ESTIMATED POPULATION SIZES FOR EXOTIC UNGULATES IN THE UNITED STATES
(Decker, 1978; Mungall, 1978 © Wildlife Management Institute. Used by permission.)

Species	Main locations	Population size
Fallow deer (*Dama dama*)	Southern and southcentral states	3,200
Barbary sheep (*Ammotragus lervia*)	New Mexico and Texas	3,800
Axis deer (*Axis axis*)	Texas	7,000
Sika deer (*Cervus nippon*)	Maryland	5,700
Blackbuck (*Antilope cervicapra*)	Texas	7,000
Sambar (*Cervus unicolor*)	Florida and California	50–190
Persian ibex (*Capra aegagrus*)	New Mexico	200
Gemsbok (*Oryx gazella*)	New Mexico	250
Thar (*Hemitragus jemlahicus*)	California	150

Endangerment in Native Habitat When the first Indian blackbuck antelope (*Antilope cervicapra*) were released on a Texas ranch in 1932, the owners hoped that they would supply viewing pleasure for themselves and their guests. Eventually, perhaps the ranchers hoped to hunt these handsome little bovids. Tens of thousands of blackbucks then thrived in India. The Indian population has since plummeted from an estimated 80,000 in 1947 to one-tenth that number in 1964. As the losses continued in India, the numbers of blackbuck in Texas grew from over 1,000 in the year 1955 to more than 4,000 in the year 1966 and over 7,300 by the year 1974 (Mungall, 1978).

Even though the blackbuck antelope was originally introduced for other reasons, it now serves as an example of a new justification for exotic introduction. Most endangered species are threatened with extinction through habitat loss. If similar habitat types are available in more secure regions of the world, should some of the species not be introduced? This justification is more compelling than the others, particularly when the only other alternatives are certain extinction or relegation to captive breeding in zoos.

Yet many specialists who have studied exotic introductions closely remain just as opposed to exotic introductions designed to save endangered species as they are to all other justifications. The reason is that the ecological effects of any exotic animal population are likely to be both complex and variable, hence impossible to predict with any real certainty. Courtenay (1978:241), for example, regards the introduction of exotics that are endangered as "merely a ploy to introduce a new species and is still biological roulette." Pressure to introduce endangered species will probably mount as their numbers decline toward extinction. The question will then be whether the benefit to the endangered species is sufficient to offset the hazards to native species and crops and to the health of wildlife, livestock, and humans.

Costs of Exotic Introductions

The problems caused by exotic introductions have been reviewed by de Vos and Petrides (1967) and by Laycock (1966). Some exotics such as the European rabbit in Australia, the black rat throughout much of the world, and the starling in North America have inflicted substantial losses upon agricultural crops. Other species outcompete native wildlife. Examples include American gray squirrels (*Sciurus carolinensis*), who compete with native red squirrels (*Sciurus vulgaris*) in England and the mynah (*Acridotheres tristis*), which is reducing native birds in Oceania through displacement.

Aggressive, adaptable competitors such as the starling pose severe threats for native species. In most cases, though, the native species have the "home court" advantage, as did most of the native New Zealand birds described in the next section. Competition may then act to limit the population of the exotic species, perhaps by confining it only to certain habitat types.

Introduced species can bring damaging parasites and diseases with them. As mentioned in Chapter 8, the accidental introduction into Hawaii of mosquitoes provided a disease vector for the island birds. Although livestock clearly differ from wildlife, they have been introduced into new environments throughout the world and are therefore exotic species. Blackhead, a serious infectious disease of wild turkeys, arrived in North America with a shipment of domesticated chickens. Duck virus enteritis, mentioned later in this chapter, apparently made its way into the United States by way of domestic ducks. Fowl cholera first appeared in North America in the late nineteenth century, presumably introduced, like DVE, through shipments of domestic waterfowl.

Another argument raised against the introduction of exotics is that their management may divert important resources away from management of native wildlife (Craighead and Dasmann, 1966). This argument applies particularly to those exotics that might be introduced onto public land.

If an exotic introduction reaches very high numbers, it can cause habitat damage. The nutria in the United States is a familiar example, as is the red deer (*Cervus elaphus*) in New Zealand. Both increased to the point of local and permanent alteration of vegetation. Expensive and time-consuming control measures became necessary.

New Zealand—Where Exotics Really Earned a Bad Name

Whenever opponents of exotic introductions want to find examples supporting their position, they need only look to New Zealand. In retrospect, that mountainous, island country was one of the worst places on earth to practice large-scale exotic introductions. Its flora and fauna developed far from continental influences. The only native mammals were a couple of species of bats whose ancestors had somehow managed the long trip. New Zealand's vegetation evolved in the absence of mammalian herbivores and thus developed no resistance to either grazing or browsing. Much of the island is mountainous and rugged, making control of expanding populations very difficult. Finally, the soil is quite subject to erosion whenever the vegetation is disturbed, as precipitation sometimes exceeds 7,620 millimeters annually (Howard, 1967).

But when the English colonized New Zealand, beginning in the 1840s, they sorely missed many of the animals that they had known in England, Europe, and elsewhere. They proceeded to introduce more than 200 species of wild vertebrates, 91 of which became permanently established (see Table 9-4). Six species of mammals were feral forms of livestock and three rodents were self-introductions that had escaped from ships. At least 29 of the mammalian species are considered serious pests today (Howard, 1967; Laycock, 1966), especially the red deer, the feral pig, and the feral goat. The government grants no protection to the more troublesome exotics; it not only encourages sport hunting but also employs dozens of professional hunters to try to keep exotic populations from reaching even higher numbers.

Those unfamiliar with the situation in New Zealand may wonder why the government has not introduced large predators, such as the cougar, to help control exotic herbivores. To begin with, introduction of an exotic predator to control an exotic prey is itself an especially hazardous undertaking, as in the case of the mongoose introduced into Jamaica to control rats. Besides, the domestic sheep industry is very important in New Zealand and would surely look unfavorably on the introduction of large carnivores. Finally, New Zealand has suffered such damage from exotics that its government will never again have an open policy of exotic introductions, including predatory species.

About the only good to come out of New Zealand's tragic experiences with exotics has been in the knowledge gained about interspecies competition and about population biology. The success rate for establishing exotic

TABLE 9-4.
SPECIES OF EXOTIC VERTEBRATES INTRODUCED INTO NEW ZEALAND
(Wodzicki, 1965 © Academic Press, Inc. Used by permission.)

Class	No. of species introduced	No. of species established	% Established
Fishes	15	13	86.7
Amphibians	5	3	60
Reptiles	3	0	0
Birds	138	43	31.1
Mammals*	54	32	59.3

*Includes 6 domesticated species—cat, cattle, sheep, goat, pig, and horse, which have established feral populations. Of these, feral horses, sheep, cattle, and goats are restricted or else are only locally common. Feral pigs and cats remain widespread and abundant.

mammals (59.3 percent) was nearly twice that for exotic birds (31.1 percent). In addition, exotic mammals have become established equally in both disturbed and undisturbed (late-successional) habitats (see Figure 9-2). Three-fourths of the exotic birds that survived are found exclusively in disturbed habitats and all the remaining 25 percent occur in both undisturbed and disturbed areas. Wodzicki (1965) explained both the differences in success rates and the habitat use of surviving exotics as an outcome of competition with native species. As New Zealand had virtually no native mammals, the exotic mammals were able to survive in large numbers and to occupy all available habitat. Exotic birds, on the other hand, had to compete with an established avifauna, which had the competitive advantage in natural or undisturbed habitats.

Large mammals are especially vulnerable to extinctions. Hunting mortality generally tends to be additive (see Chapter 7) for large mammals, and the energetic requirements combined with limited dispersal abilities require large reserves to sustain populations (see Chapter 8). As a consequence, biologists rarely are able to use techniques which require large-scale destructive sampling, particularly of the females. Since the large and destructive populations of large mammals in New Zealand were afforded no protection, the opportunity for rigorous dissection arose. Large mammal populations throughout the world may eventually benefit from the knowledge acquired from the New Zealand exotics.

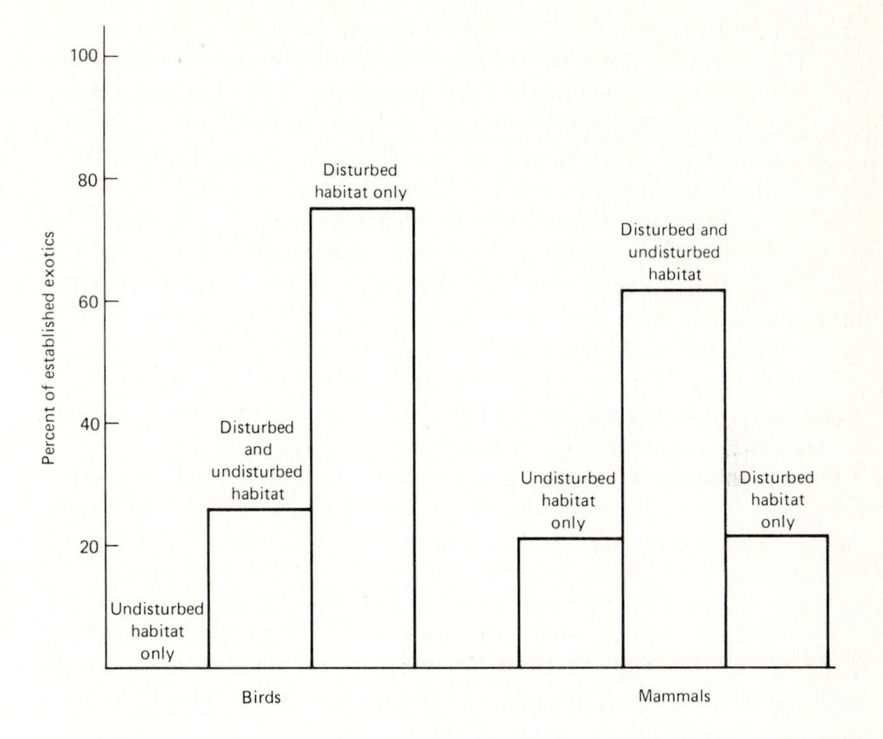

FIGURE 9-2
The occurrence of exotic birds and mammals in New Zealand in relation to habitat type. (*Wodzicki, 1965* © Academic Press, Inc. Used by permission.)

Some Exotics Are Riskier than Others

Fortunately, exotic introductions will probably never again be carried out on the scale that they were in the nineteenth century. The usual justification for exotic introductions are weak, particularly when weighed against the potential and well-documented hazards. Nevertheless, some species have characteristics which are clearly very hazardous and thus special care should be taken to protect against their introduction. Species which are *r*-strategists, for example, are always high risks and so are species with broad niche breadths, such as starlings and mongooses. Species with social behavior such as territoriality, which at times tends to regulate population size, are probably safer than are species limited purely by food supply.

Other risk factors are associated with introduction sites. Species introduced into open terrain would prove much easier to control than species introduced into deep forests or rugged, remote mountains. Introductions onto islands are far more hazardous than introductions onto continents. Risks are summarized in Table 9-5 and discussed in greater detail in Riney (1967) and Teer (1979). Species fitting the descriptions in Table 9-5 should, as a general rule, never be introduced. But even those that (by those simple criteria) rate as low risks can pose environmental hazards.

Although biologists may be able to use ecological and reproductive data to define the riskiest species of proposed exotics, the fact remains that species appearing to be far less hazardous may still pose serious ecological threats. Even the most ideal and detailed investigations can never guarantee the outcome of an attempted introduction. The range of ecological interactions between the exotic species and the new environment are simply too complex and too variable to be subject to precise analysis. Predictions still remain more of an art than a science.

Bruce Coblentz of Oregon State University, a specialist on exotic ungulates, considers exotic introductions to be one of the "big three" categories of human impacts on natural environments. Unlike the other two categories, pollutants and inappropriate resource use, exotics may be far more permanent. Pollution and inappropriate resource use can be stopped and in most cases their effects will eventually subside. Yet once an exotic species becomes firmly established, it is virtually impossible to eradicate. This unique feature should give pause to anyone contemplating an exotic introduction for whatever reason.

DISEASES AND PARASITES

More than half a century ago Leopold (1933) concluded that the effects of diseases upon wildlife had probably been radically underestimated. They

TABLE 9-5.
CHARACTERISTICS OF A SPECIES AND THE HABITAT INTO WHICH IT IS INTRODUCED THAT RELATE TO ENVIRONMENTAL RISK

Higher risk	Lower risk
r-strategist	*k*-strategist
Wide niche breadth	Limited niche breadth
Gregarious, nonterritorial	Solitary, territorial
Introduced on to oceanic island	Introduced onto continent
Introduced into mountains or deep, extensive forests	Introduced into open grassland or savanna

still are, and for several reasons. Although ecologists have developed rigorous theories and elaborate models to explain relationships between competitors or between prey and predators, they have devoted far less energy toward explaining relationships between host and disease. Another reason for the tendency to underestimate effects of diseases and parasites is the simple fact that obviously sickened wild animals are only rarely seen. Besides, wild animals usually cannot be rounded up and treated the way livestock can. Finally, many wildlife managers have been reluctant to look too closely for diseases or parasites for fear that their discovery might imply poor management practices.

Ecologically, infectious parasites and diseases can be placed into three categories: endemic, introduced, and those transmissible among species. The usual assumption is that diseases and parasites act as simply another form of density-dependent mortality, are compensatory, and hence are of little concern. This assumption might prove to be correct in the majority of cases, particularly among diseases that have long been endemic. But the exceptions can prove disastrous.

The very management practices that boost wildlife populations can also increase the chances of infectious diseases and may greatly raise levels of mortality. As natural habitat becomes scarcer, wildlife refuges and other habitat islands may harbor more and more animals, especially as habitat quality improves. Infectious diseases and parasites may be thwarted to some extent because infected animals may find it difficult to immigrate to habitat islands. But when they do, the high populations maintained through habitat management may prove highly vulnerable. Wildlife managers therefore can ill afford to ignore the effects of diseases. Until knowledge of the interplay between wildlife and their infectious diseases is far improved, both the timing as well as the effects of disease will remain largely unpredictable.

This section makes no attempt to catalogue all known wildlife parasites and diseases. Instead it describes examples of particularly troublesome diseases from an ecological standpoint. Treatment of individual wild animals showing clinical signs of disease is usually impractical and may also be undesirable. So prevention and control through manipulation of habitats and populations may be the most sensible approaches.

Fowl Cholera and Duck Virus Enteritis

Being both migratory and highly gregarious, waterfowl are especially susceptible to infectious diseases. Two of the more serious diseases, fowl cholera and duck virus enteritis, were both introduced into North America from Europe by way of domestic ducks. Since North American waterfowl have little if any resistance to these "new" diseases, their potential effects on duck and goose populations are very great.

Fowl cholera is caused by the bacteria *Pasteurella multocida.* The disease was first discovered among domestic fowl in Europe and has been found almost worldwide since then, presumably spread in shipments of domestic birds. Cholera appeared in North American waterfowl in 1944, with outbreaks in the Texas panhandle and the San Francisco Bay area (Jensen and Williams, 1964). Although the disease is not widespread, it can cause extremely high local mortality. Once an outbreak of fowl cholera begins, the effects may be reduced by raising water levels (Jensen and Williams, 1964). The higher water levels may dilute the infective materials and allow the birds to spread out over a larger area, thereby reducing crowding.

Duck virus enteritis (DVE) is a more recent arrival. This highly infectious disease was first diagnosed in the Netherlands in 1923 and first appeared in North America among domestic ducks in 1967 (Bellrose, 1976). The first American cases occurred on Long Island and within two years, wild waterfowl along the east coast were infected. The first large outbreak occurred among an especially dense concentration of waterfowl at Lake Andes National Wildlife Refuge (SD) in January 1973. By the time the outbreak was over, more than 40,000 ducks and geese had died. Survivors exposed to the disease subsequently dispersed over much of North America, establishing the virus widely.

Duck virus enteritis is caused by a herpes virus. Like other herpes viruses, it can retreat suddenly into the central nervous system of infected animals, leaving them without apparent symptoms. The virus may remain inactive throughout the individual animal's life, or it may erupt without warning months or even years later, leaving the animal critically ill and highly infective.

Heavy concentrations of ducks and geese seem to invite outbreaks of infectious diseases. Since modern management practices tend to concentrate waterfowl, disease problems will very likely become more severe. Perhaps some forms of social stress brought on by excessive crowding make the birds more susceptible, thereby triggering the outbreaks. Or perhaps those outbreaks that take place where numbers are highest are more noticeable and simply attract the most attention. Whatever the relationship between crowding and infectious diseases, the problem will ultimately constitute an important challenge to waterfowl managers in the years to come.

Meningeal Worm

Some parasites and diseases affect more than one species of host. Those that do seldom affect all species equally, and this imbalance can pose serious problems for managers. An example is the brain or meningeal worm (*Parelaphostrongylus tenuis*), a helminth which normally infects white-tailed deer without apparent harm. Eggs of this parasite pass out with the feces of infected deer. From there the larvae hatch and penetrate the foot of any one of several species of land snail, which serve as intermediate hosts. The snails in turn inhabit the vegetation, and the browsing deer inadvertently swallow infected snails, thereby completing the life cycle.

Although harmless to white-tailed deer, meningeal worms are deadly to moose, caribou (*Rangifer tarandus*), reindeer, elk, and mule deer. The clinical signs of this disease have been described as "moose sickness" and "neurologic disease," in which the accidental host exhibits listlessness, restlessness, and eventually paralysis and death (Anderson and Prestwood, 1981). Reindeer introduced onto a large island in Georgian Bay, Ontario, became infected within a single summer in the presence of white-tailed deer. The complete failure of this introduction was attributed to meningeal worm (Anderson, 1971).

Managers face some tough decisions in dealing with this parasite. Habitat disturbance throughout much of North America generally favors white-tailed deer over other cervids. Should management practices or land use changes increase the deer populations, the areas in which these increases occur become less suited for elk, moose, or caribou as parasite loads build up. In some cases, other cervids may survive only because of local habitat separation during the critical warmer months when the snails are active.

Another management problem results from deliberate transplants of infected white-tailed deer into new areas, thereby spreading the parasite wherever suitable snails occur. In Oklahoma, for example, the state's present deer population was largely established with deer transplanted from five different sites. At least one of these sites is now known to have harbored meningeal worm (Kocan et al., 1982).

Some disturbing evidence exists that meningeal worm is increasing its geographic range as white-tailed deer become more abundant. If it becomes established in the Rocky Mountain region, the effects could be quite drastic on mule deer, moose, and elk.

To sum up, infectious diseases and parasites, as with competition and predation, can have significant effects upon wildlife populations. Introductions of these agents along with exotic animals present especially severe threats. Problems may also arise when human activity brings about habitat changes that result in one species' intrusion into the geographic range of another relative, bringing an infectious agent with it. Finally, wildlife management practices themselves, by concentrating wild animals in very high numbers, may contribute both to the likelihood of disease outbreaks and to their severity.

Botulism and Lead Shot Poisoning

Infectious diseases seem to affect waterfowl populations more than those of other groups of game animals. Yet some noninfectious diseases pose significant threats as well. Over the years, the greatest losses have been to botulism, particularly in the western states. Botulism is a severe form of poisoning brought on by metabolism in the anaerobic bacteria *Clostridium botulinum.* The bacteria is quite common, and it normally exists in a dormant and harmless state. Trouble arises, though, when lowered water levels and high summer temperatures turn feeding areas into stagnant pools. With oxygen levels greatly reduced, these pools become ideal habitat for the anaerobic *Clostridium.* The bacteria multiply rapidly, exotoxin levels rise, and ducks feeding in the pools ingest the toxins and die.

Botulism can largely be avoided by controlling water levels so as to minimize the occurrence of stagnant conditions. Aquatic invertebrates may play a significant role in botulism outbreaks, either by concentrating the toxin in forms palatable to waterfowl (Jensen and Williams, 1964) or by furnishing an attractive medium for bacterial growth through their own death and decomposition (Bellrose, 1976). Water level control is not always practical as a control measure. Managers must sometimes use scare devices to keep waterfowl away from especially dangerous areas.

A problem of more recent origin is lead shot poisoning. Years of hunting have left the bottoms of some heavily hunted feeding areas littered with spent shot. Bellrose (1964) estimated that as many as 1,400 shot pellets may be deposited in feeding areas for every duck killed by gunfire. These shot pellets typically fall into the very best feeding areas, as these are the most popular hunting spots. Ducks feeding in these locations often swallow lead pellets. A heavy metal, lead can prove toxic to waterfowl, even at fairly low levels. Experiments have shown that ingestion of one no. 6 shot pellet can increase mortality by about 22 percent and that 4 such pellets cause an increase of 41 percent (Bellrose, 1959). A conservative estimate is that 2 to 3 percent of the fall and winter waterfowl populations dies each year of lead shot poisoning (Bellrose, 1976).

After thorough testing of various substitutes for lead shot, the U.S. Fish and Wildlife Service began to require hunters to use steel shot in the more heavily hunted areas. The intent is to reduce the incidence of lead poisoning by reducing the amount of shot pellets deposited in feeding areas.

WETLANDS

A wetland is any area of land covered by water for at least part of the year. Unlike lakes, wetlands are relatively shallow, with depths generally not exceeding 2 meters unless emergent aquatic vegetation is present (Cowardin et al., 1979; Horwitz, 1978). Hence estuaries, swamps, marshes, and bottomlands are all classed as wetlands.

Wetlands received a marked increase in attention from conservationists in the 1970s. This interest resulted from two problems. First, wetlands were disappearing at an ever increasing rate as more lands were converted to more profitable short-term uses. Second, wetlands are among the most productive of all ecosystems, supporting rich plant life and the fauna dependent upon it. Losses of wetlands quickly translate into losses of waterfowl, muskrats, various shore and marsh birds, and many other species.

Although the precise extent of wetland losses is difficult to determine, some estimates illustrate the magnitude of the problem. The United States has lost at least 35 to 40 percent of its original wetlands (Shaw and Fredine, 1956; Bellrose, 1976), with losses of certain freshwater types running as high as 90 percent (Horwitz, 1978). Wetlands are lost to filling, dredging, and flooding and are also susceptible to certain kinds of pollution. Greatest losses occur when wetlands are converted to farmlands at an estimated rate of half a million hectares (1,250,000 acres) annually (Allen, 1981). Wetland declines are closely pegged to rising prices for crops such as soybeans, cotton, wheat, corn, tobacco, peanuts, and feed grains.

The effects of wetland losses upon wildlife have been documented most thoroughly for the waterfowl. In the northern prairies of the United States, wetland losses since the late nineteenth century have reduced annual waterfowl production by an estimated 6 million annually (National Academy of Sciences, 1970).

Reductions in wetlands cause losses of more than just ducks and geese. Besides producing waterfowl, wetlands help control floods by moderating the effects of runoff. During heavy, prolonged rains, wetlands act as giant sponges to absorb water that might otherwise contribute to serious flooding (see Figure 9-3). The effects of drought may also be reduced simply by preserving wetlands, which retain water for longer periods than would similar areas channelized, drained, or filled.

The high productivity of wetlands makes them useful for disposing of several kinds of organic pollution. They have been described as solar-powered, self-maintained tertiary treatment systems for sewage and industrial wastes (Odum, 1979). Growth of plants is so rapid that they absorb enormous amounts of nitrogen, phosphorous, heavy metals, certain chemicals, pathogens, and waste water (Kadlec and Kadlec, 1979). Unlike artificial waste treatment systems, wetlands provide their services free.

Protection of Wetlands

Despite the considerable long-term values of wetlands, short-term economic expectations continue to favor their development and subsequent

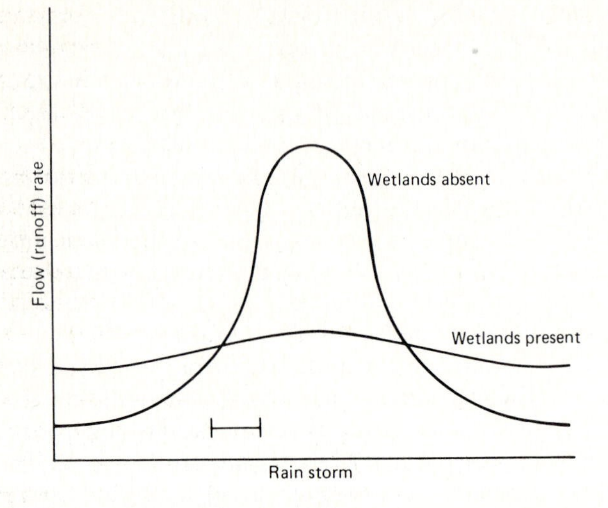

FIGURE 9-3
The effects of wetland upon runoff. (*Odum, 1979* © American Water Resources Association. Used by permission.)

loss. Since loss of wetlands means a loss of wildlife and of flood and pollution control, it is in the public interest to have wetlands protected whenever possible. Government programs are usually required to curb losses of this type of habitat.

The authority of the U.S. government to regulate development of wetlands received an important boost with passage of the Federal Water Pollution Control Act of 1972. Section 404 of the act expanded the legal authority of the Army Corps of Engineers from "all navigable waters" to "all waters of the U.S." It also authorized the Corps to regulate discharge of materials into these waters in accordance with Environmental Protection Agency (EPA) guidelines. The EPA can, in collaboration with the Secretary of the Army, prohibit activities with adverse or unacceptable effects upon wildlife and fisheries, recreation areas, shellfish beds, or municipal water supplies (McCormick, 1978). The Corps has developed policy that no alteration of wetlands will be approved unless shown to be in the public interest (Horwitz, 1978).

America's dwindling wetlands received more help through two executive orders from President Carter on May 24, 1977. The first (E.O. 11988, "Floodplain Management") discouraged filling along floodplains through regulation and licensing of developers. The second (E.O. 11990, "Protection of Wetlands") directed federal agencies to protect and preserve wetlands on federal lands (McCormick, 1978).

Prairie Potholes

The northern prairies of the United States and Canada were once dotted with thousands of small, shallow glacial lakes called "potholes." Every spring these potholes became favored breeding grounds for ducks and other aquatic and semiaquatic birds. Although prairie potholes make up only about 10 percent of North America's waterfowl breeding grounds, they may produce more than half the continent's ducks (Smith et al., 1964).

When the first inventory of America's prairie potholes was made in 1964, biologists estimated that just over 1 million hectares (2.5 million acres), probably less than half the original area, remained (National Academy of Sciences, 1970). The 4 years following the survey saw annual pothole losses in Minnesota, North Dakota, and South Dakota, averaging 12,500 hectares (31,250 acres) (Horwitz, 1978). Virtually all these losses occurred when potholes were drained and filled for growing crops. High crop prices in the face of international grain markets will likely cause further losses.

The U.S. Fish and Wildlife Service has attempted to offset these losses through establishment of Waterfowl Production Areas (WPAs) to supplement the National Wildlife Refuge System. Purchased with duck stamp revenues, these WPAs totaled nearly 400,000 hectares (1 million acres) by 1980 in North Dakota alone (Weller, 1981). Private organizations such as the Nature Conservancy and The Audubon Society also acquire tracts of wetlands. Sometimes they deed such tracts over to federal or state agencies. At other times they retain sanctuaries or refuges, managing these tracts themselves.

Bottomland Hardwoods

Another vanishing wetland type is the bottomland hardwood ecosystem in the southern United States. Bottomland hardwoods occur on the rich alluvial soils of the Mississippi Delta. These soils, deposited over centuries by the river, grow enormous hardwood trees. The resulting deep shade, combined with periodic flooding, keeps understory vegetation fairly open. Deer, turkey, squirrel, and other wildlife thrive on the rich mast crops.

Yet bottomland hardwood habitat, threatened first by logging and more recently by expanding agriculture, is greatly reduced and fast disappearing. Soaring soybean prices during the 1960s and 1970s, in particular, prompted clearing of these lowland hardwood forests. Concerned about these losses, the U.S. Fish and Wildlife Service surveyed bottomland hardwoods along the Mississippi Delta of six states—Missouri, Kentucky, Arkansas, Tennessee, Louisiana, and Mississippi. They learned that of the estimated 9.6 million hectares (24 million acres) of delta originally forested, only 4.7 million (11.8 million acres) remained by 1937, a reduction of over 50 percent. From 1937 to 1977, another 2.6 million hectares (6.5 million acres) were cleared. By 1995, only about 1.5 million hectares (3.8 million acres) are expected to contain bottomland hardwood forests, a mere 16 percent of the original area (see Figure 9-4).

The Mississippi Delta contained only about 100,000 hectares (250,000 acres) of soybeans in 1937. More and more bottomland was cleared as soybean prices rose, bringing the total to 3 million hectares (7.5 million acres) by 1972, a 15-fold increase (U.S. Fish and Wildlife Service, n.d.). Missouri was especially hard hit, losing an estimated 96 percent of its original bottomland hardwood forests between 1870 and 1975 (Korte and Frederickson, 1977).

Thus the extent of bottomland hardwood depletion exceeds even that for prairie potholes. In addition, the bottomlands require centuries of natural succession to achieve late-successional stages (see Chapter 2), making them far harder to restore. If some of the soybean fields along the Delta were abandoned tomorrow to nature's reclamation, no American living today would ever see mature bottomland forests there.

FIGURE 9-4
Loss of bottomland hardwoods. (*U.S. Fish and Wildlife Service, n.d.*)

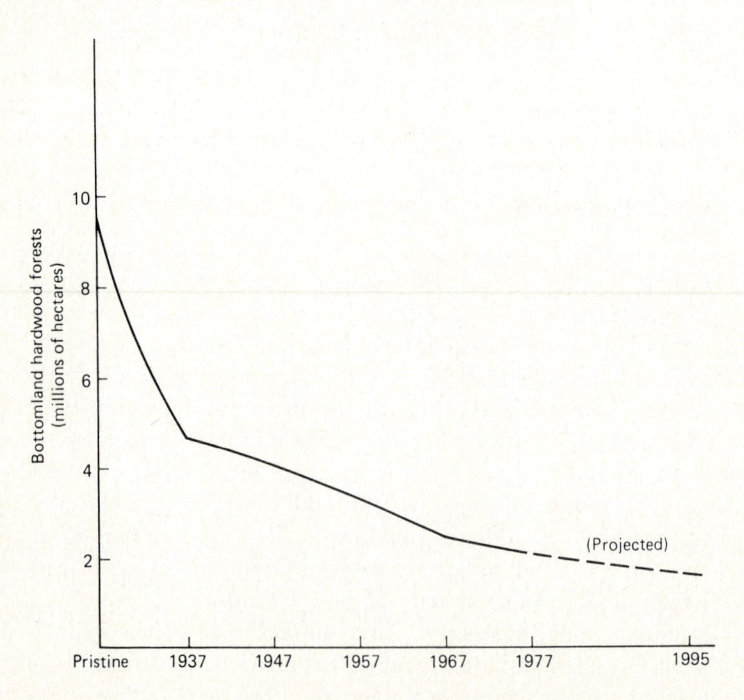

Management of Wetlands

If any appreciable areas of wetlands are to survive, they must have some form of protection from development. One quite common way of preserving wetlands is through public acquisition and subsequent management as a park or wildlife refuge. Yet public acquisition often generates serious political opposition. Sometimes the opposition is on philosophical grounds, with opponents contending that federal or state governments should not own additional land. A more common and persistent problem with public acquisition is that it removes tracts of land from county tax rolls. This objection has been met in part through revenue-sharing programs. Small portions of the original purchase price plus a larger percentage of income from crops, timber, or oil and gas leases, is allocated to county governments (National Academy of Sciences, 1970). Thus, although not taxed directly, the public agency can compensate for some of the lost revenue through revenue sharing.

Once acquired, whether through public or private acquisition, wetlands may be manipulated to achieve specific management objectives. Many marshes, for example, can be manipulated by means of water-level control and control of grazing (Weller, 1978, 1981).

Water-level control structures make it possible to drain marshes periodically. Draining can reduce excessive muskrat populations, increase rates of nutrient turnover, and rid marshes of carp (*Cyprinus carpio*), which tend to muddy marsh waters, thus reducing sunlight and hampering productivity. Many important species of marsh plants, such as many species of cattail (*Typha* spp.) and arrowhead (*Sagittaria* spp.), thrive only in shallow waters. Some semiaquatic plants germinate best under lowered water conditions (Weller, 1981). Conversely, water-level control also makes it possible to raise water levels to manipulate the marsh in the opposite direction.

Although wetlands are quite productive, heavy pressures from herbivores can alter the character of a marsh. Among the wild grazers, one of the most fecund is the muskrat. Adult females can produce up to four litters of six or seven young per adult female annually. If unchecked, such rapid population growth leads to an "eat-out," in which virtually all the emergent vegetation is removed, opening up the marsh. Trapping pressure can help control muskrat populations, as can systematic lowering of water levels.

Edge seems to be especially important in maintaining wetland wildlife diversity. Vegetative structure seems to be more important than taxonomic composition. Maximum interspersion, along with a ratio of open water to emergent vegetation of from 1 to 1 to 2 to 1, is probably a good all-around combination (Weller, 1978).

SUMMARY

This chapter introduces four special problems in wildlife management. The first two, predator control and exotic introductions, have been controversial for the past 50 years. The third problem, infectious diseases and parasites, promises to increase in severity in the decades to come. Wetlands, some of the most productive and rapidly disappearing of wildlife habitats, need special conservation efforts.

Most predator control has been undertaken to protect livestock. Since federal programs in predator control began in 1915, emphasis has shifted away from the larger species of carnivores and toward the coyote. Many control measures have been used, the most controversial being the use of poisons and steel leg-hold traps. In recent years, predator control research has taken two new routes: methods intended to be more selective, and methods that reduce or alleviate predator losses without killing the predators. More selective methods include aerial gunning, the use of toxic collars, and extension trapping. Nonlethal alternatives include aversive conditioning, use of guard dogs, and predator insurance.

Early in this century, predator control was widely used in game management. Increased understanding of predation, together with restored game abundance, led to a gradual decline of this practice. During the 1970s researchers learned that short-term, intensive predator control can lead to greater game production. If certain conditions of habitat and harvest can be met, predator control, on a much more refined basis, may once again see application in game management.

Exotic introductions have been justified on a variety of grounds including esthetics, economic incentives, sport hunting, and most recently, as a means of saving species endangered in their native habitats. Despite disappointments and even some ecological disasters, advocates continue to encourage exotic introductions, although even such benign exotics as the ring-necked pheasant still have adverse side effects.

Oceanic islands are especially vulnerable to the ecological consequences of exotic introductions. New Zealand makes a good case study because colonists there introduced more than 200 species of wild vertebrates. Slightly less than half the species became established, but of those that did, about one-third became serious pests.

The esthetic argument in favor of exotic introductions has waned considerably in the past decades, but the others, to varying degrees, remain. Biologists can effectively screen candidate species well enough to eliminate the very worst ecological offenders such as starlings and mongooses. But even the most *K*-selected endangered species could cause ecological problems, and biologists cannot predict the long-term consequences of any exotic species.

Infectious diseases fall into three ecological categories: endemic, introduced, and those affecting more than one species of host. Most endemic diseases usually act in a density-dependent way and seldom pose serious management problems. When diseases or parasites are first introduced into a region, the host population has little natural immunity and no acquired immunity. The effects of such diseases, such as fowl cholera or duck virus enteritis, can therefore be devastating. Parasites or diseases affecting more than one species of host seldom affect all of them equally. An example is the meningeal worm, a parasite harmless to white-tailed deer but deadly to other species of cervids. Land use changes or management practices that favor white-tailed deer will likely reduce other cervid populations wherever meningeal worm occurs.

Ecologists and wildlife managers know little about the ecological relationships between infectious diseases and their hosts. This shortcoming is made more serious by the fact that human activities in general and wildlife management in particular are likely to increase the effects of infectious diseases and parasites upon wildlife populations.

Some noninfectious diseases such as botulism and lead shot poisoning can have serious effects on waterfowl populations. The most effective treatments are directed at the habitat. Botulism can be controlled by minimizing stagnant water conditions and by keeping wildlife away from such poisoned areas. Lead shot poisoning can be reduced by requiring hunters to use steel shot in heavily hunted areas.

Wetlands are highly productive wildlife habitats that also play important roles in flood and pollution control. Between one-third and one-half of the natural wetlands in the United States have been drained or filled, primarily for agriculture. Direct protection, often through federal acquisition, is a common prerequisite for preservation of wetlands. Two case studies include prairie potholes and bottomland hardwoods.

Freshwater marshes also require manipulation, usually through water-level control, grazing control, or both, to regulate aquatic vegetation and control populations of carp and muskrats. Like many terrestrial habitats, marshes can be improved through increasing interspersion, in this case between open water and emergent vegetation.

REFERENCES

Allen, D. L. 1981. Private lands as wildlife habitat. In R. Dumke, G. Burger, & J. Marsh (Eds.), *Wildlife management on private lands.* Madison, Wisc.: Wisconsin Chapter of the Wildlife Society.

Anderson, R. 1971. Neurologic disease in reindeer (*Rangifer tarandus*) introduced into Ontario. *Canadian J. Zool.* 49: 159–166.

Anderson, R. 1972. The ecological relationships of meningeal worm and native cervids in North America. *J. Wildl. Dis.* 8: 304–310.

Anderson, R., & A. Prestwood. 1981. Lungworms. In W. Davidson et al. (Eds.), Diseases and parasites of white-tailed deer. *Misc. Publ. No. 7,* Tallahassee, Fla.: Tall Timbers Research Station.

Armstrong, W., & D. Harmel. 1981. Exotic mammals competing with the natives. *Texas Parks and Wildlife Magazine.* February 1981.

Beasom, S. 1974a. Intensive short-term predator control as a game management tool. *Trans. N. Am. Wild. and Natur. Resour. Conf.* 39: 230–240.

Beasom, S. 1974b. Relationships between predator removal and white-tailed deer net productivity. *J. Wildl. Manage.* 38: 854–859.

Bellrose, F. 1959. Lead poisoning as a mortality factor in waterfowl populations. *Ill. Nat. Hist. Surv. Bull.* 27: 235–288.

Bellrose, F. 1964. Spent shot and lead poisoning. In J. Linduska (Ed.), *Waterfowl tomorrow.* Washington, D.C.: USDI Fish and Wildlife Service.

Bellrose, F. 1976. *Ducks, geese, and swans of North America.* Harrisburg, Pa.: Stackpole Books.

Cain, S., J. Kadlec, D. Allen, R. Cooley, M. Hornocker, A. Leopold, & F. Wagner. 1972. Predator control—1971. *Report to the Council on Environmental Quality and the Department of the Interior by the Advisory Committee on Predator Control.* Ann Arbor, Mich.: University of Michigan Press.

Connolly, G. 1978. Predators and predator control. In J. Schmidt and D. Gilbert (Eds.), *Big game of North America.* Harrisburg, Pa.: Stackpole Books.

Connolly, G., R. Griffiths, Jr., & P. Savarie. 1978. Toxic collar for control of sheep-killing coyotes: A progress report. *Proc. 8th Vert. Pest Control Conf.* 8: 197–205.

Connolly, G., & W. Longhurst. 1975. The effects of control on coyote populations. University of California, Davis. *Div. Agr. Sci. Bull. 1872.*

Connolly, G., R. Timm, W. Howard, & W. Longhurst. 1976. Sheep killing behavior of captive coyotes. *J. Wildl. Manage.* 40: 400–407.

Cowardin, L., Virginia Carter, F. Golet, & E. Lakoe. 1979. *Classification of wetlands and deepwater habitats of the United States.* Washington, D.C.: Office of Biological Services, U.S. Fish and Wildlife Services. FWS/OBS 79-31.

Courtenay, W., Jr. 1978. The introduction of exotic organisms. In H. Brokaw (Ed.), *Wildlife and America.* Washington, D.C.: Council on Environmental Quality.

Craighead, F., & R. Dasmann. 1966. *Exotic big game or public lands.* Washington, D. C.: U.S. Dept. Interior, Bureau of Land Management.

Decker, E. 1978. Exotics. In J. Schmidt and D. Gilbert (Eds.), *Big game of North America.* Harrisburg, Pa.: Stackpole Books.

De Vos, A., & G. Petrides. 1967. Biological effects caused by terrestrial vertebrates introduced into non-native environments. *10th Technical Meeting International Union for Conservation of Nature and Natural Resources,* IUCN Pub. New Series No. 9.

Griffiths, R. Jr., G. Connolly, R. Burns, & R. Sterner. 1978. Coyotes, sheep, and lithium chloride. *Proc. 8th Vert. Pest Control Conf.* University of California, Davis. 8: 190–196.

Gustavson, C., J. Garcia, W. Hankins, & K. Rusiniak. 1974. Coyote predation control by aversive conditioning. *Science* 184: 581–583.

Gustavson C., D. Kelly, M. Sweeney, & J. Garcia. 1976. Prey lithium aversions. I: Coyotes and wolves. *Behavioral Biology* 17(1): 61–72.

Henne, D. 1977. Domestic sheep mortality on a western Montana ranch. In R. Phillips and C. Jonkel (Eds.), *Proceedings of the 1975 predator symposium.* Missoula, Mont.: Montana Forest and Conservation Experiment Station. University of Montana.

Horwitz, Elinor L. 1978. *Our national wetlands: an interagency task force report.* Coordinated by the Council on Environmental Quality, Washington, D.C.

Howard, W. 1967. Ecological changes in New Zealand due to introduced mammals. *10th Technical Meeting International Union for Conservation of Nature and Natural Resources.* IUCN Pub. New Series No. 9, pp. 219–240.

Jensen, W., & C. Williams. 1964. Botulism and fowl cholera. In J. Linduska (Ed.), *Waterfowl tomorrow.* Washington, D.C.: U.S.D.I. Fish and Wildlife Service.

Kadlec, R., & J. Kadlec. 1979. Wetlands and water quality. In P. Greeson, J. Clark, and Judith Clark (Eds.), *Wetland functions and values: the state of our understanding,* Minneapolis, Minn.: American Water Resources Association.

Kie, J., M. White, & F. Knowlton. 1979. Effects of coyote predation on population dynamics of white-tailed deer. *Proc. First Welder Wildlife Foundation Symp.* 1: 65–82.

Knowlton, F. 1972. Preliminary interpretations of coyote population mechanics with some management implications. *J. Wildl. Manage.* 36: 369–382.

Kocan, A., M. Shaw, K. Waldrup, & G. Kubat. 1982. Distribution of *Parelaphostrongylus tenuis* (Nematoda: Metastrongyloidea) in white-tailed deer from Oklahoma. *J. Wildl. Dis.* 18: 457–460.

Korte, P., & L. Frederickson. 1977. Loss of Missouri's lowland hardwood ecosystem. *Trans. N. Am. Wildl. and Natur. Resour. Conf.* 42: 31–41.

Laycock, G. 1966. *The alien animals.* Garden City, N.Y.: The Natural History Press.

Lehner, P. 1976. Coyote behavior: implications for management. *Wildl. Soc. Bull.* 4: 120–126.

Leopold, A. 1933. *Game management.* New York: Charles Scribner's Son's.

Leopold, A. 1949. *A Sand County almanac.* London: Oxford University Press, Inc.

Leopold, A., L. Sowls, & D. Spencer. 1947. A survey of over-populated deer ranges in the United States. *J. Wildl. Manage.* 11: 162–177.

Linhart, S., R. Sterner, T. Carrigan, and D. Henne. 1979. Komondor guard dogs reduce sheep losses to coyotes: a preliminary evaluation. *J. Range Manage.* 32: 238–241.

McCormick, Jack. 1978. Ecology and the regulation of freshwater wetlands. In R. E. Good, D. F. Whigham, and R. L. Simpson (Eds.), *Freshwater wetlands, ecological processes and management potential.* New York: Academic Press.

Mungall, E. 1978. *The Indian blackbuck antelope: a Texas view.* Kleberg Studies in Natural Resources. College Station, Tex.: The Caesar Kleberg Research Program in Wildlife Ecology and the Department of Wildlife Sciences, Texas A & M University.

National Academy of Sciences. 1970. Land use and wildlife resources. *National Res. Council Comm. on Agric. Land Use and Wildl. Resour.*

Odum, E. P. 1979. The values of wetlands: a hierarchial approach. In P. E. Greeson, J. R. Clark, and Judith E. Clark (Eds.), *Wetland functions and values: the state of our understanding.* Minneapolis, Minn.: American Water Resources Association.

Peek, J. 1980. Natural regulation of ungulates (What constitutes a real wilderness?). *Wildl. Soc. Bull.* 8: 217–227.

Riney, T. 1967. Ungulate introductions as a special source of research opportunities. *10th Technical Meeting International Union for Conservation of Nature and Natural Resources,* IUCN Pub. New Series No. 9. pp. 241–254.

Savarie, P., & R. Sterner. 1979. Evaluation of toxic collars for selective control of coyotes that attack sheep. *J. Wildl. Manage.* 43: 780–783.

Shaw, S., & C. Fredine. 1956. *Wetlands of the United States.* Washington, D.C.: U.S. Fish and Wildlife Service Circular 39.

Smith, A. G., J. H. Stroudt, & J. B. Gollop. 1964. Prairie potholes and marshes. In J. P. Linduska (Ed.), *Waterfowl tomorrow.* Washington, D.C.: U.S.D.I. Bureau of Sport Fisheries and Wildlife.

Teer, J. 1979. Introduction of exotic animals. In R. Teague and E. Decker (Eds.), *Wildlife conservation: principles and practice.* Washington, D.C.: The Wildlife Society.

Terrill, C. 1977. Livestock losses to predators in western states. In R. Phillips and C. Jonkel (Eds.), *Proceedings of the 1975 predator symposium.* Missoula, Mont.: Montana Forest and Conservation Experiment Station. University of Montana.

Trautman, C., L. Frederickson, & A. Carter. 1974. Relationship of red foxes and other predators to populations of ring-necked pheasants and other prey, South Dakota. *Trans. N. Am. Wildl. and Natur. Resour. Conf.* 39: 241–255.

U.S. Fish and Wildlife Service. 1978. *Report on predator damage in the west.* (draft) Washington, D.C.: U.S. Department of the Interior, Fish and Wildlife Service.

U.S. Fish and Wildlife Service, n.d. *Bottomland hardwoods along the Mississippi: a disappearing resource.* (pamphlet)

Vance, D., & R. Westemeier. 1979. Interactions of pheasants and prairie chickens in Illinois. *Wildl. Soc. Bull.* 7: 221–225.

Weller, M. W. 1978. Management of freshwater marshes for wildlife. In R. E. Good, D. F. Whigham, and R. L. Simpson (Eds.), *Freshwater wetlands, ecological processes and management potential.* New York: Academic Press.

Weller, M. W. 1981. *Freshwater marshes: ecology and wildlife management.* Minneapolis, Minn.: University of Minnesota Press.

Wodzicki, K. 1965. The status of some exotic vertebrates in the ecology of New Zealand. In H. Baker and G. Stebbins (Eds.), *The genetics of colonizing species. Proc. 1st IUBS Symp. General Biol.* New York: Academic Press.

Wood, G., & R. Barrett. 1979. Status of wild pigs in the United States. *Wildl. Soc. Bull.* 7: 237–246.

CHAPTER 10

WILDLIFE AND LAND USE

INTRODUCTION

Biological reserves, parks, and other areas of complete protection offer the habitat diversity necessary to ensure survival of most wild species. Yet only about 1 percent of the earth's land area enjoys such complete protection from human disturbance. Most of the other 99 percent is or may be used for production of food, timber, pulpwood, or other raw materials. What about wildlife on these areas? Although they cannot, in general, sustain the sorts of diversity found on more protected areas, they can meet habitat requirements for a good many species. With sufficient imagination and long-term planning, managers of commercial forests, fields, and pastures can integrate wildlife management into their plans.

The most persistent problem stems from the fact that compared with crops, livestock, timber, and pulpwood, wildlife is relative inept at paying its own way. In most western nations, this discrepancy is magnified by the tradition of public ownership of wildlife, even on private lands. Although laws usually protect game animals and endangered species from direct destruction, they do little to thwart the assault on wildlife habitat. Landowners, therefore, have both the legal right and the economic incentive to convert natural wildlife habitat to more profitable land use.

As a general rule, the more intensive the land use practice, the less suitable the land becomes for wildlife. Intensive land use means conversion of once diverse habitats into areas growing only one or two crops. Such drastic reduction in vegetative diversity renders areas unsuitable for virtually all wild species. In addition, extensive monoculture is, with rare exception, itself an unnatural occurrence. Maintenance of monocultures usually requires large-scale application of pesticides, herbicides, fungicides, and fertilizers. Such activities have high environmental costs, use enormous levels of energy, and render areas of intensive land use even less suitable for wildlife.

Despite steady trends toward more intensive land use practices, there remains some room for optimism. There is a growing realization that long-term health may be more important than short-term profits. This means that sites only marginally suited for growing large monocultures will, in the long run, prove to be unprofitable, just as they did during the dust bowl days of the 1930s. The demand for hunting on private land is resulting in more and more hunting leases. These leases furnish landowners with some financial incentives to maintain wildlife habitat on at least parts of their lands. Finally, there will undoubtedly be more emphasis in the future on land use planning. Like hunting leases, land use planning in the United States seemed at first to cut against the grain of the American character. Free hunting was a great tradition, a priceless pastime too sacred for the marketplace. Land use planning smacks of distant, faceless bureaucracy, more suited to totalitarian states than to constitutional democracies. Yet both practices are inevitable, brought about by the pressures of rising human populations and soaring energy costs.

COMMERCIAL FORESTS

Most commercial forests are in the temperate zone. Consequently, much of what is known about the effects of commercial forestry upon wildlife has come from those regions. The rapid deforestation of the tropics, as shown in Chapters 8 and 11, poses a greater environmental threat worldwide. The majority of tropical deforestation, though, is due to the clearing of land for agriculture and pasture. Clearing of native forests for conversion to intensive monoculture, such as for *Eucalyptus,* is still a relatively small-scale endeavor. The Global 2000 Report to the President of the United States predicts that because the technology of commercial forestry remains largely unproven in tropical climes, investment capital will not be forthcoming for large-scale plantings before the 1990s (Barney, 1980). Therefore, this review of commercial forestry is limited to the temperate regions of North America.

Timber and Game in the Southeast

Nowhere in North America is commercial forestry practiced more intensively than in the southern forests in the southeastern United States. Conditions of climate, soil, and topography are ideal for rapid timber growth and ease of harvest. Site preparation costs and other expenses associated with stand regeneration are all minimal.

Yet there is also an abundant demand for game in the same region. Nearly 400,000 white-tailed deer are taken annually, along with nearly 2 million squirrels, 15 million bobwhite quail, and over 16 million mourning doves, plus millions of other game animals (Table 10-1). The southeast's 5 million licensed hunters have expressed a clear willingness to pay for their sport. One survey (Horvath, 1974, as cited by Halls, 1975) found hunters willing to pay an average of $61 per day to hunt big game, $49 per day for waterfowl, and $39 per day to hunt small game. Given the total hunting-days afield for the region, the total amount rises to $3.9 billion, a considerable sum.

The demands for both game and high-yield intensive forestry have generated some management guidelines from the U.S. Forest Service and other concerned agencies. For sustaining game on commercial forests the general recommendations are:

1 Use periodic thinnings to limit stand density, admitting more sunlight to the forest floor, thereby growing more plant food for wildlife.
2 Use prescribed burns to thin out forest understory, permitting rapid nutrient cycling and promoting richer plant growth.
3 Keep cutting units as small as possible to maximize edge.
4 Maximize cutting interval (rotation) to ensure stand diversity.
5 Maintain stands of mixed species to meet game habitat requirements.

The first two guidelines are compatible with intensive silviculture practices. Unfortunately, the last three are not. Pressures for maximum yields can be derived from short cutting intervals on even-aged monocultures. Furthermore, the cost of using heavy harvesting equipment is

TABLE 10-1
ANNUAL HARVEST (CA. 1970) OF SOUTHERN FOREST GAME SPECIES
(Halls, 1975 © Wildlife Management Institute. Used by permission.)

Game	Annual harvest
White-tailed deer	396,000
Squirrels	1,942,000
Rabbits	1,177,000
Bobwhite quail	15,000,000
Mourning dove	16,532,000
Eastern wild turkey	55,900
Ruffed grouse	10,430
American woodcock	145,000
Ducks	2,323,000

considerable, thus furnishing a strong incentive to clearcut in large units for the sake of efficiency.

As the demand for hunting leases grows, so will the economic support for maintaining game habitat. By the 1970s leasing fees of more than $1 to $2 per acre annually had become common for southeastern forests and on some bottomlands were yielding $2.50 to $6 per acre per year (Halls, 1975).

Income from hunting leases may, in many cases, exceed the amounts lost by practicing less intensive forestry. One study in Missouri estimated that the costs of maintaining game habitat within commercial forests ranged from $.10 to $.54 per acre per year. Maintaining permanent forest openings to allow growth of food for wildlife is expensive, ranging up to $100 per acre per year for the plots themselves (Halls, 1975).

Maintaining an optimum balance between game and timber production has been the mission of managers of the Piedmont National Wildlife Refuge in Georgia. When the U.S. Fish and Wildlife Service obtained the 13,200-hectare (33,000-acre) tract in the 1930s, game was severely depleted and the soil was eroding seriously. The service worked closely with state and other federal agencies and by 1970 had achieved an estimated 80 percent of the optimum levels of timber and game harvests. By that time the average annual yield was 380 deer and 7 million board feet of timber, plus unspecified numbers of other game animals. Hunting leases brought from $1.50 to $2 per acre as compared with an average yield of $10 per acre per year for timber. The staff at the Piedmont NWR considered the optimum mix to be 75 percent of the forest acreage in pine, 20 percent in hardwoods, and 5 percent in permanent openings (Bureau of Sport Fisheries and Wildlife, 1970). Thus the Piedmont refuge has shown that commercial forestry and game management can be successfully coordinated on southeastern forests.

Another approach toward managing game on commercial forests has been the "featured species concept," begun by the U.S. Forest Service in 1971. The concept recognized two basic facts of forest wildlife: (1) Not all sites within a given forest offer the same habitat potential, and (2) not all forest species have the same habitat requirements. It follows, then, that no tract of forest can be managed equally well for all species. In principle the featured species concept permitted forest managers to manage individual forest compartments for those species best adapted to conditions there.

Once the featured sites have been selected, the Forest Service, working with detailed data on the species' habitat requirements, works out management options, usually in two stages. The first stage develops options consistent with even-aged silviculture. These options involve stand size, rotation lengths, site preparation and regeneration, intermediate thinnings, and prescribed burns (Holbrook, 1974). When a species' habitat requirements cannot be met through silvicultural manipulations, the service uses a second set of options designed to improve habitat. These may include creation of permanent forest openings or direct manipulation of cover or food production.

The featured species concept is not limited to game animals. It is applied to threatened or endangered species such as the red-cockaded woodpecker and the bald eagle. Even songbirds are included (see Table 10-2). Figure 10-1 is a hypothetical example of just how the featured species concept can be applied to a particular area.

Incentives for Private Landowners

Most schemes for preserving wildlife on commercial forests have been developed by agencies such as the Forest Service for application to large tracts of publicly owned land. They can also be applied to large tracts of forest owned by private industry. But the majority of commercial timberland in the United States falls into neither category. Some 59 percent of this land is privately owned, though not by the forest industry. Landowners are free to choose what to do with their timber.

When compared in strictly economic terms, publicly owned wildlife fares poorly against privately owned timber. A private landowner can usually reap larger and faster profits through intensive harvests, often to the detriment of wildlife. What incentives can such a landowner be offered to ensure maintenance of wildlife habitat? Larger landowners may receive direct financial compensation through hunting leases, just as does the

TABLE 10-2
USING THE FEATURED SPECIES CONCEPT TO ADAPT MANAGEMENT OPTIONS TO SPECIES
(Holbrook, 1974 © The Wildlife Society, Used by permission.)

Featured species	Rotation age (yrs)	Stand size (ha)
Game		
Eastern wild turkey	Hardwoods 80	4–20 for hardwoods
	Pines 50–80	4–40 for pines
Gray squirrel	Hardwoods 70–100	4–20
Bobwhite quail	Pines 50–80	4–40
Endangered or threatened		
Red-cockaded woodpecker	Pines 70–80	4–40
Bald eagle	Leave nest & perch trees	N/A
	Do not disturb during nesting	
Songbirds	Hardwoods & pines 80–100	4–80
	Maximize vertical diversity	

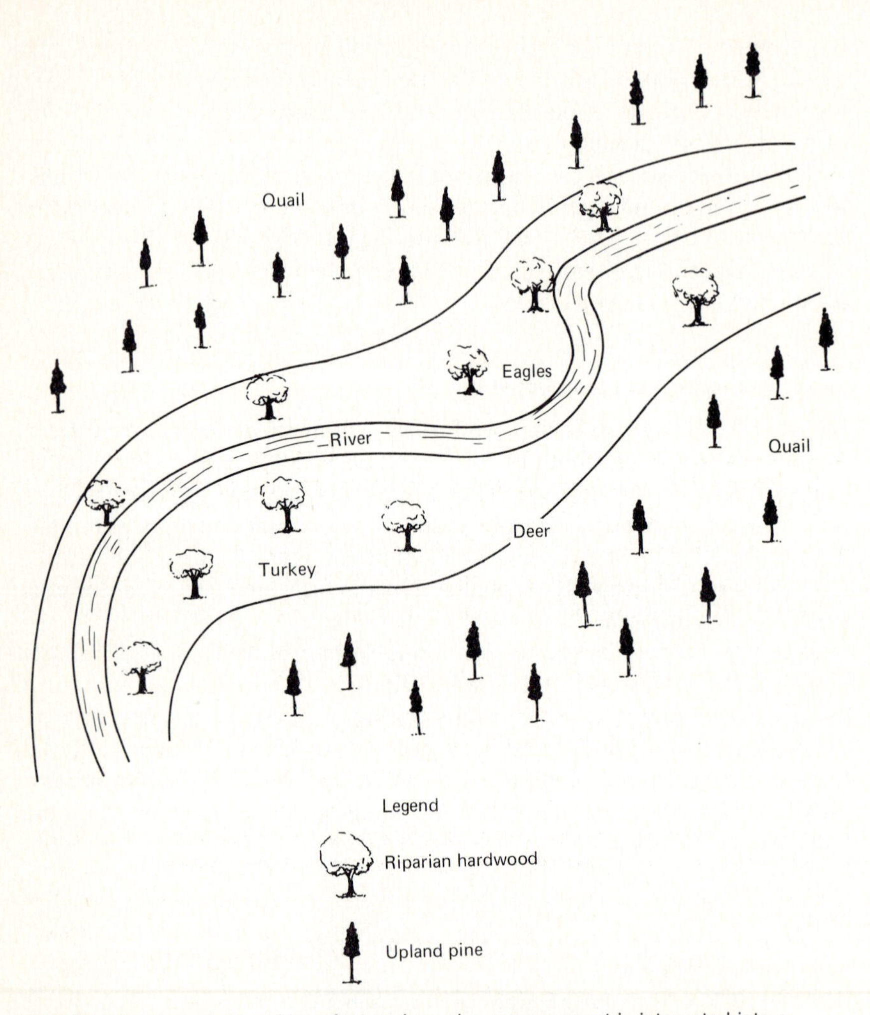

FIGURE 10-1 An example of how featured species management is integrated into a forested area. (*Modified from Holbrook, 1974* © The Wildlife Society. Used by permission.)

Piedmont Wildlife Refuge. Amounts received for leases will seldom, if ever, equal the income from timber, but they may serve to offset the costs of preserving snags, openings, or other habitat needs.

Smaller landowners especially need additional incentives. Shaw and Gansner (1975) maintained that any system of incentives should have as objectives (1) long-term retention of woodlands, and (2) adequate maintenance of those woodlands. If public funds, particularly those derived from hunting license sales are used, then it follows that a third objective should be public access.

One category of incentives is the tax break. States may offer tax reductions to landowners willing to keep their property forested. This alone will not ensure optimum wildlife habitat, but it will serve to discourage conversion from forest to open fields or pastures. Connecticut, for example, offers renewable 10-year contracts to landowners with 10 or more forested hectares (25 acres). These contracts allow tax assessment on the basis of "current use value" rather than "highest and best use value." By the mid-1970s some 4,600 landowners had enrolled nearly 200,000 hectares (half a million acres) in the program (Shaw and Gansner, 1975).

Other incentives come as cost-share programs, in which a public agency pays the landowner part of the cost of preserving forests or wildlife. Such programs are consistent philosophically and legally with the American tradition of public ownership of wildlife. In theory, the agency simply pays the landowner for the difference between the amount of money to be made by retaining wildlife habitat and the amount to be made by converting the land to more profitable and intensive uses. In practice, of course, the method is both troublesome and expensive. The Agricultural Stabilization and Conservation Service (ASCS) of the U.S. Department of Agriculture has used such programs for timber and more recently for wildlife. The latter has thus far been used so rarely that it has done little good (Shaw and Gansner, 1975) but still offers possibilities for the future.

Wildlife Diversity on Commercial Forests

Commercial logging operations can have major impacts upon wildlife habitat. Webb (1977) and Webb et al. (1977) measured the effects of four different intensities of logging on avian populations and diversity in the northern hardwood forests. The researchers compared 4 levels of cutting: 25 percent, 50 percent, 75 percent, and 100 percent of all merchantable timber on blocks of more than 200 hectares (500 acres) each. (A cut of 100 percent of the merchantable timber is not the same as a clearcut because trees not useful for logs are left standing.) They left another stand untouched to serve as a control. Then they monitored avian populations and diversity for 10 years after the cuts were made, as natural regeneration proceeded.

The results were somewhat surprising. The total numbers of individual birds counted changed little and the differences showed no clear pattern or trend (Table 10-3). Species richness showed a steady increase with cutting intensities. Species diversity was significantly higher on areas following 75 percent and 100 percent cuts than on the control area.

Not all species were affected equally. Of the species most intensively studied, 8 showed population increases with logging, 7 experienced decreases, and 11 showed no real change. Species in this last category

TABLE 10-3
SONGBIRD DATA FROM FIVE STUDY SITES FOR 10 YEARS FOLLOWING LOGGING
(Webb, 1977 © Wildlife Management Institute. Used by permission.)

	Area				
	Control	**25%**	**50%**	**75%**	**100%**
No. of individuals	1,237	1,204	1,129	1,287	1,149
Total no. of species	37	43	45	46	45
$\bar{x}$ no. of species/yr	25.3	25.6	25.9	28.1	28.3

apparently have quite wide habitat tolerance and cannot be managed through habitat manipulation.

There was yet another surprise. Of the 15 species showing definite population changes after the cuttings, all tended to return to precutting population levels within 10 years. Thus, those species responding positively to cutting within the first 4 years tended to decline sharply during the next 6. Likewise, species adversely affected by logging began to recover within a few years after the cuts.

Several conclusions can be drawn from Webb's studies. First, they could show no adverse effects of logging on avian diversity (indeed, there was a slight increase). Logging per se, then, could not be shown as harmful to most wildlife. Webb (1977) also suggested that a traditional view of wildlife managers, that frequent and light cuts in small stands are best, may be questionable. Here it is important to recall a distinction emphasized in Chapters 7 and 8. The traditional wisdom seeks to maximize edge, thus favoring those species which are edge-dependent. Among these are most of our forest game species. When the management objective is maximum productivity of edge-dependent game, then the conventional wisdom still holds. It becomes questionable, as Webb has shown, when the objective is to maintain maximum species diversity, at least for northern hardwood forests. Only further research in other forest types will determine the extent to which Webb's findings can be applied elsewhere.

A quite different problem is what to do with interior or "area-sensitive" species. Among birds, these often include long-distance migrants that winter in the tropics. When they move north to breed in the spring, these species select nesting sites only in large areas of contiguous habitat. These birds are declining wherever development is converting large-scale forests into progressively smaller fragments.

Why do some birds such as the wood thrush (*Hylocichla mustelina*), red-eyed vireo (*Vireo olivaceous*), scarlet tanager (*Piranga olivacea*), and ovenbird need large tracts of woodland for nesting? One reason seems to be that many area-sensitive birds nest on or near the ground. As fragments become smaller, disturbance and trampling by humans and livestock usually increase, reducing nesting success. Nest predation by small mammals, crows (*Corvus brachyrhynchos*), and other animals probably increases, as does nest parasitism by cowbirds (*Molothrus ater*). Finally, year-round resident birds may have significant competitive advantages in foraging and rates of reproduction and dispersal (Whitcomb, 1977). In short, as more areas are disturbed and disrupted, the edge-sensitive *r*-selected species win out over the interior species.

Several species of long-distance migrants have declined in the eastern United States as forests have been fragmented. The densities of nesting pairs of long-distance migrants were measured on five different areas in the 1940s and 1950s. By the 1970s the densities on these same sites declined by an average of 53.6 percent (Robbins, 1979).

Table 10-4 gives some preliminary estimates of the minimum areas of contiguous forest needed to sustain various species of area-sensitive birds. To obtain these estimates, Robbins determined the occurrence of species on 500 breeding bird survey stops in areas with varying sizes of forest fragments. He then plotted the percent occurrence of each species against the size of the forest fragments. The resulting estimates represent thresholds beyond which increases in the size of the forest fragments do not cause significant increases in frequency of occurrence.

If these area-sensitive birds are to be maintained on commercial forests, the stand sizes of even-aged, mixed stands still have to be kept rather large. The recommendation is obviously inconsistent with the usual prescription for maximizing game (typically edge-dependent) production. The proximity of stands of different ages is also an important consideration. If adjacent stands are close to the same age and development, their combined areas may act like that of a single, larger stand. Perhaps the best spatial arrangement is depicted in Figure 10-2. Here adjacent stands are close to the same age, and cutting one every few years in a clockwise direction would keep the age intervals close indefinitely. An ideal arrangement would include a natural area in the center of the stands. This area would never be cut and could provide snags and nesting holes typical of "overmature" forests. In contrast, a checkerboard pattern would be least suited to the needs of area-sensitive species (see Figure 10-2). This scheme would favor edge-dependent wildlife over area-sensitive ones, to the considerable detriment of the latter.

Another way of providing for the needs of area-sensitive species is through the use of habitat corridors, in essence, the same strategy as described in Chapter 8 for the preservation of biological reserves. The size of the area is not the only important consideration. A small fragment, if

TABLE 10-4
PRELIMINARY ESTIMATES OF MINIMUM FOREST AREAS REQUIRED TO SUSTAIN POPULATIONS OF AREA-SENSITIVE FOREST BIRDS
(Modified from Robbins, 1979)

Species	Critical size (ha)
Red-bellied woodpecker	4
Hairy woodpecker	4
Eastern wood pewee	4
Blue jay	4
Tufted titmouse	4
Great crested flycatcher	10
Carolina wren	10
Acadian flycatcher	30
Pine warbler	30
Kentucky warbler	30
Hooded warbler	30
Red-shouldered hawk	100
Wood thrush	100
Yellow-throated vireo	100
Red-eyed vireo	100
Prothonotary warbler	100
N. Parula warbler	100
Louisiana waterthrush	100
Scarlet tanager	100
Summer tanager	100
Black and white warbler	300
Worm-eating warbler	300
Ovenbird	2,650

connected by a corridor of habitat to larger ones, will typically support more interior species than will a similar-sized fragment without a corridor.

The value of such corridors has been clearly demonstrated in the eastern United States. In Maryland, for example, researchers found a small forest fragment of 13.7 hectares (34 acres) which contained more interior species than expected (MacClintock et al., 1977). Closer examination revealed that the fragment was connected by a narrow habitat corridor to another forest of 156 hectares (390 acres). This second forest led, in turn, to another forest of over 3,900 hectares (9,750 acres) by way of several more corridors. Even such area-sensitive species as the ovenbird persisted within the 13.7-hectare tract, although at a lower density than in the larger fragments. Those ovenbirds establishing territories within the smallest fragment did so only in the interior of the fragment, avoiding otherwise

20–30 years	10–20 years	0–10 years
30–40 years	Natural area	70–80 years
40–50 yesrs	50–60 years	60–70 years

60–70 years	30–40 years	70–80 years
10–20 years	Disturbed area	20–30 years
50–60 years	0–10 years	40–50 years

FIGURE 10-2
Two extreme examples of spatial relationships among forest cuts in terms of their effects upon area-sensitive species. (*Robbins, 1979.*)

suitable sites near the edge. The researchers speculated that ovenbirds, and perhaps other interior species as well, practice strict habitat selection that restricts use to forest interiors.

Even selective logging shifts the balance away from interior species toward edge species. Whitcomb et al. (1977) compared avian diversity on two old growth tracts of about 14 hectares (35 acres), each located on exceptionally fertile soil. One tract was selectively logged for white oak and the other remained intact. The resulting changes for both edge and interior species appear in Table 10-5. Like Webb and his associates, these researchers found that avian species richness actually increased following selective logging. The balance shifted away from canopy-dependent species and those which forage or nest near the forest floor (characteristics of interior animals). The habitat changes favored instead those edge species well-adapted to shrubs and smaller trees. As the canopy was opened, shrubs and smaller trees underwent more rapid growth.

An important, pioneering approach to integrating wildlife habitat needs into forest management was developed for the Blue Mountains of Oregon

TABLE 10-5
EFFECTS OF SELECTIVE LOGGING OF WHITE OAK UPON FOREST INTERIOR AND EDGE SPECIES IN MARYLAND
(Modified from Whitcomb et al., 1977 © National Audubon Society. Used by permission.) Numbers are territorial males.

Species	Unlogged 1975	Unlogged 1976	Logged 1975	Logged 1976
Interior species				
Red-eyed vireo	36	30.5	22.5	28
Ovenbird	9.5	7	1.5	1
Wood thrush	38	21	23	24
Yellow-throated vireo	0.5	1	2.5	2
Cardinal	9.5	8	13	11
Downy woodpecker	5	6	6	8
Kentucky warbler	4	3	7	5
Hooded warbler	0	1	11	14.5
Edge species				
Common flicker	+*	1	2	3
Gray catbird	0	0	1	+*
Common yellowthroat	0	0	2	+*
Indigo bunting	0	0	2	+*
Rufous-sided towhee	2	0	7	6
White-eyed vireo	0	0	3.5	1
Mourning dove	0	0	2	2
Baltimore oriole	0	0	1	0
Brown thrasher	0	0	+*	+*

*+ = present.

and Washington. Forest managers there needed a system to help them predict the impact of logging and related activities upon wildlife populations. A team of biologists and natural resource specialists (Thomas et al., 1979) devised such a system of habitat data for all 378 species of native vertebrates.

This Blue Mountain system summarizes the effects of successional changes on wildlife for each major habitat type. Since logging and other forest practices predictably alter successional stages, managers can use this system to anticipate the effects of any management option on groups of species or on individual species. In addition, the system allows calculation of a versatility index for each vertebrate species. The index is based on the number of habitat types and the number of successional stages within those habitat types that the species uses for feeding and reproduction. Forest managers can thus predict which species will thrive following logging and which will decline. Special management programs may be planned to ensure survival of species with low versatility indexes long before their situations become critical. The basic approach used for the Blue Mountains may eventually be adopted for use in many other managed forests.

This section reviews three distinctly different management objectives for wildlife on commercial forests:

1 Maximum production of game species
2 Maximum diversity of wildlife, including nongame
3 Preservation of interior or area-sensitive species

The three objectives are not completely compatible. The first requires maintenance of small stands for maximum edge, while the second seems to be attainable by logging in large blocks at more frequent intervals. Preservation of interior species can best be accomplished through management of large stands, adjacent to stands of similar age, connected to still larger stands with corridors of natural forest habitat. All three objectives do have one thing in common, though. They can be met only when at least parts of the forest are allowed to regenerate naturally. Large-scale monocultures provide quality habitat only for the pest species that plague them.

WILDLIFE ON CROPLANDS

Of all human activities, none have had more impact upon wildlife than agriculture. This generalization is true for the United States (McConnell and Harmon, 1976) and for the developing world (Myers, 1979). The specific effects vary enormously depending upon geographic location and types of agricultural practices. Some species almost always benefit from the disturbance and simplification of habitat that results from farming, but most do not. One generalization stands out clearly: the suitability of farmland for wildlife declines as intensity of agriculture increases. The trends are toward more intensive farming involving large-scale monocultures; cultivation of former fence rows, weed patches, gallery forests, small woodlots, and windbreaks; heavy energy and capital investment, and greater dependence upon fertilizers, pesticides, and herbicides.

Game animals decline sharply with more intensive agriculture. For example, between 1939 and 1974, one Illinois farm lost 100 percent of its prairie chickens, 78 percent of its bobwhite quail, and 96 percent of its cottontail rabbits (see Table 10-6). These changes followed shifts in land use, including a 51 percent reduction in the total number of fields (thus reducing interspersion drastically), a total loss of hay fields, and a 91 percent decline in pasture (see Table 10-7). Meanwhile, the area devoted to row crops rose by 268 percent. These changes were typical for farms in the midwest.

TABLE 10-6
CHANGES IN GAME POPULATIONS ON A 1,117-HECTARE FARM IN ILLINOIS, 1939–1974
(Vance, 1976 © The Wildlife Society. Used by permission.)

Species	1939	1974	% Decline
Displaying male prairie chickens	131	0	100
Bobwhite quail	226	49	78.3
Cottontail rabbits (index)	57	2	96.0

Intensive agriculture also reduces species diversity. Comparing avian surveys from 1906 to 1909 with surveys half a century later, the Illinois Natural History Survey found a sharp drop in species evenness. An equal number (17) of species increased as decreased during the 50 years of intensifying agriculture, while 6 others remained about the same. However, in 1909, 18 species comprised 70 percent of the state's breeding bird population. By 1959, 70 percent of the population was made up of only 9 species (Burger, 1978).

Large agricultural monocultures, in turn, are highly vulnerable to insect pests. Modern agriculture thus depends upon application of chemical pesticides. Both persistent pesticides such as the chlorinated hydrocarbons (DDT, dieldrin, chlordane, heptachlor, and related insecticides) and nonpersistent pesticides such as parathion, malathion, azodrin, and phosphamidon (organophosphates), plus sevin, baygon, and pyrolan (carbamates) pose direct threats to wildlife. Pesticides also damage wildlife indirectly through reducing the insect populations upon which many birds depend.

The actions of persistent pesticides differ substantially from those of nonpersistent ones. Because they are stable compounds, persistent pesti-

TABLE 10-7
CHANGES OF A 1,117-HECTARE FARM IN ILLINOIS, 1939–1974
(After Vance, 1976 © The Wildlife Society. Used by permission.)

	1939		1974	
Land use	No. fields	% of area	No. fields	% of area
Row crops	78	28.4	80	76.1
Hay fields	77	36.0	0	0
Mixed herbaceous	40	8.2	21	11.2
Pasture	32	11.1	6	1.0
Woodlots	9	1.5	9	1.3

cides can remain in soil or water for indefinite periods. Eventually, the chemicals work their way into food chains and are picked up by animals. They are stored in animal tissues, particularly fat and liver, generally for the life span of the individual. Concentrations increase up the food chains, a process known as "biological magnification." Higher-level predators can accumulate very high levels, as explained in Chapter 8 for peregrine falcons. The best-documented long-term effect of persistent pesticides is the thinning of eggshells, resulting in declines in reproductive success. This effect may also affect avian behavior, leading to abandonment of young, thus further depressing survival rates.

Nonpersistent pesticides are free of these problems, but their use has brought on new problems. Their hazards result not from gradual accumulation but through acute toxicity. Some of these pesticides, including ethyl parathion, phosphamihon, and axodrin, are extremely toxic to birds (Risebrough, 1978). Large-scale use of these chemicals has resulted in deaths of millions of songbirds.

Land Retirement Programs

Large surpluses can depress prices. Attempting to stabilize crop prices, the U.S. government began in 1956 a program known as the Soil Bank. Under this program, the federal government subsidized farmers for taking land out of production. Between 16 and 24 million hectares (40 to 60 million acres) were idled (Harmon and Nelson, 1973). The result was a bonanza for wildlife. Pheasant populations in South Dakota, for example, rose as more and more hectares were enrolled in the Soil Bank (see Figure 10-3).

The last Soil Bank contracts expired in 1969. A similar program, the Cropland Adjustment Program (CAP) arose in its place. In Indiana, four times more game was taken on CAP farms than on nonparticipating farms. The cost per unit of hunter effort dropped from $5.12 to $1.82 on CAP farms (Machan and Feldt, 1972). Both Soil Bank and CAP programs involved long-term contracts of 5 to 10 years, avoiding the costs of writing annual contracts and allowing time for natural succession to form protective vegetative cover. Two other programs, Diverted Acres and Set-Aside, attempt to retire land on only an annual basis (see Table 10-8), a practice far less beneficial to wildlife and soil conservation (Harmon and Nelson, 1973).

During the 1970s the growth of international grain markets, a rising balance of payments deficit, and general economic instability led to a decline in land retirement programs in the United States. But by the early 1980s crop surpluses had again begun to depress crop prices. The Reagan administration began in 1983 its own version of land retirement, called payment in kind (PIK). Under this arrangement, farmers who agreed to

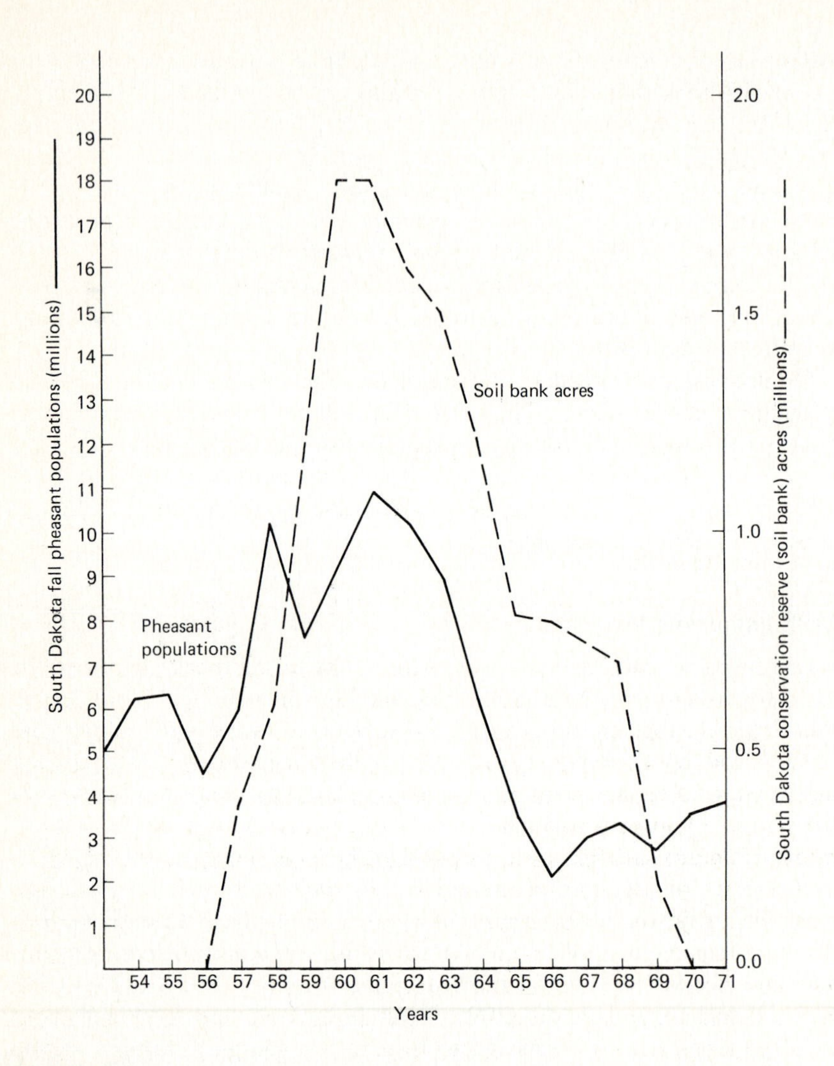

FIGURE 10-3
Effects of land retirement upon ring-necked pheasant populations. (*Erickson and Wiebe, 1973* © The Wildlife Society. Used by permission.)

withdraw acreage from production were given compensation in the form of government-owned surplus grain. The PIK program was designed to reduce large government surpluses while providing the means to pay farmers for reducing production. As both surpluses and production receded, the crop prices were expected to rise. Unfortunately, the PIK program was authorized only for one year, thus affording little opportunity for wildlife habitat restoration or improved soil conservation.

TABLE 10-8
A COMPARISON OF THE TWO BASIC APPROACHES TO LAND RETIREMENT. The Soil Bank and Cropland Adjustment Program involved 5- or 10-year contracts, retiring for the most part whole farms. Set-Aside and Diverted Acres operated through annual contracts and usually involved partial retirement of farms (information based upon Harmon and Nelson, 1973 © The Wildlife Society. Used by permission.)

	5- to 10-year contracts	Annual contracts
Soil conservation	Generally effective	Less effective
Wildlife production	Generally good	Generally poor
Crop rotations	Easier planning	More erratic planning
Program costs	Lower	Higher
Local economy	Loss of purchasing power	Less reduction in purchasing power
Political support	Weaker	Stronger

Management Options

Given the land use trends toward more intensive agriculture, the outlook for wildlife on most private farmlands is poor. The international grain market as well as that for other crops is likely to keep the economic pressure on toward high-yield production at least on prime farmlands. State wildlife agencies are generally retreating from management programs on private lands, concentrating instead upon more intensive wildlife management on state game management areas and other public lands. Are there any realistic solutions, or is farmland always to contain only those species so well-adapted to monoculture that they become serious pests?

States often favor voluntary programs intended to coax landowners into setting aside a few hectares here and there for the benefit of wildlife. Acres for Wildlife is one such scheme. Landowners willing to enroll at least .4 hectares (1 acre) of land agree to prevent destruction of cover and avoid most forms of mowing and spraying. In exchange for their cooperation the landowners receive a certificate of appreciation from the state wildlife department, plus signs for display. Such programs are inexpensive and politically acceptable, and are probably worth continuing. The problem is they fail to entice enough landowners to have a really substantial impact. The economic pressures against Acres for Wildlife are simply too great.

Attempting to counter these economic pressures, a few states have developed programs which compensate landowners directly for establishing habitat. Nebraska, for example, increased license fees and issued new $7.50 habitat stamps to hunters. Proceeds were then made available to farmers at varying rates per acre, depending upon the type of habitat or cover preserved. South Dakota has had a similar program, with the state paying farmers to keep land out of production (Harmon, 1979).

Of course not all farmland is of equal value. Bearing this in mind, some authorities have advocated wildlife management primarily upon marginal lands, leaving prime agricultural lands solely for growing crops (Burger, 1978). This strategy has important advantages. Since marginal lands cannot produce as much as prime lands, the costs of keeping them out of production are much lower. Moreover, marginal lands may have received less intensive use to begin with, so that natural cover on some of them is already present. Finally, good wildlife management and good soil conservation on marginal lands are quite compatible. During exceptionally dry years, losses of crops and soil on marginal lands are greater than on prime farmlands. Many marginal lands should not be planted at all. They may be reasonably productive during periods of high to moderate rains. But when dry cycles come, as they typically do in several consecutive years, crops fail and the exposed soils blow away. This happened on a large scale in the United States with the "dust bowl" and in the Soviet Union with development of the "virgin lands" (Eckholm, 1976). It can happen wherever short-term economic pressures tempt farmers to intrude into marginal lands.

Whatever option is chosen for wildlife management on private farmlands, it must be economically competitive. Normally, of course, wildlife, a publicly owned resource, is not competitive. Hence, the question becomes whether or not wildlife enthusiasts, hunters and nonhunters alike, can or will pay the price. Perhaps user-pay programs such as those described for Nebraska will serve as models for the future. Hunting leases may also bring in enough revenue to help offset the losses incurred by landowners who take land out of production. Leases may be especially appealing because of the lower overhead compared with other forms of land use. Perhaps, too, the general public will ultimately decide to pay either for taking land out of production, or by granting tax incentives to cooperating landowners. It seems likely, though, that marginal farmlands will play a crucial role in wildlife management.

Then, too, even prime farmlands have sites that are marginal. Narrow strips of riparian vegetation along streams often cost too much to clear. Preserving them aids wildlife and helps retain soil. Windbreaks also serve both the aims of wildlife and soil conservation. Their maintenance, however, requires that strips of land be kept out of production. Most owners of prime farmland may not object to retaining riparian woodlands or windbreaks during ordinary times. But when crop prices rise sharply, as they have for soybeans in the 1960s and 1970s, farmers feel the temptation to expand their cultivation. Larger tractors, combines, and other implements are difficult to maneuver and farmers may see windbreaks and riparian vegetation as troublesome obstacles to increased power and efficiency. So out come the windbreaks and down come riparian woods. Soil and wildlife both suffer as a consequence.

So the same principle applied to marginal lands in general can be used for certain sites on prime farmland. When crop prices rise, some form of compensation should be made available to landowners to reduce the temptation to cultivate every hectare. Again, the compensation could be in the form of direct subsidies, tax breaks, or hunting leases. The causes of dwindling wildlife on farmlands are economic and will require economic solutions.

Just as with timber crops, the size and shape of harvest units can drastically affect edge, cover, and hence habitat quality. One method proposed to improve wildlife habitat on farms is a shift toward triangular fields (Powers, 1979). Triangular fields (see Figure 10-4) provide significantly more edge per unit area than do conventional square or rectangular fields. One practical drawback to this scheme, however, is the difficulty in maneuvering large farm machinery in triangular corners. This problem could be reduced by rounding the edges, thus providing even more cover.

FIGURE 10-4
Triangular field arrangement designed to maximize edge. (*Powers, 1979* © The Wildlife Society. Used by permission.)

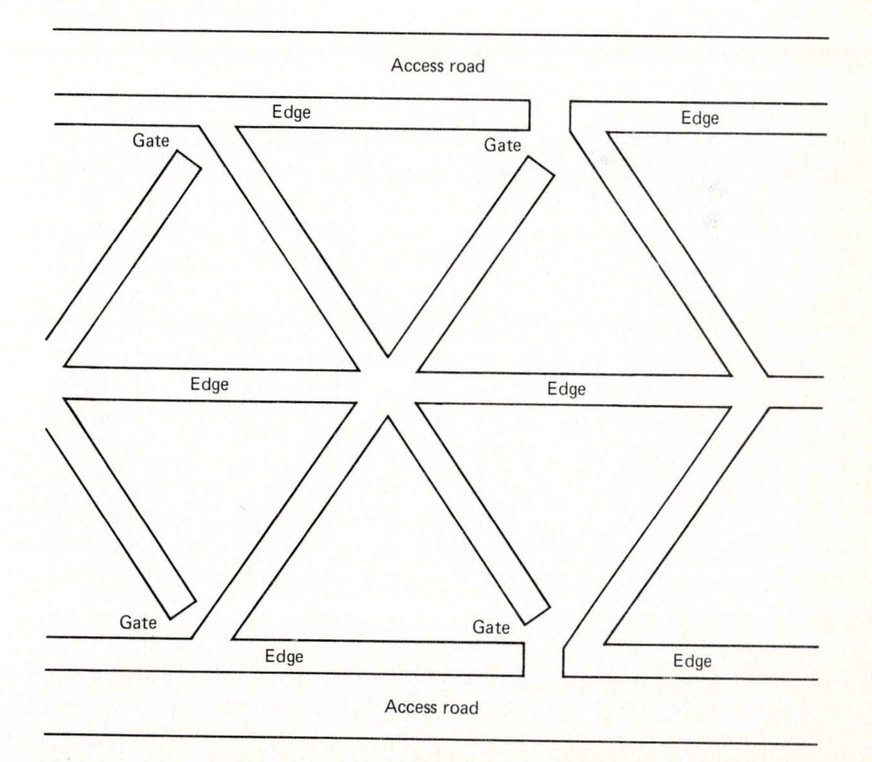

The more cover and edge, of course, the more land is taken out of production. Once again, it seems likely that farmers would have to be given some pecuniary incentive for reducing land under cultivation.

Landowners also need tough, effective trespass laws. Without such laws, a landowner's "reward" for maintaining wildlife habitat may be a greater rate of trespass by hunters. Incidents leading to property damage, vandalism, and even liability for accidents, can act as strong disincentives for landowners who might otherwise be willing to devote some of their land to wildlife habitat. States with the greatest success with wildlife management on private lands have generally been those with effective trespass laws (Burger, 1978; Burger and Teer, 1981).

RANGELANDS AND WILDLIFE

Next to agriculture, grazing of domestic livestock has had more effect upon wildlife habitat than has any other activity. Grazing is concentrated, as a general rule, on lands too arid to support intensive farming. To understand and interpret the effects of grazing upon wildlife, it is necessary first to understand the effects of both fire and grazing upon rangeland.

Fire

Although the specific effects of fire vary among different types of grasslands, some generalizations can nonetheless be made. Fire, like other forms of disturbance, commonly favors forbs over grasses. It can be devastating to ground-nesting birds when it occurs during the nesting season. Insect populations are temporarily reduced, thereby suppressing food for insectivorous wildlife. Most small rodent populations decline with the occurrence of fire. Fire usually increases the nutrient content of vegetation, as new postfire growth takes advantage of the increased pool of nutrients. Whether fire increases or decreases total plant production depends upon the amount of rainfall. More arid regions tend to experience a reduction in productivity after fires, while moister ones tend to produce more vegetation (DeVos, 1969).

Fire may also keep large areas of grassland or savanna in subclimax conditions. Should human activity then exclude fire, resulting successional advance may alter drastically the structure of the ecosystem. The extent of this alteration depends, of course, upon how the plant and animal life had adapted to natural patterns of fire.

Patterns of fire are often drastically altered once humans begin to raise livestock. In many parts of the world, fires are deliberately set in a periodic attempt to stimulate new and more nutritious forage. Heavy grazing, though, may so reduce the amount of organic material that it diminishes the chances of fire.

Grazing

Differences in grassland ecosystems, grazing patterns, and grazing intensities make broad generalizations about the effects of grazing upon wildlife populations virtually impossible. Nonetheless, one generalization remains: wildlife populations and diversity decline under prolonged and heavy grazing pressure. This is especially true of "overgrazing," or sustaining more livestock on a range than it can support. Overgrazing can lead to permanent changes in a grassland's vegetation, the most common of which is the invasion of shrub or brushland into areas previously dominated by grasses and forbs. As livestock deplete the more palatable grasses, less palatable brush such as creosote (*Larrea tridentata*), mesquite (*Prosopis* spp.), and sagebrush (*Artemisia* spp.) become established. Species of small mammals also change (see Table 10-9).

As grassland gives way to brushland, carrying capacity drops for both livestock and many species of wildlife. More arid grasslands may even be transformed into deserts, as parts of the Mediterranean and Middle East have been, through sustained overgrazing. As ground cover diminishes, soil erosion increases, further reducing the land's productivity.

Regions vary considerably in their vulnerability to livestock grazing. Communities long subjected to grazing by wild herbivores are usually more resistant to livestock than are communities never exposed to significant levels of grazing. The *Bouteloua* Province, named for the dominant grass species east of the Rocky Mountains, for example, long supported huge herds of bison. Dominant grasses there became resistant to grazing pressures so that the replacement of bison by cattle had little real effect on the plant community.

In contrast, the more arid *Agropyron* Province west of the Rockies, never sustained the large populations of bison. The native plants proved to be far less able to withstand grazing once cattle were introduced, and the dominant plants were soon destroyed and largely replaced by winter annuals that were themselves exotic introductions. Cattle thus profoundly

TABLE 10-9
POPULATION CHANGES AMONG NORTH AMERICAN WILDLIFE SPECIES WITH INCREASING GRAZING PRESSURES
(After DeVos, 1969 © Academic Press, Inc. Used by permission.)

Species decreasing	Species increasing
Cottontail rabbit	Jackrabbit
Cotton rat	Kangaroo rat
Meadow mice	Prairie dog
Harvest mice	Ground squirrel
	Wood rat

changed the composition of the *Agropyron* Province but not the *Bouteloua* Province (Mack and Thompson, 1982).

Competition between Livestock and Wildlife

Ranchers are often concerned about wild herbivores competing with their livestock. As in the East African ungulates, two or more species of herbivores coexisting with little or no competition, may together be considerably more productive than a single species. This greater productivity applies, of course, only if sufficient habitat variety is available to offer varied diets. Improved pastures or other practices which greatly simplify extensive areas will support only livestock.

Among North American big game, elk are most commonly viewed as competitors with livestock, especially cattle (Nelson, 1982). Elk have quite wide foraging niches, being capable grazers as well as browsers. They are adaptable and wide-ranging, often coming down the mountains and into winter pastures to take hay put there for cattle. Relationships between cattle and elk were investigated extensively in the Missouri River breaks of north central Montana (Mackie, 1970). There food habitat overlapped extensively during spring and fall. In winter, competition was avoided only because cattle were kept away from elk ranges. Summer competition was minimal, as elk switched their diets to forbs from late May through the summer. The lack of serious competition in this case was in part due to the small elk population. Should either elk or cattle numbers be increased, greater competition should be expected during those seasons in which food habits overlap.

Figure 10-5 shows how herbivore food habits can be partitioned. If a domestic species takes grasses and forbs (A through C), while a wild herbivore takes forbs and shrubs (B through D), then competition is limited to forbs (B through C). Elk feed all the way from A through D. Pronghorns, on the other hand, typically confine their foraging to forbs and browse (Schwartz and Nagy, 1976; Schwartz et al., 1977). They will therefore not normally compete with cattle, which favor grasses, although they may compete with domestic sheep. A rough approximation of the extent of dietary overlap is shown in Figure 10-6. By proper planning, a

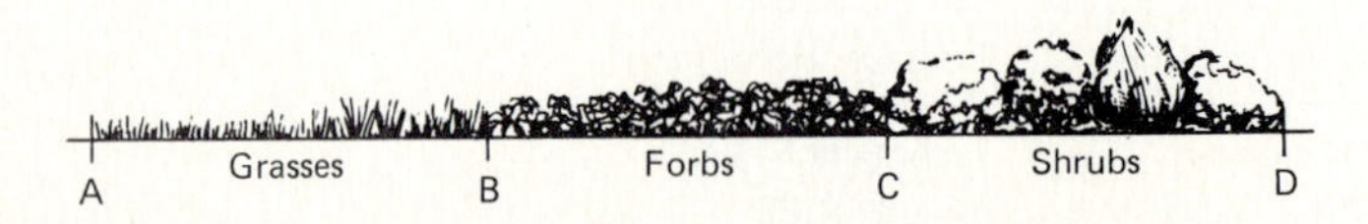

FIGURE 10-5
Range of forage types. (*Modified from National Academy of Sciences, 1970.*)

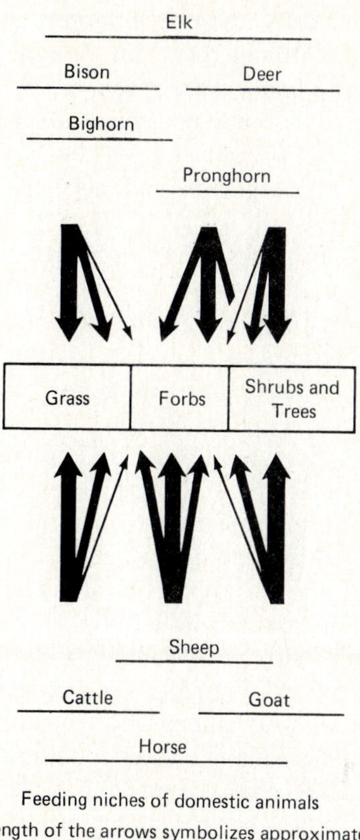

Note: Strength of the arrows symbolizes approximate feeding preferences. Potential for competition is greatest when two species have similar preferences.

FIGURE 10-6
Forage competition between various wild and domestic herbivores. (*Wagner, 1978.*)

rancher may successfully combine wild and domestic species with minimal dietary overlap and hence no significant competition.

Management of Wildlife on Ranges

The primary problems on most rangelands is overgrazing, a practice that reduces carrying capacity for wild and domestic species alike. One survey (Box et al., 1977) of public rangelands in the United States revealed that about 75 percent of these lands were producing forage at less than half their potential. This pattern of range use works to no one's long-term benefit. Range conditions on many private lands may be even worse, though no comparable surveys are available (Wagner, 1977).

Elimination of overgrazing, then, is the first order of business in improving rangelands for wildlife. On public ranges, reductions in grazing pressure can be achieved by tighter regulations governing grazing leases. The Bureau of Land Management has gradually been reducing the livestock carried on its lands since about 1918, while the U.S. Forest Service lands have experienced decreasing grazing pressure since the early 1960s (Wagner, 1978). Still many ranges under the jurisdiction of both agencies still suffer from overgrazing and they have a long way to go before any real recovery occurs.

Once grazing pressure slacks off, most ranges begin to recover through natural succession. This recovery tends to boost overall production and helps restore the natural condition of the communities.

Where sustained overgrazing has led to brushland invasions, range recovery may take a very long time. Range managers have sought ways to destroy brush or shrub to speed up recovery. At stake are some 100 million hectares (250 million acres) of brushland in the United States alone (National Academy of Sciences, 1970). Various mechanical treatments, fire, and herbicides have been used to get rid of brush. Usually these treatments are followed by seeding treated sites to grasses.

Just how these brush control programs affect wildlife habitat is unclear. Several studies have shown that wild herbivores use treated sites intensively, although most of them have failed to show any real increase in population size (Wagner, 1978). Thus, such control programs may merely shift the distribution of wild animals by concentrating them on treated areas rather than actually boosting wild populations. The spatial pattern of treatment is most likely a critical element. If brush is cleared in long strips so as to maximize edge between remaining brushland and restored range, both livestock and wildlife can benefit (Box and Powell, 1965).

As recovery of rangeland proceeds, managers can decide which species of wild and domestic herbivore they wish to feature. If properly selected, these species may actually help to balance forage composition. Pronghorns, by selecting forbs, can help shift range conditions back toward grasses. Cattle do just the opposite, eating grasses and pushing floral composition toward forbs. The two species together, then, can produce more biomass than can either species alone, just as pronghorn and bison once did.

Grazing should be carefully regulated in and around wetlands. Livestock are attracted to ponds and other sources of surface water and will consume large amounts of forage around them. They also cause compaction and erosion. Since cattle seldom wander more than a kilometer from surface water, placement of ponds has been recommended as a means of controlling their movements (Mackie, 1970; Nelson, 1982). Sometimes it becomes necessary to fence sensitive portions of ponds and other wetlands to discourage adverse effects of livestock.

Just as on agricultural lands, wildlife on rangelands need the naturally occurring strips of woody vegetation that grow along the banks of rivers and streams. Even narrow strips can preserve a wide variety of wild species. One investigation in Iowa measured breeding bird diversity against width of riparian zones along streams (Stauffer and Best, 1980). Results (see Table 10-10) showed that as little as 3.5 meters of natural woody vegetation can support up to 25 avian species. Preservation of these narrow riparian zones keeps little land out of production, helps conserve soil, functions as windbreaks, and maintains wildlife diversity.

WAITING FOR THE LAND ETHIC

Regardless of the type of land use, the effects upon wildlife can be summarized by the following principles:

1 As land use intensifies, the land becomes less suitable as wildlife habitat.

2 Intensification of land use results from economic incentives directed toward higher yields and greater efficiency.

3 Wildlife habitat has rarely been able to compete effectively with other forms of land use in strictly economic terms. This "market failure" results from the following: a tradition of public ownership of wildlife even when it occurs on private land, the fact that few species of wildlife have demonstrable economic value, and a deeply rooted belief that free hunting is the right of American sport hunters.

Aldo Leopold recognized all these principles decades ago and believed that conservation programs based only on economic self-interest were "hopelessly lopsided" (Leopold, 1949). He proposed "the land ethic" to counter the damaging practices resulting from the pursuit of short-term

TABLE 10-10
EFFECTS OF WIDTH OF RIPARIAN VEGETATION UPON BREEDING BIRD DIVERSITY IN IOWA
Widths are for one side of the stream (Stauffer and Best, 1980 © The Wildlife Society. Used by permission.)

Width (m)	No. breeding bird species
2.5	10–15
3.5	20–25
4.5	25–30
5.5	27–32

profits. Under such an ethic, abuses of the land would become socially unacceptable, even contemptible, thus serving as a strong deterrent.

Leopold was among the first to recognize land use problems and to describe them consisely. But the fact that his writings of long ago ring true today is less a tribute to Leopold's visionary abilities than it is an indictment of the lack of progress in wildlife conservation. Very little has changed, especially on private lands where habitat conditions have grown steadily worse.

There is even now no widely accepted land ethic. Land is still regarded as mere property, and property rights, including rights to abuse and destroy, remain sacred. Perhaps a widespread catastrophe such as a resurgence of the dust bowl will have to occur before the land ethic takes root. Environmental education in the public schools may also help, but the effects of such programs are long in coming.

While waiting for the land ethic, many wildlife professionals continue to promote economic incentives for wildlife management on private lands. These people are neither ignorant of nor indifferent to Leopold's land ethic. As with so many conservationists, they have become pragmatic, and the successes of some hunting leases and shooting preserves look good when compared with extensive soybean and wheat fields, overgrazed rangelands, and pine monocultures.

Yet many noneconomic incentives may prove, in the short run, anyway, to be at least as important as economic ones. Effective and enforceable trespass laws, as noted earlier, are essential. Even hunting leases, though offering obvious economic incentives, may prove more valuable to landowners for noneconomic reasons. Many landowners, especially absentee ones, have found that lessees work hard to protect their investments and in so doing, help protect the land itself.

It seems very unlikely that the land ethic will arise solely from the enlightened self-interest of landowners. Individually, landowners benefit from maximum short-term economic practices and pay little for the long-term consequences. Most of the consequences are absorbed, directly or indirectly, by government. Disposal of toxic wastes, for example, aroused little concern for effects upon wildlife; it became a serious issue only because of effects on human health and property values. Those who established the toxic waste dumps often escaped the full consequences of their actions after the federal government, through the EPA Superfund, agreed to pay cleanup costs.

Overgrazing and farming of marginal lands can be profitable to landowners in the short run, but costly in terms of erosion and loss of wildlife and other resource values in the long run. Yet the remedies offered for these short-sighted practices through the Soil Conservation Service, the U.S. Fish and Wildlife Service, and their state counterparts are paid for by the public.

Foresters can point with pride to their high-yield pine monocultures developed at the exclusion of native hardwoods. Meanwhile, hardwood timber remains in such high demand that much of it must be imported. Such imports contribute to America's balance of payments deficit and thus have a negative impact on the economy as a whole.

In the spring of 1983 severe floods ravaged much of the United States. Most people attributed the floods to simply too much rainfall. Someday they will realize that removal of forests and the draining and filling of wetlands, done largely on private lands for the economic benefit of landowners, contributed substantially to the destruction. As usual, disaster relief, much of it in the form of federal money, came to the aid of victims.

There is as yet no land ethic because landowners can continue to reap the economic benefits of short-term gains while passing the longer-term costs of their actions to society as a whole. The land ethic will emerge when the general public finally realizes just who benefits from and who pays for ecologically unsound land use. Ironically, the land ethic will most likely come about more for economic reasons than for ecological ones. In the final analysis that irony will matter little. The important issue will be the establishment, long overdue, of the land ethic.

SUMMARY

This chapter reviews wildlife management on lands devoted primarily to other uses. In general, the more intensively land is used for timber, crops, or livestock, the less suitable it becomes for wildlife habitat. Economic incentives have led to more intensive production and, as a result, both game and nongame populations have declined. Wildlife competes poorly in strictly economic terms, in part because of the tradition of public ownership. Nevertheless, most wildlife managers, especially those concerned with private lands, advocate remedies which offer at least some economic incentives for landowners.

Contrary to popular belief, timber harvesting itself has little impact upon most wildlife populations, at least in the temperate zone. The most serious problems in terms of wildlife habitat on commercial forests include widespread conversion of native forests to monocultures and the loss of old stands of "overmature" timber. Wildlife management on such forests should then seek to minimize the extent of conifer monocultures, particularly on marginal sites, and to preserve some old growth hardwood and pine. Monetary losses from reduced timber production could be offset through hunting leases, tax incentives, and cost sharing.

Area-sensitive species pose special problems for commercial forests. These species, mostly birds that migrate each winter to the tropics, need large, contiguous areas of forest for nesting. Quite unlike more familiar game species, area-sensitive species usually decline with increased edge. To sustain such species, managers need to keep adjacent stands of forest trees as nearly to the same age as possible. They also need to maintain corridors of natural habitat between tracts of forests.

Wildlife on croplands has shown steady and consistent declines as American agriculture has become more mechanized and more intensive. Prime agricultural lands will probably continue to produce high yields of food and fiber and will thus remain unsuitable for wildlife. Most authorities recommend developing wildlife habitat primarily on marginal croplands, where economic competition is less severe.

Land retirement programs, implemented to stabilize crop prices, can have beneficial effects on wildlife. Such idle lands, particularly if they are idled for several consecutive years, can develop the necessary cover and food for many wildlife species.

State wildlife agencies have tried various programs to aid wildlife on croplands. Voluntary programs, in which landowners agree to take some land out of production, cost little but affect very little land. A few states have required hunters to purchase habitat stamps. The funds raised through the sale of such stamps are then used to pay landowners for each unit area of land preserved as wildlife habitat.

The greatest problem for wildlife on rangeland is the reduced productivity caused by prolonged overgrazing. Accordingly, reduction of grazing pressure is in the best interest of both wildlife and livestock and is the first step for wildlife management. If prolonged overgrazing has led to brushland invasion, then brush control may be necessary to restore the habitat. Ideally, brush control should be done in long strips so as to maximize edge.

Competition between livestock and wild herbivores can be reduced or avoided by matching domestic animals with wild species that have different food habits. If done correctly, the combined productivity of both game and livestock can be boosted well beyond that of either one separately. Another important step in managing wildlife habitat on rangelands is adequate protection of both wetlands and riparian habitat from excessive damage by livestock.

As a remedy to ecologically damaging land use practices, Leopold proposed development of the land ethic. Once adopted by society, a land ethic would discourage irresponsible land use through social pressure. The land ethic has yet to become effectively implemented because society as a whole does not recognize the long-term social, economic, and ecological costs incurred by the public for ecologically damaging land use.

REFERENCES

Barney, G. O. 1980. *The global 2000 report to the president of the U.S. vol. 1, the summary report.* Washington, D.C.

Box, T. W., D. D. Dwyer, & F. H. Wagner. 1977. *The public rangelands and their management—a report to the president's council on environmental quality.* Washington, D.C.: Council on Environmental Quality.

Box, T. W., and J. Powell. 1965. Brush management techniques for improved forage values in south Texas. *Trans. N. Amer. Wildl. and Natur. Resour. Conf.* 30: 285–295.

Bureau of Sport Fisheries and Wildlife. 1970. *Piedmont National Wildlife Refuge, Jones and Jasper County, Georgia: even-aged forestry management.* Washington, D. C.: U.S. Dept. of Interior, Fish and Wildlife Service. RL-417-R.

Burger, G. V. 1978. Agriculture and wildlife. In H. P. Brokaw (Ed.), *Wildlife and America.* Washington, D. C.: Council on Environmental Quality.

Burger, G., & J. Teer. 1981. Economic and socioeconomic issues affecting wildlife management of private lands. In R. Dumke, G. Burger and J. March (Eds.), *Wildlife management on private lands.* Madison, Wisc.: Wisconsin Chapter of the Wildlife Society.

DeVos, A. 1969. Ecological conditions affecting the production of wild herbivorous mammals on grasslands. *Adv. in Ecol. Res.* 6: 137–183.

Eckholm, Erik P. 1976. *Losing ground: environmental stress and world food prospects.* New York: W. W. Norton & Co.

Erickson, R. E., & J. E. Wiebe. 1973. Pheasants, economics, and land retirement programs in South Dakota. *Wildl. Soc. Bull.* 1: 22–27.

Halls, L. K. 1975. Economic feasibility of including game habitats in timber management systems. *Trans. N. Am. Wildl. and Natur. Resour. Conf.* 40: 168–176.

Harmon, K. W. 1979. Private land wildlife—a new program is needed. In R. D. Teague and E. Decker (Eds.), *Wildlife conservation: principles and practice.* Washington, D.C.: The Wildlife Society.

Harmon, K. W., & M. M. Nelson. 1973. Wildlife and soil considerations in land retirement programs. *Wildl. Soc. Bull.* 1: 28–38.

Holbrook, H. L. 1974. A system for wildlife habitat management on southern National Forests. *Wildl. Soc. Bull.* 2: 119–123.

Horvath, J. C. 1974. *Economic survey of wildlife recreation (detailed analyses).* Atlanta, Georgia: Env. Research Group, Georgia State University.

MacClintock, Lucy, R. F. Whitcomb, & B. L. Whitcomb. 1977. Island biogeography and "habitat islands" of eastern forests II. Evidence for the value of corridors and maximization of isolation in preservation of biotic diversity. *Am. Birds.* 31: 6–16.

Machan, W. J., & R. D. Feldt. 1972. Hunting results on cropland adjustment program land in northwestern Indiana. *J. Wildl. Manage.* 36: 192–195.

Mack, R., & J. Thompson. 1982. Evolution in steppe with a few large, hooved mammals. *Am. Nat.* 119: 757–773.

Mackie, R. J. 1970. Range ecology and relations of mule deer, elk, and cattle in the Missouri River breaks, Montana. *Wildl. Monogr. 20.*

McConnell, C. A., & K. W. Harmon. 1976. *Agricultural effects on wildlife in America—a brief history.* Paper presented at the 31st annual meeting of the Soil Conservation Society of America. Minneapolis, Minnesota.

Myers, N. 1979. *The sinking ark.* New York: Pergamon Press.

National Academy of Sciences. 1970. *Land use and wildlife resources.* Washington, D.C.: National Res. Council Comm. on Agric. Land Use and Wildl. Resour.

Nelson, J. R. 1982. Relationships of elk and other large herbivores. In J. Thomas and D. Toweill (Eds.), *Elk of North America: ecology and management.* Washington, D.C.: Wildlife Management Institute, and Harrisburg, Pa.: Stackpole Books.

Powers, J. E. 1979. Planning for an optimal mix of agricultural and wildlife land use. *J. Wildl. Manage.* 43: 493–502.

Risebrough, R. W. 1978. Pesticides and other toxicants. In H. P. Brokaw (Ed.), *Wildlife and America.* Washington, D.C.: Council on Environmental Quality.

Robbins, C. S. 1979. Effects of fragmentation on bird populations. *Management of north-central and northeastern forests for nongame birds.* Gen. Tech. Rep. NC-51 USDA For. Serv. N.C. For Exp. Sta., St. Paul, Minn.

Schwartz, C. C., & J. G. Nagy. 1976. Pronghorn diets relative to forage availability in northeastern Colorado. *J. Wildl. Manage.* 40: 469–478.

Schwartz, C. C., J. G. Nagy, & R. W. Rice. 1977. Pronghorn dietary quality relative to forage availability and other ruminants in Colorado. *J. Wildl. Manag.* 41: 161–168.

Shaw, S. P., & D. A. Gansner. 1975. Incentives to enhance timber and wildlife management on private forest lands. *Trans. N. Amer. Wildl. and Natur. Resour. Conf.* 40: 177–185.

Stauffer, D. F., & L. B. Best. 1980. Habitat selection by birds of riparian communities: evaluating effects of habitat alterations. *J. Wildl. Manage.* 44: 1–14.

Thomas, J., P. Miller, C. Maser, R. Anderson, and B. Carter. 1979. Plant communities and successional stages. In J. Thomas (Ed.), *Wildlife habitats in managed forests: the Blue Mountains of Oregon and Washington.* Forest Service, Washington, D.C., Agricultural Handbook No. 553.

Vance, D. Russell. 1976. Changes in land use and wildlife populations in southeastern Illinois. *Wildl. Soc. Bull.* 4(1): 11–15.

Wagner, F. H. 1977. Western rangeland: troubled American resource. *Trans. N. Am. Wildl. and Natur. Resour. Conf.* 43: 453–461.

Wagner, F. H. 1978. Livestock grazing and the livestock industry. In H. P. Brokaw (Ed.), *Wildlife and America.* Washington, D.C.: Council on Environmental Quality.

Webb, W. L. 1977. Songbird management in a northern hardwood forest. *Trans. N. Am. Wildl. and Natur. Resour. Conf.* 42: 438–448.

Webb, W. L., D. F. Behrend, & B. Satsorn. 1977. Effect of logging on songbird populations in a northern hardwood forest. *Wildl. Monogr. 55.*

Whitcomb, R. F. 1977. Island biogeography and "habitat islands" of eastern forests I. Introduction. *Am. Birds.* 31: 3–5.

Whitcomb, B. L., R. F. Whitcomb, & D. Bystrak. 1977. Island biogeography and "habitat islands" of eastern forests III. Long-term turnover and effects of selective logging on the avifauna of forest fragments. *Am. Birds.* 31: 17–23.

CHAPTER 11

INTERNATIONAL WILDLIFE CONSERVATION

The twentieth century has seen the earth grow smaller, more interdependent, and far more populous. Sovereign states rely increasingly upon international commerce for raw materials as well as finished products. Economic development or economic disaster in one part of the world effects other parts, making the global community truly interdependent economically. It follows, then, that nations so linked together have a common responsibility in ensuring survival for the world's wildlife.

So intertwined are the peoples of the earth, that seemingly trivial actions in one nation can have major ecological consequences in others. For example, Norman Myers has reported that Central America's rain forests are being turned into North America's hamburgers. Fast-food franchises grew at an unprecedented rate in the United States during the 1970s. Beef prices rose sharply, in part due to increased demand, but also because of rises in the prices of feed grains used to fatten cattle. The highly competitive fast-food business quite naturally sought less expensive sources of ground beef and found them in imported beef. This new market spurred ranchers in Central American countries to increase cattle production as exports to the United States exceeded the 100,000-ton mark annually. Since 1960, the pasture lands of Central America have been increased by 67 percent; most of the clearing has been of rain forests (Myers, 1981).

This chapter recognizes the interdependence of modern people and the resulting need to cooperate internationally in conserving wildlife. To date, the greatest efforts have been directed at international trade in wildlife and products derived from wildlife. Other, longer-term efforts are aimed at establishing a system of biological reserves in each of the world's 193 biogeographical provinces. Both topics are covered here, and the chapter ends with some predictions for international conservation into the twenty-first century.

WILDLIFE MANAGEMENT IN DIFFERENT REGIONS

Wildlife Management in Europe

The roots of modern wildlife management are largely European. For centuries Europeans, like their counterparts throughout the world, lived as hunter-gatherers. But during the late middle ages, hunting gradually shifted from subsistence to sport. Increasing agriculture along with unchecked hunting led to regional extinctions of several large mammals such as the European bison (*Bison bonasus*), the wild horse (*Equus prezalsky*), and the wild oxen (*Bos taurus*) (Leopold, 1936a).

Neighboring nobles, unhappy over the decline of game animals on their lands, began to establish treaties among themselves by the fifteenth century in Germany (Leopold, 1936a,b). These nobles started some of the earliest wildlife management in Europe through preservation of food-bearing trees and shrubs, local predator control, and restrictions on harvest along with severe penalties for poaching. Sport hunting, particularly for big game such as red deer and roe deer (*Capreolus capreolus*) become the exclusive privilege of the nobility.

In many forested regions of Europe, intensive game management has, since the early nineteenth century, been integrated into intensive forest management. This merger has not been an easy one since intensive levels of these two forms of management are largely incompatible. Spruce monocultures furnish almost nothing in the way of wildlife food, so supplemental feeding is often necessary. Deer frequently impair regeneration of forest trees by eating the seedlings, making careful culling necessary to keep deer populations within the limits that managed forests can maintain, even with supplemental feeding.

Hunting is no longer reserved for the nobility, but it is very strictly regulated through licensing requirements, rigid codes of conduct, and stringent management responsibilities placed upon the hunters. Applicants for hunting licenses must pass a series of strenuous examinations on game management, life histories of game species, and hunting regulations and traditions, plus ballistics and firearm safety. Prospective hunters in Czechoslovakia, for example, must undergo 60 hours of classroom instruction, followed by a one-year apprenticeship in game management before they can even take the license exam. Then they must pass exams in each of seven subjects before being granted a license (Newman, 1979). Violations of either traditional codes of conduct or game laws are quite serious matters in most European countries and can result in permanent revocation of a violator's license.

European hunters assume much of the responsibility for game management. Various private hunting clubs or sports organizations purchase leases for hunting on private lands. There they are expected to practice habitat improvement, supplemental feeding, predator and disease control, and game censuses. Often they must estimate harvest quotas and see to it that the quotas are filled. Sometimes, hunters play active roles in wildlife research, as through Finland's permanent observer system (Salo, 1976).

Several wildlife specialists from the United States have compared European game management with American game management (Gottschalk, 1972; Leopold, 1936b; Newman, 1979; Taber, 1961; Wolfe, 1966). In general they have found a slight trend in the American system toward that of European countries, a trend attributed mainly to increasing pressures in the United States for lands on which to hunt. The European system provides far more land per hunter because of its strict licensing requirements. One study found that Polish hunters had an average of about 13 times as much land on which to hunt as did American hunters (Taber, 1961). The more laissez faire American system of issuing hunting licenses cannot distribute hunting pressure evenly, nor can it limit the total number of hunters except for special hunts on restricted areas.

Nevertheless, the European system of game management has severe drawbacks. The first drawback is the artificiality or loss of naturalness inherent in game management. Heavy reliance on predator control, supplemental feeding, and, in the case of big game, systematic culling greatly reduces what most Americans or Canadians would regard as quality of the hunt. Moreover, most European countries are far less egalitarian than Canada or the United States, so the privilege of hunting reserved for only a few would be less acceptable in North America. After evaluating the European and American approaches to game management, Wolfe (1966) concluded that the complete German hunting system would be politically unacceptable in the United States. He did advocate modifications toward the European systems through requiring more testing for license applicants and increasing hunting leases on private lands.

Wildlife Management in the Developing World

The United States, Canada, Europe, Japan, and Australia face abundant problems in conserving their native wildlife. But far greater threats exist to wildlife survival in the so-called developing countries of Asia, Africa, and Latin America. Human population growth in these less developed countries is much higher than in the developed ones (see Table 11-1). The

TABLE 11-1.
HUMAN POPULATION STATUS AND GROWTH RATES FOR DIFFERENT REGIONS OF THE WORLD
(Ehrlich et al., 1977 © W. H. Freeman & Co. Used by permission.)

Region	Human population in 1976 (millions)	Annual rate of population growth (%)	Projected population by 2000 A.D.
Africa	413	2.6	815
Asia	2,287	2.0	3,612
Latin America	326	2.8	606
Oceania	22	1.8	33
USSR	257	0.9	314
Europe	476	0.6	540
U.S. & Canada	239	0.8	294
World	4,020	1.8	6,214

annual population growth rate for Africa is about 2.6 percent, for Asia 2.0 percent, and for Latin America 2.8 percent. At these rates the populations will double in from 25 to 35 years. This trend, combined with rising expectations for higher standards of living, is placing unprecedented stress upon entire ecosystems and the wildlife therein. Developed nations must share responsibility for such stresses. Their heavy demands for cheap raw materials are adding substantially to the rate of destruction.

Developing nations are almost entirely tropical. The tropics contain perhaps three-quarters of all species on earth (Myers, 1979), arranged in the world's most intricate and complex ecosystems. Scientists have barely begun to understand even the most rudimentary workings of tropical communities, but one characteristic is certain: for all their complexity, tropical ecosystems, particularly moist forests, are among the most fragile. Many of these forests occur on soils unable to hold nutrients. Undisturbed forests retain sufficient nutrients in their abundant vegetation, quickly recycling from dead plants into live ones. But once the moist forest cover is stripped away for pasture, farmland, or through large-scale clear-cutting, rainfall washes nutrients away. Once disturbed by human activity on a large scale, tropical forests are slow to recover, if they can recover at all.

The problems in wildlife conservation in Belize (formerly British Honduras) are typical of those in a good many developing nations. Habitat loss is severe, principally because of land clearing for agriculture and a population growing at about 3.5 percent annually, making Belize the fastest growing nation in all of Central and South America (Ehrlich et al., 1977). Although laws exist to protect various categories of wildlife, enforcement is hampered by lack of adequate funds and trained personnel

(Frost, 1977). Thus, commercially valuable wildlife, such as the jaguar, is declining. Also declining are the paca (*Cuniculus paca*), the manatee (*Trichechus manatus*), and even the nine-banded armadillo, which are hunted mainly for food. Frost (1977) found that the Belize government has enjoyed some success through public education programs, but he warned that improvements in personnel, basic biological data, and liasons between resource management departments were critically needed.

A few developing nations have been more fortunate. Botswana contains large areas too dry for agriculture and livestock raising, together with a relatively sparse human population density. Because the semiarid climate limits other types of food production, wild animals remain important dietary supplements for Botswanans. Wildlife management is new in Botswana, but already about 18 percent of the nation's area, including representatives of all major habitat types, have been designated as national parks or game reserves. Game harvests are carefully controlled through strict quota systems; these seem to furnish adequate subsistence hunting for tribe members and also permit a limited amount of sport hunting. Direct and indirect revenues generated through wildlife-related activities totaled over $10 million in 1973 (Butynski and Von Richter, 1975).

Despite their biological wealth, developing nations lack the financial resources to pay for wildlife management. Unchecked human population growth, accompanied as always by chronic poverty, is leading to more and more clearing for agriculture, even where soil conditions permit only 1 to 2 years of production before becoming sterile. Meanwhile, the developed nations continue their seemingly inexhaustible demand for raw materials at the lowest possible prices. This contributing role of the developed nations, combined with the fact that developing nations are often too poor to pay for conservation, means that the developed world will have to bear much of the costs for wildlife management in the developing world. Perhaps the greatest challenge for wildlife conservation during the last two decades of the twentieth century will be to ensure the survival of tropical wildlife in representative areas of protected natural habitat.

INTERNATIONAL TRADE IN WILDLIFE

Part of the justification for passing the Lacey Act in 1900 was the need to regulate imports of wildlife to the United States. Thirteen years after passage of the Lacey Act, importation of wild bird plumes was outlawed by the Wilson Tariff Act. Both acts were later amended to prohibit importation of any wildlife taken illegally in the country of origin (King, 1978).

Despite these and other conservation efforts, the international trade in wildlife continues to climb. Legal imports of wildlife products into the United States rose from 4 million in 1973 to 187 million in 1978 (Ehrlich and Ehrlich, 1981). Even assuming that some of this staggering increase resulted from improved record keeping, one would have to conclude that international trade is on the rise. Commercial trade is second only to habitat loss as the leading threat to wildlife.

The Endangered Species Acts of 1969 and 1973 sharply curbed legal imports of officially listed endangered species to the United States. For some species of spotted cats, the numbers imported before 1969 were considerable. Skins from some 1,300 cheetahs, 9,600 leopards, 13,500 jaguars, and 129,000 ocelots (*Felis pardalis*) legally entered the United States in 1968 (King, 1978).

Consumers and Producers

As a general rule, consumers of wildlife are the developed nations with ample supplies of hard currency. Western Europe, North America, and Japan together consume most of the world's international wildlife trade. Lifestyles, affluence, and leisure time within developed nations create most of the demand for wildlife products and live animals from other parts of the world.

Most of the producers are tropical, developing nations. Many governments of such nations encourage trade in wildlife as a source of foreign exchange. If properly regulated, such a practice is philosophically similar to that of state and provincial wildlife agencies in the developed countries that encourage sustained yields of game animals by sport hunters. Unfortunately, though, developing nations rarely have the resources to regulate wildlife harvests. Should the wealthier consumer nations, where most of the demand for wildlife and their products originates, help developing nations prevent excessive exploitation? This question is more than merely philosophical; it remains at the very core of international cooperation in regulating the wildlife trade.

The CITES Treaty

The Convention on International Trade in Endangered Species (CITES) began in 1973 (see Chapter 8) as a formal treaty to regulate international commerce in species threatened with extinction. Nations signing CITES agree to establish their own scientific authorities to regulate exports and to abide by the treaty's conditions under international law.

Any wild species in question can be listed on one of three CITES appendixes. Those listed under Appendix I, the most serious category, are regarded as "endangered." Their movement internationally requires an import permit from the country of destination as well as an export permit from the country of origin. Practically speaking, commercial trade in such species is, between signatories, virtually nil. Appendix II is a less serious classification, corresponding to "threatened." Species listed on Appendix II must have an export permit from the nation of origin. Appendix III is reserved for those species believed by the nation having jurisdiction over them to be threatened. Thus any country may list one of its own species on Appendix III, allowing that nation to regulate trade in that species.

Whenever the question of whether or not to protect a species arises, along with it comes the problem of burden of proof. Should it lie with those who wish to exploit the species or with those who seek to protect it? In the past, the burden rested with those who wanted the animal protected. Such has not been the case with CITES. Instead, at their 1976 meeting in Berne, Switzerland, delegates decided that any errors in listing and delisting species should be made in favor of protection. The convention adopted the "Berne criteria," which, for the delisting of species, require:

1 Positive scientific evidence that the species can withstand exploitation
2 Well-documented population surveys or trend estimates showing increasing populations
3 Comprehensive trade analysis (Johnson, 1980).

Now the burden of proof rested with those who would advocate exploitation and, predictably, controversy errupted.

At the root of the controversy is the same protectionist versus user conflict that arose between John Muir and Gifford Pinchot (see Chapter 1). Wildlife management agencies in the United States, entrenched in the "user" point of view, reacted sharply to imposition of the Berne criteria. North American species, including the widespread and presumably common bobcat and river otter (*Lutra canadensis*), were listed on Appendix II. Such action was not only philosophically counter to agency tradition, it also intruded on jurisdiction previously confined to the agencies themselves. State wildlife departments found themselves at odds with the Scientific Authority, creating within the United States another federal–state confrontation.

Yet CITES has not proved to be the catastrophe to traditional wildlife management that its opponents feared. The listing of bobcat and otter on Appendix II prompted a flurry of research on the status of the two species and on the effects of trapping pressures upon them. Much of this research was overdue, in part because both species are quite difficult to census, and there simply had not been enough incentive to invest the necessary time and effort. Such "emergency" research should, over the long run, afford state agencies a better basis for regulating harvest of bobcat, otter, and other furbearers. Moreover, the Berne criteria, while difficult, are not impossible to meet. Results of detailed reports on the American alligator met the criteria, and the species has been changed from Appendix I to Appendix II.

Should whole taxa, rather than individual species, be listed on CITES appendixes? Certain groups, including crocodilians, spotted cats, and lutrines, have been blanket-listed for two basic reasons. First, the products derived from one member of a group can be extremely difficult to distinguish from others. Customs inspectors and wildlife officials may thus be unable to distinguish a truly endangered species from a similar but more common one. Second, a sort of "domino effect" may set in if morphologically similar species subject to exploitation are listed individually. Importers will then shift their purchasing to closely related unlisted species, placing more pressure on them. Eventually the unlisted species may have to be listed as their populations decline.

The leopard presents a special problem. Threatened in some portions of its range, it remains quite common in parts of Africa. Realizing that the animal remains abundant throughout their jurisdictions, some African wildlife officials have voiced objections, contending that the developed world was presumptive in placing the leopard on Appendix I (Teer and Swank, 1978). Such a listing was viewed as an intrusion on their internal affairs, a claim much like that voiced by their American counterparts concerning the bobcat.

Any effective long-range strategy in regulating international trade in wildlife will have to strike a balance between protection and use. Complete protection of nonendangered species, favored by some people and organizations, could destroy an important utilitarian justification for preserving the very habitat on which wildlife depends. CITES is getting the message, especially from many African nations, that if wildlife is to survive it must be able to pay at least some of its own way (IUCN, 1981).

A certain amount of conflict between consumer and producer nations is inevitable in any international effort to regulate trade in wildlife. Consumer nations feel that too much regulatory responsibility has been left up to their enforcement officials at ports of entry. The real slaughter, they contend, occurs in the remote regions of developing or producer nations. Producer nations counter that consumer nations are at fault through their reluctance to enforce import laws vigorously. Time and experience should help work out many of these difficulties. But as King (1978) has pointed out, the Lacey Act and other such measures have shown that control of trade at the consumer end is both more feasible and more critical.

Another controversy concerns the role of nongovernmental organizations (NGOs). At the 1979 CITES meeting in Costa Rica, NGOs were well-organized and worked hard at lobbying delegates. (Under Convention rules, NGOs are free to attend meetings, though they cannot vote. Voting remains exclusively the right of governmental delegations.) Some 55 NGOs attended the Costa Rica meeting. Roughly three-quarters of them represented conservation or environmental groups and the remainder represented trade interests. The conservation NGOs had a real impact on the meeting, bringing about the defeat of a United States–backed proposal to weaken the Berne criteria (Barber, 1979). Not everyone agrees on either the wisdom or the effectiveness of NGOs in influencing CITES policy (see opposing views by Talbot, 1979, and Johnson, 1980), but it seems likely that NGOs have become a permanent part of the process.

Day-to-day functions for CITES are carried out by its secretariat, consisting of two scientists, one attorney, and a secretarial staff based in Switzerland. CITES is both larger and more effective than other conservation conventions, and its growth and effectiveness have been largely attributed to the secretariat (IUCN, 1981). Functioning as a global nerve center for regulating the wildlife trade, the secretariat accumulates much information of value to CITES and to law enforcement agencies.

Despite its progress and accomplishments, CITES does have its limitations. Not all nations have yet signed, including some key ones in international trade. Both producer and consumer nations have been reluctant to enforce laws to the limits, perhaps because there is not yet enough public pressure for them to do so. But CITES does at least substantially slow the traffic in endangered species, even though it cannot hope to stop it completely. It also functions as an important focal point for international cooperation in wildlife conservation. The framework has been built and an international management network created. Wildlife conservation is a cause that no nation has publicly rejected and thus serves as an important starting point for international cooperation on broader issues as well.

HABITAT DESTRUCTION

The CITES Treaty was never intended to preserve wildlife habitat. Other conservation approaches are needed to ensure preservation of adequate reserves of habitat. Ideally, a system of reserves representing all the world's 193 biogeographical provinces should be developed. Such a system would preserve intact ecosystems of all major types. Just as with international trade in wildlife, a certain amount of international cooperation and coordination is essential. Standards must be set up and maintained, even between nations not otherwise on the best of terms.

An International Classification for Habitat

Botanists and biogeographers have long been classifying the world's major habitat types. Refinements during the 1970s (Dasmann, 1973a, 1973b; Udvardy, 1975) had a pragmatic motive; namely, to conserve representative samples of the world's vanishing wild areas. These new classification schemes integrated both the distribution of species and the distribution of ecosystem units. They attempted to standardize ecological associations through a hierarchical classification. The one presented here is from Udvardy (1975).

Realms In the global classification of habitat types, the highest taxon is the realm. There are 8 biogeographical realms (see Figure 11-1):

1 Nearctic realm
2 Palaearctic realm
3 Africotropical realm
4 Indomalayan realm
5 Oceanic realm
6 Australian realm
7 Antarctic realm
8 Neotropical realm.

Biogeographical Provinces Realms are divided into biogeographical provinces, which correspond roughly to floral and faunal regions of older classifications. Each realm has its own set of provinces. The Nearctic realm, for example, has 22 biogeographical provinces (see Table 11-2), the Palaearctic 44, the Africotropical 29, the Indomalayan 27, the Oceanic 7, the Australian 13, the Antarctic 4, and the Neotropical 47, for a total of 193.

Principal Biome Types The third and final taxon of this classification system is in a sense the most general. Ecologists have long recognized that similar climate-plant-animal associations can occur in different regions of the world. These general associations are called "biomes," and under the Udvardy classification 14 principal biome types are recognized:

1 Tropical humid forest
2 Subtropical and temperate rain forest or woodlands
3 Temperate needle-leaf forests or woodlands
4 Tropical dry or deciduous forests (including monsoon forests) or woodlands
5 Temperate broad-leafed forests or woodlands, and subpolar deciduous thickets
6 Evergreen sclerophyllous forests, scrubs or woodlands
7 Warm deserts and semideserts
8 Cold winter (continental) deserts and semideserts
9 Tundra communities and barren arctic desert
10 Tropical grasslands and savannas
11 Temperate grasslands
12 Mixed mountain and highland systems with complex zonation

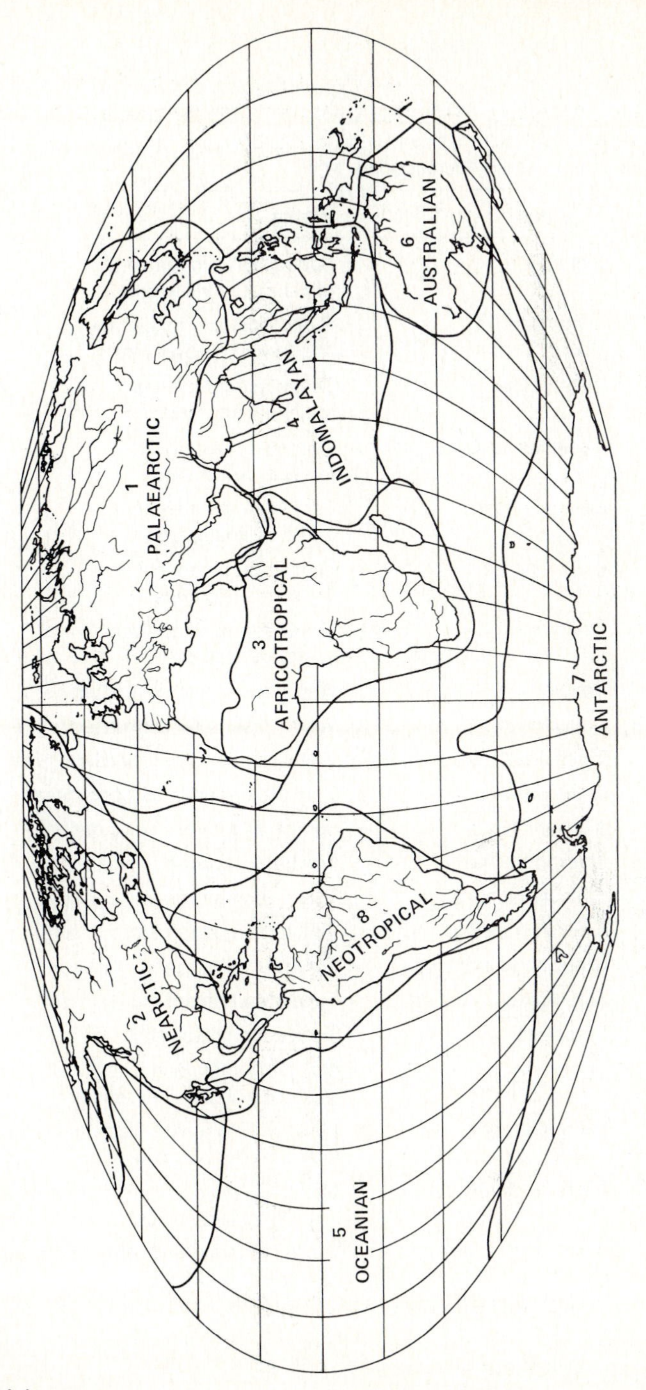

FIGURE 11-1
The biogeographical realms of the world. (*IUCN, 1980* © International Union for Conservation of Nature and Natural Resources. Used by permission.)

TABLE 11-2.
THE BIOGEOGRAPHICAL PROVINCES AND BIOMES OF THE NEARCTIC
(After Udvardy, 1975 © International Union for Conservation of Nature and Natural Resources. Used by permission.)

No.	Province	Biome
1.1.2	Sitkan	Subtropical and temperate rain forest or woodlands
1.2.2	Oregonian	Subtropical and temperate rain forest or woodlands
1.3.3	Yukon taiga	Temperate needle-leaf forest or woodlands
1.4.3	Canadian taiga	Temperate needle-leaf forest or woodlands
1.5.5	Eastern forest	Temperate broad-leafed forest or woodlands
1.6.5	Austroriparian	Temperate broad-leafed forest or woodlands
1.7.6	Californian	Evergreen sclerophyllous forest, scrub, woodland
1.8.7	Sonoran	Warm desert and semidesert
1.9.7	Chihuahuan	Warm desert and semidesert
1.10.7	Tamaulipan	Warm desert and semidesert
1.11.8	Great Basin	Cold winter desert and semidesert
1.12.9	Aleutian Island	Tundra and barren arctic desert
1.13.9	Alaskan tundra	Tundra and barren arctic desert
1.14.9	Canadian tundra	Tundra and barren arctic desert
1.15.9	Arctic Archipelago	Tundra and barren arctic desert
1.16.9	Greenland tundra	Tundra and barren arctic desert
1.17.9	Arctic desert & ice cap	Tundra and barren arctic desert
1.18.11	Grasslands	Temperate grassland
1.19.12	Rocky Mountains	Mixed mountain and highland systems with complex zonation
1.20.12	Sierra-Cascade	Mixed mountain and highland systems with complex zonation
1.21.12	Madrean-Cordilleran	Mixed mountain and highland systems with complex zonation
1.22.14	Great Lakes	Mixed mountain and highland systems with complex zonation

13 Mixed island systems
14 Lake systems.

All three taxa, realms, provinces, and biomes are combined to furnish both regional and ecological descriptions for any natural area on earth. Wood Buffalo National Park in Canada, for example, would be classified 1.4.3 (Nearctic, Canadian taiga, temperate needle-leaf forest or woodlands). The Wia Wia Nature Reserve in Surinam is 8.4.1 (Neotropical realm, Guyanan, tropical humid forest).

Classification of Protected Natural Areas

Throughout the world, natural biological communities continue to disappear. Inevitably, so do the wild vertebrates living in them. The only long-term hope for the world's wildlife is adequate protection of representative samples of these vanishing natural communities. Every year more species are found only in zoos or protected natural areas. It is essential, then, to understand how these protected areas are classified. Many of these places were established for certain types of outdoor recreation, for preservation of geological treasures, or for reasons other than the preservation of wildlife habitat. Nevertheless, the protection they are granted assures preservation of at least limited patches of wildlife habitat.

The United Nations passed a resolution in 1959 recognizing national parks and equivalent reserves as important in the wise use of natural resources. Since then the United Nations, in counsel with the International Union for Conservation of Nature and Natural Resources, has developed international criteria for these areas and maintained lists of national parks and equivalent reserves meeting those criteria. Three major classes of protected natural areas are recognized in the UN–IUCN guidelines (IUCN, 1980):

1 *Nature reserves,* including both strict nature reserves and managed nature reserves, in which development is excluded and access by the public severely restricted or, in some cases, entirely forbidden.

2 *National parks and equivalent reserves,* in which protection is conferred by the "highest competent authority," meaning the national or federal government. The public is allowed to visit these areas, and a certain limited amount of construction of roads and facilities is permitted.

3 *Biosphere reserves* are set aside primarily for education, research, and conservation of representative ecosystems under guidelines from the United Nations Educational, Scientific, and Cultural Organization (UNESCO). Some biosphere reserves are also national parks, provincial parks, or other types of reserves.

National nature reserves, national parks and equivalent reserves, and biogeographic provinces must all meet the same three basic criteria. To be included on the UN–IUCN list an area must:

1 Have adequate protection under statue
2 Have adequate de facto protection
3 Be at least 1,000 hectares in size, with exceptions made for islands of smaller size
4 In general, prohibit all forms of exploitation including hunting, fishing, lumbering, mining, and public works construction as well as agricultural and pastoral practices.

Table 11-3 presents the extent of national parks and other reserves for a few selected nations. Note that developing nations such as Brazil and Nepal have protected far smaller proportions of their land area than have the United States and Canada.

The Biosphere Reserve Programme

One such system, the UNESCO Biosphere Reserve Programme, is already well under way. Unlike many older national parks and nature reserves, the biosphere reserves are selected primarily on the basis of representativeness, rather than on uniqueness. Thus the first concern is whether or not an area under consideration really represents the biogeographic province in

TABLE 11-3.
THE EXTENT OF NATIONAL PARKS, EQUIVALENT RESERVES, NATURE RESERVES, AND BIOSPHERE RESERVES FOR SELECTED NATIONS.
Figures compiled from IUCN (1980 © International Union for Conservation of Nature and Natural Resources. Used by permission.)

			Total area (km²) (no. of units)		
% Land area protected	Nation	Land area (km²)	National parks and equivalent reserves	Nature reserves	Biosphere reserves
0.1	China	9,560,990	0	8,820(26)	0
3.8	U.S.A.	9,363,425	273,361(93)	0	86,179(33)
1.4	Canada	9,221,016	129,757(28)	101(2)	581(2)
0.7	Brazil	8,511,178	55,649(20)	1,409(7)	0
1.9	Nigeria	970,994	5,309(1)	13,152(4)	5(1)
6.8	Tanzania	937,062	42,803(11)	20,474(6)	0
18.2.	Botswana	574,980	37,700(3)	66,693(6)	0
0.5	France	550,787	2,788(7)	219(5)	0
5.2	Japan	379.927	19,919(26)	0	0
3.3	Nepal	140,619	3,991(4)	588(3)	0
7.3	Costa Rica	49,826	2,517(10)	1,115(7)	0

which it occurs. Scenic splendors, such as giant waterfalls or an unusually high mountain, are granted less importance. Their presence might even be detrimental, as they could render an area nonrepresentative of the province as a whole, hence make the area unfit for inclusion.

By late in 1979, tbe world had a total of 162 biosphere reserves in 76 of the 193 biogeographical provinces (IUCN, 1980). Not only were more than half the provinces unprotected, but the distribution of reserves was extremely uneven (see Figure 11-2). Much more remains to be done. The United States has at least 28 officially designated biosphere reserves (see Table 11-4). Notice that several are well-known national parks, and others are administered by the U.S. Forest Service, the Bureau of Land Management, and the U.S. Fish and Wildlife Service.

Major Management Problems

Aside from the problem of acquiring more biosphere reserves in regions where they are most needed, several other problems are inherent in the management of all natural reserves. The first and foremost problem is avoidance of extinctions. For many species, particularly on larger reserves, extinction will remain unlikely. But for extinction-prone species, such as the larger carnivores, area-sensitive birds, and commercially tempting animals such as rhinoceroses, the odds favor extinction. In some cases, populations of extremely extinction-prone species may periodically require reintroductions, a tactic that requires preservation of populations elsewhere. Species yielding high profits to poachers will have to be carefully protected by well-paid and well-trained guards, while international efforts continue to retard market sources abroad.

Another problem is maintenance of representative successional stages. How patchy should the reserve be? Should managers deliberately disturb sites within reserves and if so, by what means, how frequently, and in what size units? These questions can be answered only through further ecological research to determine the characteristics of natural disturbance. Managers can then seek to mimic such disturbances to aid preservation of natural diversity.

Should biosphere reserves seek to maximize species diversity or should they concentrate on saving the more extinction-prone species? This crucial question, addressed briefly in Chapter 8, will prove troublesome. The overall aim of biosphere reserves is to preserve genetic diversity, an objective which offers no clear-cut answer.

Biosphere reserves are also established to provide international cooperative research. Far more scientific support is needed to help justify, improve, and strengthen reserve management. Insular ecology, reviewed in Chapter 8, provides an exciting and promising set of guidelines. But

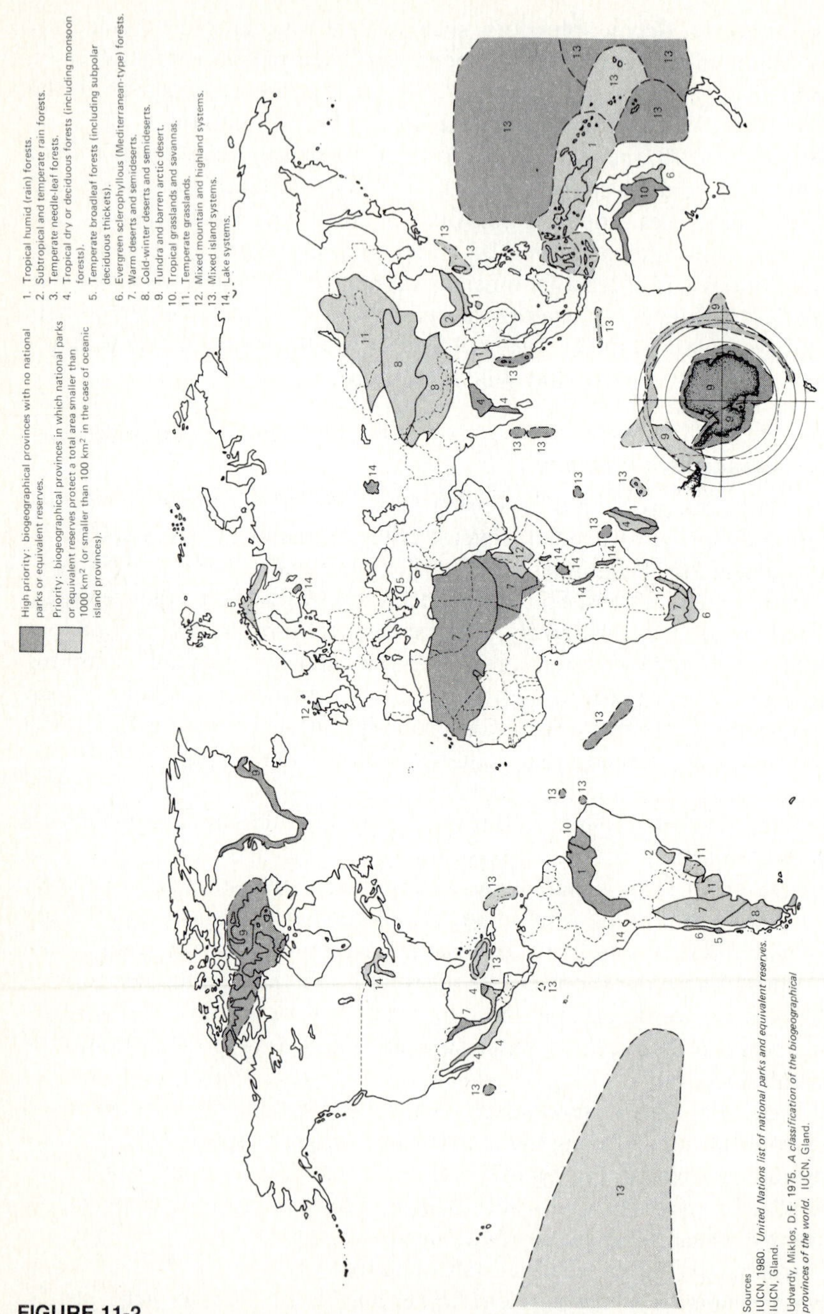

FIGURE 11-2
Global needs for additional biological reserves, national parks, or other protected areas. (*IUCN, 1980* © International Union for Conservation of Nature and Natural Resources. Used by permission.)

TABLE 11-4.
SOME BIOSPHERE RESERVES WITHIN THE UNITED STATES
(Modified from Franklin, 1979)

Name and location	Biogeographic province	Size (ha)	Administering agency
Noatack National Arctic Range, Alaska	Alaska tundra	3,000,000	Bureau of Land Management
Aleutian Island National Wildlife Refuge, Alaska	Aleutian Island	1,100,000	Fish and Wildlife Service
Channel Islands National Monument, Calif.	Californian	7,440	National Park Service
San Dimas Exp. Range, Calif.	Californian	6,947	Forest Service
San Joaquin Exp. Range, Calif.	Californian	1,861	Forest Service
Big Bend National Park, Tex.	Chihuahuan	286,600	National Park Service
Jornada Exp. Range, N. Mex.	Chihuahuan	77,000	Agricultural Research Service
Coweeta Exp. Forest, N.C.	Eastern forest (south)	2,300	Forest Service
Great Smoky Mts. National Park, Tenn. & N.C.	Eastern forest (south)	207,500	National Park Service
Hubbard Brook Exp. Forest, N.H.	Eastern forest (northeast)	3,057	Forest Service
Everglades National Park, Fla.	Everglades	566,800	National Park Service
Central Plains Exp. Sta., Colo.	Grasslands (short grass)	6,280	Agriculture Research Service
Desert Exp. Range, Utah	Great Basin (south)	22,513	Forest Service
Luguilla Exp. Forest, P.R.	Greater Antillian	11,300	Forest Service
Virgin Islands Nat. Park, V.I.	Lesser Antillian	6,130	National Park Service
Cascade Head Exp. Forest, Oreg.	Oregonian	7,051	Forest Service
Olympic National Park, Wash.	Oregonian	362,850	National Park Service
Coram Exp. Forest, Mont.	Rocky Mountain (north)	2,984	Forest Service
Glacier National Park, Mont.	Rocky Mountain (north)	410,000	National Park Service

Yellowstone National Park, Wyo., Idaho, and Mont.	Rocky Mountain (north)	900,000	National Park Service
Fraser Exp. Forest, Colo.	Rocky Mountain (south)	9,300	Forest Service
Rocky Mountain National Park, Colo.	Rocky Mountain (south)	106,160	National Park Service
H.J. Andrews Exp. Forest, Oreg.	Sierra-Cascade (north)	6,050	Forest Service
Three Sisters Wilderness, Oreg.	Sierra-Cascade (north)	80,900	Forest Service
Sequoia-Kings Canyon National Parks, Calif.	Sierra-Cascade (south)	342,754	Forest Service
Staislaus Exp. Forest, Calif.	Sierra-Cascade (south)	683	National Park Service
Oregon Pipe Cactus National Monument, Ariz.	Sonoran (typical)	134,000	National Park Service
Mt. McKinley National Park, Alaska	Yukon taiga	784,900	National Park Service

these guidelines need more refinement and greater empirical support to help convince governments of the importance of reserves to long-term human needs. Many questions remain unanswered. How sensitive are different ecosystems to human disturbance? How resilient are they when disturbed? What really constitutes an effective population size and how does that size vary between species? Are the general size guidelines for reserves realistic and do they vary between realm or biome? What is the role of dispersal and how does it affect the vulnerability of species?

Four immediate priorities have been recommended for the UNESCO program (Di Castri and Loope, 1979):

1 Encourage nations to create their own network of biosphere reserves

2 Encourage cooperation among countries with reserves representing similar ecosystem types, including exchange of personnel and information

3 Continue to maintain full-fledged support of such international organizations such as IUCN, UNEP, and FAO

4 Provide funds for assisting developing nations in establishing reserves.

Funding for Biosphere Reserves

Obviously, a global network of biosphere reserves will be costly, both for initial purchase and for management. UNESCO provides guidelines and coordination but no direct funding. In most cases the costs of developing reserves must be borne by the nation in which the reserve occurs. This has not been a problem for the developed world but remains beyond the reach of many developing countries where reserves are badly needed.

Just as with regulating international trade in wildlife, the question again arises as to who benefits and who pays. Should the developed nations help defray costs of establishing and maintaining reserves in developing nations? Some have argued that developed nations should pay on ethical grounds, that much habitat destruction in developing countries results from demands inside developed ones. The "hamburger connection" reported by Myers (1981) is one such case in point. Another line of reasoning is more pragmatic. It stands that unless developed nations bear the costs, biosphere reserves simply will not be established in many developing countries.

Either justification supports the need for developed nations to pay more for biosphere reserves. This need can be more easily accepted when genetic resources are viewed, as they should be, as useful to all humankind.

The World Conservation Strategy (IUCN, 1980) accepted the need for assistance to developing nations. The strategy proposed several mechanisms to implement such assistance. One method would consist of an expanded version of the already established International Board for Plant Genetic Resources (IBPGR), a Rome-based organization dedicated to preserving genetic resources of food crops. The board has already classified 10 high-priority regions for preserving crop genetic materials, including some on-site protection. Another suggestion from the World Conservation Strategy was that industries and other commercial enterprises using particular species should sponsor establishment of reserves; so should industrial and developmental projects relying upon naturally occurring compounds. In both cases, sponsorship could work to long-term advantage for industry by ensuring availability of materials in the future.

Myers (1979) proposed one-way transfers of funds (grants economics) to developing nations to help establish and maintain reserves. Wealthier nations could, for example, subsidize reserves in critical areas such as tropical rain forests, perhaps matching small amounts of funding from within the developing countries. Global taxes derived from exploitation of deep ocean manganese could furnish funds for reserves. Myers calculated that a .1 percent ad valorum tax on internationally traded oil would, within about 20 months, provide enough money to establish 150 biological reserves and maintain them for 20 years.

A different tack could be taken by developing nations by imposing a user tax on development. Brazil already uses taxes from harvest of commercially grown *Eucalyptus* to help fund national parks and other reserves. This method provides a steady source of income and assures that conservation is integrated into international development.

INTERNATIONAL WILDLIFE CONSERVATION IN THE FUTURE

By the year 2000, the world population will have reached an estimated 6.4 billion. Approximately 90 percent of the increase from 1975 levels will occur in developing countries (Barney, 1980). Such changes will greatly increase the stresses on the earth's remaining wildlife. If present trends continue without bold new conservation measures, somewhere between 500,000 and 2 million species will vanish by the year 2000 (Lovejoy, 1980). Strong remedial measures are clearly needed.

The success of the CITES Treaty offers hope that international cooperation can and will work. Perhaps it is a testimony to wildlife's universal appeal that nations have been able to work together so well. In any case, there is good reason to believe that the projected extinctions can be cut substantially.

A system of biological reserves, under the guidelines of the Biosphere Reserve Programme or some similar framework, will be essential. The developed nations will have to bear the brunt of the costs, but the projected costs of such reserves are minuscule indeed when compared with overall costs of development.

Conservationists are becoming increasingly sophisticated in economic and political, as well as biological matters. Increasingly, those who seek to preserve representative samples of wild nature are realizing that conservation and development need not be mutually exclusive. This is particularly true with respect to the developing nations. Their populations will stabilize only after economic development has ensured their people of a decent, hopeful existence.

The trick will be to integrate conservation, both in the form of biological reserves and healthier, more productive environments, in the early stages of economic development. Whenever conservation is applied as a last minute consideration, it is viewed by most people as obstructionist and even misanthropic (IUCN, 1980). But once long-term environmental considerations become an established part of development, the friction will be eased.

SUMMARY

Nations of the world are increasingly dependent economically and actions on one part of the earth can have significant ecological impacts in other regions. Moreover, the parts of the world least prepared financially to cope with environmental problems, the tropical and developing nations, also contain the greatest biological diversity in some of the world's most fragile and complex ecosystems. Some of the most critical challenges for wildlife managers late in the twentieth century will be in international wildlife conservation.

This chapter begins with brief reviews of wildlife conservation in different regions. Europeans have practiced wildlife conservation in the form of intensive game management for centuries. Whenever European wildlife management is compared with that practiced in the United States and Canada, questions arise regarding whether or not North American game management will, under the pressures of increasing demand, become like that of Europe. Hunting leases and shooting preserves in the United States certainly resemble the elite and expensive hunting practices of many European nations, and hunter safety courses resemble compulsory European hunter training. On the other hand, the North American traditions of wildlife as a common resource, together with the practice of hunting opportunities for everyone, should mitigate to some extent against a drift too close to the European system.

Wildlife conservation is quite different in the developing world. High rates of human population growth, along with poverty, poor educational opportunities, and a heavy dependence on subsistence agriculture, are placing unprecedented pressure on wildlife habitat. Enforcement of laws to protect wildlife typically ranges from poor to nonexistent. Professional training in the conservation of wildlife and other natural resources is generally lacking. Finally, governments struggling just to feed their people are unlikely to place much of a premium on programs to save their wildlife. Even if they do grant wildlife a high priority, they lack adequate funding for conservation programs.

To date, the greatest international efforts have been directed at controlling trade in threatened and endangered species. The CITES Treaty has pledged all signatories to establish scientific authorities to help set up trade quotas. Periodically, CITES meets, and representatives from member nations decide which species or groups of species should be placed on Appendix I (endangered, and requiring both export and import permits) and Appendix II (threatened, and requiring only export permits). The convention has generated conflict between utilitarians and preservationists and between producing nations and consuming nations. Yet overall it has been successful, both in regulating international trade and in establishing important cooperation, even between nations not otherwise on the best of terms.

Habitat destruction remains the single greatest threat to wildlife worldwide and there is considerable need for increased international cooperation to preserve habitat. One system of habitat classification developed by the IUCN forms the basis for inventory and for establishing priorities for protection. It relies on a hierarchy of realms, biogeographical provinces, and principal biome types.

Some global network of parks, biological reserves, or other protected areas must be developed. Ideally, the objective is to preserve representative samples of each of the nation's 193 biogeographical provinces. One such program is the UNESCO Biosphere Reserve Programme, in which the United Nations furnishes guidelines and technical assistance to all countries to encourage protection of habitat. Unfortunately, the financial costs of developing and maintaining such reserves must be borne by the nation holding sovereignty over each area. As a result, a disproportionate number of reserves occur in developed nations while their establishment lags behind in poorer tropical countries, where they are needed most.

The IUCN plays a crucial role in international habitat conservation. Acting in consultation with the United Nations, the IUCN has produced strict international guidelines for nature reserves. These high standards serve as models for developing nations, many of which recognize the values of tourism and international prestige in establishing national parks and reserves.

A persistent and crucial question in international wildlife conservation concerns who should pay the costs. Developing nations pose the immediate threats to their own wildlife, their fledgling governments are often plagued with corruption and other problems. Yet much of the demand for wildlife in international trade and for the cheap beef and wood products produced in developing nations comes from developed ones. Developed nations must therefore accept a larger share of the responsibility for the deteriorating status of wildlife throughout the developing world. In addition, from a purely practical standpoint, developed nations such as the United States, Canada, Great Britain, and Japan are the only ones who can afford the high costs of conservation.

There is cause for guarded optimism. The CITES Treaty has shown that international cooperation can and will work for wildlife. International agencies, both public and private, are working steadily to improve the global status of wildlife, especially in developing nations. Various proposals have been made which, if enacted, would raise the necessary funding for a truly effective system of global habitat reserves. There will be losses, and the worldwide extinction rates, even under the best of projections, will be high. But with hard work, commitment, and a little well-placed luck, those losses can be minimized and finally contained. On a global scale, only consistent and dedicated international cooperation can ensure a future for the world's wildlife.

REFERENCES

Barber, Janet. 1979. The NGO's show their mettle. *IUCN Bull.* 10(5): 34,40.

Barney, G. O. 1980. *The global 2000 report to the president of the U.S. vol. 1: the summary report.* Washington, D.C.

Butynski, T., & W. Von Richter. 1975. Wildlife management in Botswana. *Wildl. Soc. Bull.* 3: 19–24.

Dasmann, R. F. 1973a. *A system for defining and classifying natural regions for purposes of conservation.* IUCN Occ. Paper No. 7, Morges, Switzerland.

Dasmann, R. F. 1973b. *Biotic provinces of the world.* IUCN Occ. Paper No. 9, Morges, Switzerland.

Di Castri, F., & L. Loope. 1979. Thoughts on the biosphere reserve concept and its implementation. In J. F. Franklin and S. L. Krugman (Eds.), *Selection, management and utilization of biosphere reserves.* Pacific NW Forest and Range Exp. Sta. Washington, D.C.: U.S. Dept. of Agriculture, U.S. Forest Service.

Ehrlich, P., Anne Ehrlich, & J. Holden. 1977. *Ecoscience: population, resources, environment.* San Francisco, Calif.: W. H. Freeman & Co.

Ehrlich, P., & Anne Ehrlich. 1981. *Extinction: the causes and consequences of the disappearance of species.* New York: Random House.

Franklin, J. F. 1979. The conceptual basis for selection of U.S. Biosphere Reserves and features of established areas. In J. F. Franklin and S. L. Krugman (Eds.), *Selection, management and utilization of biosphere reserves.* Pacific NW Forest and Range Exp. Sta. Washington, D.C.: U.S. Dept. of Agriculture, U.S. Forest Service.

Frost, M. 1977. Wildlife management in Belize: Program status and problems. *Wildl. Soc. Bull.* 5: 48–51.

Gottschalk, J. S. 1972. The German hunting system, West Germany. 1968. *J. Wildl. Manage.* 36: 110–118.

IUCN. 1980. *World conservation strategy.* International Union for Conservation of Nature and Natural Resources. Gland, Switzerland.

IUCN. 1981. A convention comes of age. IUCN Bull. New Series 12 (1-2): 122–123. International Union for Conservation of Nature and Natural Resources. Gland, Switzerland.

Johnson, M. K. 1980. Management involvement lacking at recent international meeting. *Wildl. Soc. Bull.* 8: 65–69.

King, F. W. 1978. The wildlife trade. In H. P. Brokaw (Ed.), *Wildlife and America.* Washington, D.C.: Council on Environmental Quality.

Leopold, A. 1936a. Deer and the dauerwald in Germany. I. History. *J. Forestry* 34: 366–375.

Leopold, A. 1936b. Deer and dauerwald in Germany. II. Ecology and policy. *J. Forestry* 34: 460–466.

Lovejoy, T. E. 1980. A projection of species extinctions. In G. O. Barney (Ed.), *The global 2000 report to the president of the U.S. vol. 1: the summary report.* Washington, D.C.

Myers, N. 1979. *The sinking ark.* New York: Pergamon Press.

Myers, N. 1981. The hamburger connection: how Central America's forests become North America's hamburgers. *Ambio* 10: 3–8.

Newman, J. R. 1979. Hunting and hunter education in Czechoslovakia. *Wildl. Soc. Bull.* 7: 155–161.

Salo, L. 1976. History of wildlife management in Finland. *Wildl. Soc. Bull.* 4: 167–174.

Taber, R. D. 1961. Wildlife administration and harvest in Poland. *J. Wildl. Manage.* 25: 353–363.

Talbot, L. 1979. Conservationists win in Costa Rica. *IUCN Bull.* 10(5): 34.

Teer, J. G., & W. G. Swank. 1978. International implications of designating species endangered or threatened. *Trans. N. Am. Wildl. and Natur. Resour. Conf.* 43: 33–41.

Udvardy, M. D. F. 1975. *A classification of the biogeographical provinces of the world.* IUCN Occ. Paper No. 18, Morges, Switzerland.

Wolfe, M. 1966. *Eine vergleichende betrachtung der jagdrechtlichen regelungen in der Bundesrepublik Deutschland und in der vereinigten staaten von Amerika.* (Summary in English) Hann, Munden.

INDEX